Petroleum Engineering

The Springer series in Petroleum Engineering promotes and expedites the dissemination of new research results and tutorial views in the field of exploration and production. The series contains monographs, lecture notes, and edited volumes. The subject focus is on upstream petroleum engineering, and coverage extends to all theoretical and applied aspects of the field. Material on traditional drilling and more modern methods such as fracking is of interest, as are topics including but not limited to:

- Exploration
- Formation evaluation (well logging)
- Drilling
- Economics
- Reservoir simulation
- Reservoir engineering
- Well engineering
- Artificial lift systems
- Facilities engineering

Contributions to the series can be made by submitting a proposal to the responsible Springer Editor, Charlotte Cross at charlotte.cross@springer.com or the Academic Series Editor, Dr. Gbenga Oluyemi g.f.oluyemi@rgu.ac.uk.

More information about this series at http://www.springer.com/series/15095

Bernardo Carreón-Calderón ·
Verónica Uribe-Vargas ·
Juan Pablo Aguayo

Thermophysical Properties of Heavy Petroleum Fluids

Springer

Bernardo Carreón-Calderón
Instituto Mexicano del Petróleo
Gerencia de Ingeniería de Yacimientos
Mexico City, Mexico

Verónica Uribe-Vargas
Instituto Mexicano del Petróleo
Gerencia de Ingeniería de Yacimientos
Mexico City, Mexico

Juan Pablo Aguayo
Instituto de Ciencias Aplicadas y Tecnología
Universidad Nacional Autónoma de México
Mexico City, Mexico

ISSN 2366-2646 ISSN 2366-2654 (electronic)
Petroleum Engineering
ISBN 978-3-030-58830-4 ISBN 978-3-030-58831-1 (eBook)
https://doi.org/10.1007/978-3-030-58831-1

This Springer imprint is published by the registered company Springer Nature Switzerland AG
The registered company address is: Gewerbestrasse 11, 6330 Cham, Switzerland

To my wife, Rosalba, and to my children,
Bernardo and Berenice
The reason for being and existing

—Bernardo Carreón-Calderón

To my husband, Manuel and to my two
children, Andrew and Atenea
For their unconditional love

—Verónica Uribe-Vargas

To my parents, José and Gloria, to my wife,
Leticia, and to my son, Fernando.
For the example you have given me.

—Juan Pablo Aguayo

Preface

The main purpose of writing this book is to present new approaches for studying thermophysical properties of petroleum in general and heavy petroleum in particular. Unlike the most books on these topics, this book provides an understanding description of conceptual and mathematical tools for the estimation of such properties, where, whenever possible, adjustments of the model parameters to experimental information is left as the last option. Thus, this book is an ambitious attempt to apply the physical chemistry concepts to property estimation of heavy petroleum fluids predictively. Rheology is an exception to this effort; however, heavy petroleum rheology is analyzed, considering the complex non-Newtonian response that heavy petroleum may present.

The overall approach here is based on theoretical developments, which are contrasted against reported experimental observations, covering the physical aspects encountered within some of the main applications in the exploitation, transportation and processing of heavy petroleum. The book is intended to be comprehensible to readers who have some previous knowledge of fundamentals of physical chemistry (thermodynamics and transport properties), and of organic-petroleum chemistry (terminology) as well as significant mathematical background (differential and integral calculus, and tensorial notation). This book is a compendium which is accessible to engineers and researchers, graduate and postgraduate students devoted to the study of petroleum physical chemistry.

In view of the modest size of the book, it is not possible to cover all possible topics and approaches that can be addressed about heavy petroleum fluids. Therefore, the choice of the topics reflects the experience and prejudgments of the authors. However, authors consider that even though they cannot claim that the issue has been entirely addressed in this book, they believe that some of its contents are considerably addressed here for investigation purposes and future developments.

Mexico City, Mexico
July 2020

Bernardo Carreón-Calderón
Verónica Uribe-Vargas
Juan Pablo Aguayo

Acknowledgements Authors express their gratitude to Instituto Mexicano del Petróleo (IMP) and SENER-CONACyT for the support in order to write this book through the 280158 research project (IMP-Y.61069). Besides, authors are very pleased to acknowledge with gratitude the support received from our colleagues Ángel González-Esquivel, Humberto Aguilar-Cisneros, and Nicolás Ruiz-Fuentes, who were the connection between research and industrial practice, and who, therefore, made the writing of this book possible. In addition, opinions and discussions about Rheology from Dr. Gabriel Ascanio, Dr. Marisol Córdova and Dr. Alberto Rosas, were fundamental to the project, and the help for rheological measurements from Karen Pérez, Brenda Becerril, Aloha Sánchez and Alexa Barrientos is gratefully acknowledged. Part of the project was realized through the support of the Instituto de Ciencias Aplicadas y Tecnología (ICAT-UNAM).

Contents

About the Authors

Bernardo Carreón-Calderón holds M.Sc. and Dr. Eng. degrees in chemical engineering from the National Autonomous University of Mexico (UNAM) and Metropolitan Autonomous University (UAM), respectively. He is the author of several articles about Phase Transitions, Hydrocarbon Thermodynamics, and Molecular Diffusion. He has worked for the Mexican Institute of Petroleum (IMP) on Oilfield Processing and Flow Assurance during last 18 years. He is also a Transport Phenomena, Separation Processes and Thermodynamics professor for National Autonomous University of Mexico during 21 years. Currently, he holds the position of full-time researcher in the Reservoir Engineering Department.

Verónica Uribe-Vargas holds Master and Doctor Degrees in chemical engineering from the National Autonomous University of Mexico (UNAM). She has worked for the Mexican Institute of Petroleum on Refining Processes and Flow Assurance and Reservoir Engineering during last 22 years. She is the author of several articles about Phase Equilibria and Hydrocarbon Thermodynamics. She has been a postgraduate staff member in the Mexican Institute of Petroleum (IMP) as well. Currently, she holds the position of full-time researcher in the Reservoir Engineering Department.

Juan Pablo Aguayo has Bachelor's and Master's degrees in chemical engineering from the Universidad Nacional Autónoma de México (UNAM) and Ph.D. from the University of Wales, Swansea (now Swansea University). Formerly he was a lecturer at universities La Salle, Simón Bolívar, Universidad Autónoma de Tlaxcala and currently lectures for undergraduate and graduate courses at UNAM, in subjects such as Energy Transfer, Transport Phenomena, Rheology and Applied Mathematics. He has worked at the Instituto Mexicano del Petróleo (IMP) as a process engineer and later as a postdoctorate. Since October 2013, he has been a researcher at Instituto de Ciencias Aplicadas y Tecnología (ICAT-UNAM).

Chapter 1
Introduction

List of Abbreviations

DOE	United States Department of Energy
UNITAR	International Group of the United Nations

Thermophysical properties of fluids are of great significance in design and operation of facilities for the exploitation, transportation and processing of petroleum fluids. The thermophysical term implies two types of physical properties: thermodynamic and transport properties. The former have their origin in the study of processes at equilibrium by the area of physics known as Thermodynamics, whereas the latter have their origins in the study of processes of non-equilibrium by the area of physics known as Transport Phenomena. The thermodynamic properties govern the feasibility for a given process to be realizable, while the transport properties determine either the size or the needed time for a given process to occur. Thus, the necessity of accurate and reliable methodologies for estimating thermodynamic and transport properties of petroleum fluids over a wide range of conditions is recognizable. Chronologically, thermodynamic properties should be analyzed before transport properties, because it does not make any sense to determine sizes and times, if such processes are not feasible. This is the treatment that will be given to the study of thermophysical properties of heavy petroleum fluids in this book: first, the thermodynamic properties from their fundamental concepts to applications will be addressed in depth, then the probably most important transport property will be addressed: the viscosity.

1.1 Characteristics of Heavy Petroleum

People are familiar with the terms oil and crude oil, which encompass most hydrocarbon mixtures at liquid state. In the oil industry, the conventional oils are those crude oils characterized by its low viscosity, which allows them to be recovered

B. Carreón-Calderón et al., *Thermophysical Properties of Heavy Petroleum Fluids*, Petroleum Engineering, https://doi.org/10.1007/978-3-030-58831-1_1

by conventional process. In recent decades, the discovery of high quality (conventional) oil has declined, and as consequence heavy oils and bitumen have begun to be exploited, becoming important resources of crude oils. Bitumens have a higher production and processing cost than conventional crude oils, but they are found with greater abundance [1].

1.1.1 Physical Analysis

The international definition of heavy oil was first discussed at the World Petroleum Congress in 1980; the United States Department of Energy (DOE) continued the work through UNITAR (an international group of the United Nations): Heavy crude oil is described as dead oil (oil without gas) with a density below 21° API and its viscosity is between 100 and 10,000 centipoise (cP) at reservoir temperature. The World Petroleum Congress subsequently adopted the definition of UNITAR, in 1987, with minor modifications. Later, Venezuela added its own definition of extra heavy oil, such as a crude oil that is below 10° API, with a viscosity of less than 10,000 mPa s. In contrast, Canadian heavy oil, which is obtained from the oil sands or carbonates, has an API gravity less than 10° and a viscosity greater than 10,000 mPa s at reservoir conditions. It is the most viscous hydrocarbon and is practically a solid at room temperature, now recognized worldwide as bitumen [2]. Figure 1.1 shows the classification of heavy, extra-heavy, and bitumen oils based on their viscosity.

The presence of light hydrocarbons or other kind of volatile chemical components may be extremely scarce in all these oils irrespective of the used classification to appoint them. Hence, in this book, the designation heavy petroleum fluids or petroleum fluids in general will refer to mixtures formed by oils according to the above conventional meanings and other identifiable components, such as light

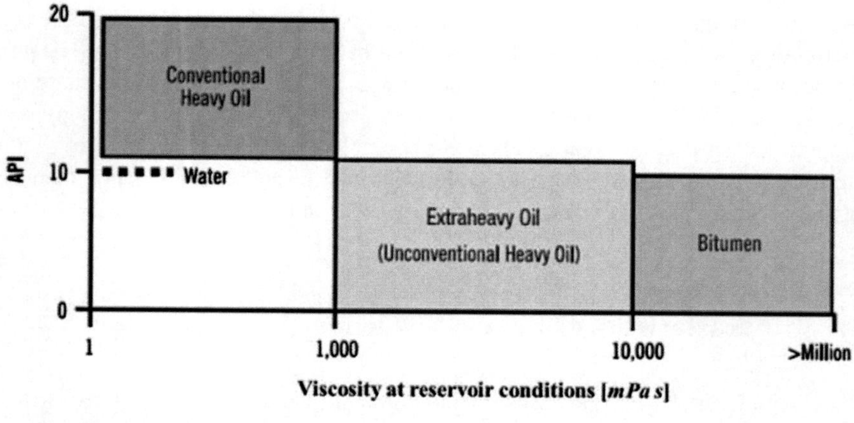

Fig. 1.1 Heavy oil and bitumen definition [2]

alkanes, cycloalkanes, and aromatic components, carbon dioxide, hydrogen sulfide, nitrogen, and water, either they are in the oils naturally or they are added as a part of some kind industrial process.

The other important definition for crude oils is related to the sulfur content: if the crude oil has more than 0.5 wt% sulfur, it is called sour; while for less than 0.5 wt% it called sweet. This distinction between sour and sweet crude oil was originated with the refinery practices employed during most of the twentieth century. At that time, sweet crude oils did not require the removal of sulfur when used for the manufacture of transport fuels. However, since the 1990s, stricter restrictions on the sulfur content of manufactured fuels have made sulfur content important [1].

1.1.2 Chemical Analysis

Heavy oils generally have relatively low proportions of volatile components with low molecular weights and fairly high proportions of lower volatility components with high molecular weights. The high molecular weight fraction of heavy oils is composed of a complex variety of different molecular and chemical types, a mixture of components (not necessarily only paraffins or asphaltenes) with high melting points and high pour points that greatly contribute to the properties of heavy oil. This contributes to its low mobility compared to conventional crude oil [3].

The main components of oil are hydrocarbons, hydrogen and carbon components that show a great variation in their molecular structure. They are generally divided into, paraffins and isoparffins, naphthenic and aromatic. The simplest hydrocarbons are a large group of chain-shaped molecules known as paraffins. Methane (CH_4) is the simplest of all hydrocarbons, and the main component in a natural gas. A paraffinic component consists of hydrocarbon segments of type C, CH, CH_2 or CH_3. Carbon atoms are connected by simple bonds. The paraffins are divided into normal paraffins (n-paraffins) and iso-paraffins (i-paraffins). In an n-paraffin, carbon atoms form linear chains, while an i-paraffin contains at least one side chain. Paraffins are also known as alkanes. Naphthenes: these components are similar to paraffins in that they are formed by the same types of hydrocarbon segments, but they contain one or more cyclic structures. The segments in the ring structures (for example, CH_2) are connected by simple bonds. Most naphthenic ring structures contain six carbon atoms, but naphthenic components with five or seven carbon atoms connected in ring structures are also common in oilfield fluids; naphthenes are also called cycloalkanes. And finally, the aromatic components are like naphthenes, aromatics in which they contain one or more cyclic structures, but carbon atoms in an aromatic component are connected by double bonds. Benzene (C_6H_6), is the simplest aromatic component. Polycyclic aromatic components with two or more ring structures are also found in oilfield fluids.

The percentage of paraffinic as one (P), naphthenic (N) and aromatic (A) components in a reservoir fluid is often called PNA distribution. Petroleum field fluids may also contain inorganic components, of which nitrogen (N_2), carbon dioxide (CO_2) and hydrogen sulfide (H_2S) are the most common. Water (H_2O) is another important component of reservoir fluids. Since water has a limited miscibility with hydrocarbons, most of the water in a reservoir is usually in a separate water zone located below the gas and oil zones [4].

1.1.3 Modelling

According to a recent survey about industrial requirements for thermodynamics and transport properties [5], there is still a necessity for reliable thermodynamic and transport property models for a wide range of systems and industries such as chemical, agrochemical, pharmaceutical, power generation, exploration and production, among others. Some conclusions from this survey are: there is still the necessity for models from a single framework, that cover all conditions of interest; the framework should be generalized so that simple equations of state can be derived for nonpolar, polar, associating, and high boiling substances and their mixtures, with sufficient accuracy for industrial purposes. The number of available models should be reduced by recommending either a unique model or a small number of models; there is a lack of models for larger molecules and complex molecules such as polar and associating ones. Besides, there is a necessity to improve the stability test, the most time-consuming element of the multiphase calculations. The simultaneous description of different thermodynamic properties as well as phase and chemical equilibrium types is of great importance. Thus, there is a necessity for combined models for predicting thermodynamic and transport properties. Finally, there is a necessity for models for predicting properties, especially transport properties of complex molecules or products. Most of these necessities are addressed in this book for the specific case of heavy petroleum fluids.

1.1.3.1 State of Art

Together with the components mentioned in the previous paragraph, in order to perform the modeling of the crude oils, they are separated into identifiable components with a number less than 7 carbons, the components with more than 7 carbons are called the heptane fraction plus (C_{7+}). On these fractions falls the responsibility of the difficulty of modeling in the thermodynamic characterization of petroleum fluids, especially when molecular structures are required [6]. Therefore, a convenient set of components and unidentifiable fractions is needed for the thermodynamic characterization of petroleum [5, 7]. An important aspect is that the components that make up the fractions do not have to be neither real nor present in the mixture. They can be well defined or constructed individual molecules [8–10].

The characterization of unidentifiable petroleum fractions, fractions of unknown composition, implies the use of measurable properties to calculate basic parameters required by state equations or similar expressions. In the case of the cubic state equations, the required parameters are the critical properties and the acentric factor. Therefore, several sets of correlations have been developed to estimate these basic parameters from measurable properties such as molecular weight and liquid density. From the point of view of calculations, a given unidentifiable fraction of oil is considered as a pure hypothetical component in this approach, the above, it allows to calculate the chemical and physical properties with greater precision. However, more computation time and experimental data are required. On the other hand, when unidentifiable oil fractions are treated mathematically as individual components, some volume properties are used to estimate the missing ones. One of the disadvantages of this approach is the use of specific correlations, which are limited to certain conditions and types of oil. This has resulted in different sets of correlations [11, 12]. Therefore, thermodynamic characterization also involves the selection of models and procedures to estimate the volume properties of unidentifiable oil fractions [13, 14]. Attempts have been made to combine the advantages of characterization at the molecular level with the use of unidentifiable fractions as individual components for the concept of group contribution [15, 16]. There, the chemical and physical properties are written in terms of functional groups (CH_3-, $-CH_2-$, etc.) either by correlations [17, 18] or by trial and error procedure [19]. In the latter case, a suggested chemical structure is adjusted so that the available experimental information matches the calculations. These procedures require much less experimental information than characterization at the molecular level. For example, molecular weight, density, true boiling point and the number of hydrocarbon families are commonly needed [20].

In addition to correlations, more rigorous approaches, such as molecular reconstructions, have been suggested. Although these methods differ from each other in the techniques used, they all seek to construct representative molecule mixtures from partial experimental data [7–9, 21]. The objective is that the physical and chemical properties of the actual mixtures, unidentifiable petroleum fractions, can be reproduced using those constructed; good results have been obtained in the case of heavy petroleum fractions [10]. A possible disadvantage of molecular reconstruction methods is that tens, hundreds, or even thousands of molecules are considered necessary. In fact, a rigorous chemical structure and composition may be required for petroleum refining purposes, but a detailed characterization is not necessary for oil production computations because the only properties required are bulk properties such as density, viscosity, etc. In the context of molecular reconstruction methods, the use of group contribution methods to estimate the physical and chemical properties of the properties of pure components may, in principle, give better results than correlations based on massive properties because molecular interactions they are inherently taken into account [22]. Therefore, the use of these methods has already been suggested to estimate the properties of the individual components of the heavy oil fractions [23]. Particular, the Joback-Reid [24] and Marrero-Gani [25] methods to estimate the Critical properties have been used for extremely heavy hydrocarbon mixtures, such as vacuum tower bottoms [26]. In addition, group contribution methods have been developed not only to estimate the critical properties of individual molecules, but also to estimate the parameters used in cubic equations of state, such as the parameters of binary interaction belonging to the classical mixing rule; [27, 28]. Between a characterization at the molecular level, and the use of unidentifiable petroleum fractions as hypothetical individual components, we have the representation by using of the group contribution concept. In this type of methods, the characterization consists not only in determining the volume properties, but also in determining the functional groups, which are structural units that form molecules, so that a volume property can be adjusted. Using these calculated molecular structures, it is possible to employ methods that require molecular structures to calculate other volume properties such as surface tension, solubility [15, 16] and so on. Typically, these methods require fewer input data than molecular reconstruction methods: true boiling point analysis (TBP) or the paraffin-naphthenic-aromatic (PNA) analysis is generally enough. In these methods, a molecular pseudostructure, a set of non-integer functional groups, is calculated for each family of PNA components, that is, from three to one pseudostructures are calculated for each fraction of indefinite oil [17–19, 29].

In this book, the chemical pseudostructure concept will be used widely. In Chap. 2, fundamental concepts of the Classical Thermodynamics will be introduced and analyzed from the fundamental relationships and concepts to the equilibria criteria in terms of pressure, temperature, and composition for multiphase and multicomponent systems. Although the focus is on heavy petroleum fluids, the issues are addressed in a general way, so that they are valid and can be used as starting point for studying systems formed by any kind fluid. Once equilibrium criteria are given, then mathematical expressions that connects pressure, volume and composition of fluids are required, the so-called equations of state. In Chaps. 3 and 4, Cubic Equations of State ($CEoS$) and the Perturbed-Chain-Statistical-Associating-Fluid-Theory equation of state ($PC-SAFT$) are presented in a general form, pointing out strengths and weaknesses for being applied to heavy petroleum fluids. The main emphasis in Chap. 3 will be on EoS/G^E mixing rules, which allows us to extend the prediction capabilities and applicability of $CEoS$ to more complex phase equilibriums, such as those in the presence of mixtures formed by asymmetric components (solvents and heavy oil). Regardless of the equation or the mixing rule, a finite number of components is required for equations of state to be used; however, petroleum is a mixture that shows a continuous distribution concerning its properties, that is, an infinite number of components forms petroleum mixtures. Thus, to turn oil into a mixture of discrete and finite number of components (pseudocomponents) that can be handled by equations of state, it is necessary to make use of distribution functions for random variables. This issue will be addressed in Chap. 5. All concepts and theoretical tools that are developed in Chaps. from 2 to 5 will be applied to real phase equilibrium problems involving heavy petroleum fluids in Chap. 6. First, numerical algorithms for such problems are derived and explained, including phase stability analysis. Then phase equilibriums of heavy petroleum fluids in the presence of gases, light hydrocarbons, water, and asphaltenes are determined and compared with experimental data. The chemical equilibrium is undertaken using the same approach in Chap. 7, but in less detail because experimental information is less abundant. Finally, in Chap. 8, methods are shown and analyzed in order to study viscosity from a Newtonian and non-Newtonian standpoint, considering also other non-Newtonian features that have an impact on the flow response and its modeling.

References

1. Gray MR (2015) Upgrading oilsands bitumen and heavy oil, 1st edn. PICA Press, Canada
2. Banerjee DK (2012) Oil sands, heavy oil, and bitumen: from recovery to refinery. PennWell Corporation, USA
3. Speight JG (2009) Enhanced recovery methods for heavy oil and tar sands. Gulf Publishing Company, USA
4. Pedersen KS, Christensen PL (2007) Phase behavior of petroleum reservoir fluids. Taylor & Francis Group, New York
5. Hendriks E, Kontogeorgis GM, Dohrn R, De Hemptinne JC, Economou IG, Ẑilnik, LF, Vesovic V (2010) Industrial requirements for thermodynamics and transport properties. Ind Eng Chem Res 49:11131
6. Habhtalab A, Mahmoodi P (2010) Vapor—liquid equilibria of asymmetrical systems using UNIFAC-NRF Group contribution activity coefficient model. Fluid Phase Equilib 289:61
7. Albahri TA (2005) Molecularly explicit characterization model (MECM) for light petroleum fractions. Ind Eng Chem Res 44:9286
8. Khorasheh F, Khaledia R, Grayb MR (1998) Computer generation of representative molecules for heavy hydrocarbon mixtures. Fuel 77:241
9. Hudebine D, Verstraete JJ (2004) Molecular reconstruction of LCO gasoils from overall petroleum analyses. Chem Eng Sci 59:4755
10. Verstraete JJ, Schnongs P, Dulot H, Hudebine D (2010) Molecular reconstruction of heavy petroleum residue fractions. Chem Eng Sci 65:304
11. Jianzhong Z, Biao Z, Suoqi Z, Renan W, Guanghua Y (1998) Simplified prediction of critical properties of nonpolar components, petroleum, and coal liquid fractions. Ind Eng Chem Res 37:2059
12. Pedersen KS, Milter J, Sørensen H (2004) Cubic equations of state applied to HT/HP and highly aromatic fluids. SPE J 186
13. Riazi MR, Al-Adwani HA, Bishara A (2004) The impact of characterization methods on properties of reservoir fluids and crude oils: options and restrictions. J Pet Sci Eng 42:195
14. Aladwani HA, Riazi MR (2010) Some guidelines for choosing a characterization method for petroleum fractions in process simulators. Chem Eng Res Des 83:160
15. Fahim MA, Elkilani AS (1992) Prediction of solubility of hydrogen in petroleum cuts using modified UNIFAC. Can J Chem Eng 70:335
16. Darwish E, A1-Sahhaf TA, Fahim MA (1995) Prediction of the surface tension of petroleum cuts using a modified UNIFAC group contribution method. Fluid Phase Equilib 105:229
17. Avaullée L, Neau E, Jaubert JN (1997) Thermodynamic modeling for petroleum fluids II. Prediction of PVT properties of oils and gases by fitting one or two parameters to the saturation pressures of reservoir fluids. Fluid Phase Equilib 139:171
18. Avaullée L, Duchet-Suchaux P, Durandeau M, Jaubert JN (2001) A new approach in correlating the oil thermodynamic properties. J Pet Sci Eng 30:43
19. Ruzicka V, Fredenslund A, Rasmussen P (1983) Representation of petroleum fractions by group contribution. Ind Eng Chem Process Des Dev 22:49
20. Mohammad R, Daubert TE (1986) Prediction of molecular-type analysis of petroleum fractions and coal liquids. Ind Eng Chem Process Des Dev 25:1009
21. SaineAye MM, Zhang N (2005) A novel methodology in transforming bulk properties of refining streams into molecular information. Chem Eng Sci 60:6702

22. Ahmada MI, Zhang N, Jobsonb M (2011) Molecular componentsbased representation of petroleum fractions. Chem Eng Res Des 89:410
23. Neau E, Jaubert JN, Rogalski M (1993) Characterization of heavy oils. Ind Eng Chem Res 32:1196
24. Joback KG, Reid RC (1987) Estimation of pure-component properties from group-contributions. Chem Eng Commun 57:233
25. Marrero J, Gani R (2001) Group-contribution based estimation of pure component properties. Fluid Phase Equilib 183–184:183
26. McFarlane RA, Gray MR, Shaw JM (2010) Evaluation of covolume mixing rules for bitumen liquid density and bubble pressure estimation. Fluid Phase Equilib 293:87
27. Jaubert JN, Privat R, Mutelet F (2010) Predicting the phase equilibria of synthetic petroleum fluids with the PPR78 approach. AIChE J 56(12):3225
28. Jaubert JN, Privat R (2010) Relationship between the binary interaction parameters (kij) of the Peng—Robinson and those of the Soave-Redlich-Kwong equations of state. Application to the definition of the PR2SRK model. Fluid Phase Equilib 295:26
29. Avaullée L, Neau E, Jaubert JN (1997) Thermodynamic modeling for petroleum fluids III. Reservoir fluid saturation pressures. A complete PVT property estimation. Application to swelling test. Fluid Phase Equilib 141:87

Chapter 2
Fundamental Concepts of Thermodynamics

List of Abbreviations

Nomenclature

A	Helmholtz free energy
f	Homogeneous function
$f(x, y)$	Legendre transformation
g	Function equivalent
G	Gibbs free energy
H	Enthalpy
n_i	Number of moles
P	Absolute pressure
P^0	Reference pressure
Q	Thermal energy
R	The universal gas constant
S	Entropy
T	Temperature
U	Internal energy
V	Total volume
W	Work
Z	Compressibility factor
α_j	Phase fraction
δ_{jk}	Kronecker delta
η	Order of homogeneity
λ	Number that increases the mass of the system
μ_i	Chemical potential
σ	Phases number

B. Carreón-Calderón et al., *Thermophysical Properties of Heavy Petroleum Fluids*, Petroleum Engineering, https://doi.org/10.1007/978-3-030-58831-1_2

Superscripts

~	Molar
0	Pure component
E	Excess property
hg	Hypothetical ideal gas
ig	Ideal gas
is	Ideal solution
R	Real

Subscripts

i	Each component
j	Subsystem

As part of the classical thermodynamics, the study of physical and chemical equilibrium starts with the fundamental laws of nature known as the first and the second laws of thermodynamics. These laws were initially stated as postulates; however, they have been verified by experience so that they have achieved the condition of axioms or laws. In what follows, the basic framework of these laws is outlined so that fundamental concepts and equations are available for the phase and chemical equilibriums involving heavy petroleum fluids to be addressed in following chapters. More detailed discussion about the topics that will be presented here can be found in specialized textbooks [1, 2].

2.1 The First Law of Thermodynamics

The first law of thermodynamics, which is also known as the energy conservation law, states that:

> Energy is not destroyed or created, but it can only be transformed from one type of energy to another.

The mathematical expression of the first law depends on the types of energy that are involved in a given transformation or change. In the case of physical or chemical equilibrium between phases, kinetic and potential energies are usually disregarded; hence, for a system where no mass is exchanged through its boundaries (closed system), the first law of thermodynamic is

$$dU = dQ - dW \tag{2.1}$$

where the internal energy U refers to the energy that the system has by itself. When the system consists of molecules, U refers to the energy that the molecules possess due to their translational, vibrational and rotational movements within the system. On the other hand, Q and W are the thermal energy and macroscopic work exchanged with the surroundings through the boundaries that set the limits of the system. Q and W are functions of the path followed by the system from one thermodynamic state to another, whereas U is a function of state; that is, its value depends only on the state of the system, and not on the path followed to reach it. A thermodynamic state of a given system is found when each of its properties have definite values.

2.2 The Second Law of Thermodynamics

The first law expresses that the energy of the universe remains constant regardless of the changes occurring within the universe itself. However, there is no restrictions from the first law associated with the way or direction for such changes to occur, and consequently, for the corresponding energy and work exchanges. It is the second law of thermodynamics precisely, which provides criteria to establish the direction of natural changes within the universe.

The natural observations that served as starting point to establish the second law are based on the study of the so-called thermal machines, which are machines that transform thermal energy to macroscopic work from cyclic processes. However, the postulates and conclusions from the second law are valid for any type of process or system. The performance of the thermal machines, on which the second law is based, are:

> It is not possible to transform completely the thermal energy absorbed from the surroundings by the system into work performed on the surroundings by the system.
>
> A system in a cyclic process needs at least two reservoirs of thermal energy at two different temperatures to perform work on the surroundings.

The previous observations led to the Clausius inequality o mathematical expression of the second law of thermodynamics for closed systems:

$$dS \geq \frac{dQ}{T} \tag{2.2}$$

which states that, for closed systems subjected to natural or spontaneous processes, the ratio of the change of thermal energy dQ to the absolute temperature T is lesser than or equal to the function of state called entropy S. According to the second law, for an isolated system ($dQ = 0$ and $dW = 0$) undergoing an irreversible transformation ($>$), the transformation will occur in natural way or spontaneously as long as the entropy of the system increases, that is, if $dS > 0$. On the other hand, if the isolated system undergoes a reversible transformation ($=$), the second law establishes that the entropy of the system remains constant. As observed, the second law is connected with the concept of irreversible and reversible transformations. During irreversible transformations, systems and their surroundings undergo permanent changes, while, during reversible transformations, systems and their surroundings can be restored to their original states. Therefore, reversible transformations are not natural processes, but ideal ones. Natural transformations are always irreversible processes.

2.3 Fundamental Expressions of Thermodynamics

Mathematical expressions for calculating and studying thermodynamic properties, and physical and chemical equilibriums between phases are obtained from the fundamental expression of thermodynamics. The fundamental expression arises from combining the first and second laws.

The fundamental expression for closed systems arises from replacing dQ in the second law with the first law, that is,

$$dU \leq TdS - PdV \tag{2.3}$$

where the expansion-compression work is the only one considered here ($dW = PdV$). P and V symbolize the absolute pressure and the total volume of the system, respectively. As can be observed, Eq. (2.3) has the mathematic form of the total differential of U with S and V as independent variables; therefore, from the definition of total differential of any function, we have that

$$\left(\frac{\partial U}{\partial S}\right)_V = T \tag{2.4}$$

and

$$\left(\frac{\partial U}{\partial V}\right)_S = -P \tag{2.5}$$

The simplicity of these partial derivatives suggests that S and V are the natural variables of U. Although Eq. (2.3) is general and valid for any closed system, it has one inconvenient for its practical use: S appears as independent variable. V can be measured or determined experimentally; however, it is not possible to determine S from experiments directly. Therefore, it is desirable to express the fundamental expression in terms of accessible variables such as pressure, temperature and volume only.

From the mathematical point of view, the Legendre transformation allows functions to be expressed in terms of their partial derivatives rather than their independent variables. In the case of the fundamental expression, it seems convenient because, according to Eqs. (2.4) and (2.5), the partial derivatives of the function U are equal to measurable variables: temperature and pressure.

The Legendre transformation of any function $f(x, y)$ is a function g equivalent to $f(x, y)$, where one independent variable is replaced by the corresponding partial derivative. Hence, to obtain g, it is necessary first to establish the partial derivatives of $f(x, y)$ from the definition of total differential

$$df = \left(\frac{\partial f}{\partial x}\right)_y dx + \left(\frac{\partial f}{\partial y}\right)_x dy = pdx + qdy \tag{2.6}$$

If the independent variable x is replaced by p then the Legendre transformation is

$$g(x, y) = f(x, y) - xp \tag{2.7}$$

Although $g(x, y)$ appears to be function of x and y, it can be shown that $g(x, y)$ is function of p and y actually, because

$$dg = df - xdp - pdx = qdy - xdp \tag{2.8}$$

As mentioned before, it is desirable to choose accessible experimental variables as independent ones for the fundamental inequality. If S is intended to be changed as independent variable, then, according to the Legendre transformation as expressed by Eq. (2.7), we can use the partial derivative given by Eq. (2.4) in order to change S for T as independent variable. Therefore, the Legendre transformation of U, which is symbolized by A, is

$$A \equiv g(T, V) = U - S\left(\frac{\partial U}{\partial S}\right)_V = U - ST \tag{2.9}$$

The energy A defines the Helmholtz free energy, which allows the fundamental expression of thermodynamics to be written as

$$dA \leq -SdT - PdV \tag{2.10}$$

From which and the from definition of total differential again, we found that

$$\left(\frac{\partial A}{\partial T}\right)_V = -S \tag{2.11}$$

and

$$\left(\frac{\partial A}{\partial V}\right)_T = -P \tag{2.12}$$

On the other hand, we may prefer pressure rather than volume as independent variable in the Legendre transformation given by Eq. (2.9). This can be carried out as illustrated above, but now the Legendre transformation of A is symbolized by G and is determined as

$$G \equiv g(T, P) = A - V\left(\frac{\partial A}{\partial V}\right)_T = U - ST + VP \tag{2.13}$$

The energy G is the Gibbs free energy and allows the fundamental expression of thermodynamics to be written as function of T and P as follows:

$$dG \leq -SdT + VdP \tag{2.14}$$

where the corresponding partial derivatives are

$$\left(\frac{\partial G}{\partial T}\right)_P = -S \tag{2.15}$$

and

$$\left(\frac{\partial G}{\partial P}\right)_T = V \tag{2.16}$$

Finally, it is worth introducing the so-called enthalpy H, because of its significance for the energy exchange in chemical reactions and processes at constant pressure. H may be seen as the Legendre transformation of U, where V is replaced by P as independent variable by means of the partial derivative given by Eq. (2.5), as follows:

$$H(S, P) = U - V\left(\frac{\partial U}{\partial V}\right)_S = U + PV \tag{2.17}$$

From which and from the fundamental expression, we arrive at

$$dH \leq TdS + VdP \tag{2.18}$$

whose associated partial derivatives are

$$\left(\frac{\partial H}{\partial S}\right)_P = T \tag{2.19}$$

and

$$\left(\frac{\partial H}{\partial P}\right)_S = V \tag{2.20}$$

U, A, G, and H are known as thermodynamic potentials because their variations from one state of the system to another establish the corresponding energetic exchanges and whether such exchanges will occur spontaneously or not.

In the study of phase and chemical equilibriums, it is important the tendency of systems to exchange both energy and mass with their surroundings. This implies that the fundamental expressions described above have to be generalized for open systems. By definition, an open system can exchange both mass and energy with its surroundings. Mathematically, this means that U depends not only on S and V, but also on the number of moles n_i of each component i in the system, that is, we have that

$$U = U(S, V, n_1, n_2, \ldots, n_C) \tag{2.21}$$

Therefore, its total differential is

$$dU = \left(\frac{\partial U}{\partial S}\right)_{V,n_i} dS + \left(\frac{\partial U}{\partial V}\right)_{S,n_i} dV + \sum_{i=1}^{C} \left(\frac{\partial U}{\partial n_i}\right)_{S,V,n_j,j\neq i} dS \tag{2.22}$$

Substitution of Eqs. (2.4) and (2.5) leads to

$$dU \leq TdS - PdV + \sum_{i=1}^{C} \mu_i dn_i \tag{2.23}$$

In which, the so-called chemical potential μ_i has been defined as

$$\mu_i \equiv \left(\frac{\partial U}{\partial n_i}\right)_{S,V,n_j,j\neq i} \tag{2.24}$$

Analogously, the fundamental expression for open systems can be written now in terms of A, G, and H in the following ways

$$dA \leq -SdT - PdV + \sum_{i=1}^{C} \mu_i dn_i \tag{2.25}$$

$$dG \leq -SdT + VdP + \sum_{i=1}^{C} \mu_i dn_i \tag{2.26}$$

$$dH \leq TdS + VdP + \sum_{i=1}^{C} \mu_i dn_i \tag{2.27}$$

Because A, G, and H are equivalent functions to U, the changes of U with respect to n_i must be the same for all the thermodynamic potentials. Hence, μ_i can be defined from any thermodynamic potential

$$\mu_i \equiv \left(\frac{\partial U}{\partial n_i}\right)_{S,V,n_j,j\neq i} = \left(\frac{\partial A}{\partial n_i}\right)_{T,V,n_j,j\neq i} = \left(\frac{\partial G}{\partial n_i}\right)_{T,P,n_j,j\neq i} = \left(\frac{\partial H}{\partial n_i}\right)_{S,P,n_j,j\neq i} \tag{2.28}$$

One of the most outstanding characteristics of all previous expressions is that they are only combinations of properties of state; therefore, they can be applied to any state of the system regardless of the path followed to reach a given thermodynamic state. Because temperature, pressure and volume are in practice the available properties that are measured and controlled, Eqs. (2.25) and (2.26) form of basis for phase and chemical equilibrium problems to be addressed.

2.4 Spontaneity and Equilibrium Criteria

In thermodynamics, the system is at equilibrium when there is not tendency to change on macroscopic scale. This means that even though molecules inside the system exchange energy and momentum with each other at molecular level, the average value of all these exchanges remains unaltered.

The spontaneity and equilibrium criteria establish whether a system suffers a spontaneous (natural) transformation or it remains at thermodynamic equilibrium, respectively. These criteria are based on the definition of irreversible and reversible processes. The irreversible ones involve the presence of finite and measurable driving forces for changes of state to occur. The change continues until the driving forces vanish achieving thus an equilibrium state. This final equilibrium state can also be reached by means of a reversible process; however, in this case, driving forces are infinitely small so that the system is always at equilibrium state with its surroundings. In other words, the reversible processes are theoretical simplifications of the natural or irreversible processes. Although the reversible process are ideal transformations, they are important because the mass and energy exchanges associated with them are always extremes (maxima or minima) in comparison with those present in irreversible transformations. Besides, from the mathematical point of view, reversible processes are simpler to outline and to solve than irreversible ones. For these reasons, reversible processes are a common starting point to analyze the natural transformations.

The first criterion of spontaneity and equilibrium arises from the second law applied to isolated system ($dQ = 0$). According to Eq. (2.2), this implies that

$$dS \geq 0 \tag{2.29}$$

Table. 2.1 Spontaneity and equilibrium criteria for closed systems

Conditions	Spontaneity	Equilibrium
$dQ = 0$ and $dW = 0$	$dS > 0$	$dS = 0$
$dT = 0$ and $dP = 0$	$dG < 0$	$dG = 0$
$dT = 0$ and $dV = 0$	$dA < 0$	$dA = 0$
$dS = 0$ and $dV = 0$	$dU < 0$	$dU = 0$
$dS = 0$ and $dP = 0$	$dH < 0$	$dH = 0$

From this expression, it is concluded that the entropy of the system increases ($dS > 0$) during a natural or irreversible transformation, whereas it remains constant ($dS = 0$) during reversible one. The former case corresponds to the spontaneity criterion, while the latter corresponds to the equilibrium criterion for isolated systems.

When temperature and pressure are set as independent variables during a given transformation, then, as explained in the previous section, G plays the role of S to find the corresponding spontaneity and equilibrium criteria. Thus, for transformations at constant pressure and temperature, Eq. (2.14) reduces to

$$dG \leq 0 \tag{2.30}$$

In this case, we have that G of the system decreases ($dG < 0$) during natural transformations, whereas it remains constant ($dG = 0$) during reversible ones. The former case corresponds to the spontaneity criterion, while the latter corresponds to the equilibrium one. As observed from these two examples, the criteria rely on the conditions under which the transformation occur. In Table 2.1, the spontaneity and equilibrium criteria for closed systems are summarized.

When the criteria from Table 2.1 are applied to a system formed by σ subsystems as depicted in Fig. 2.1, more practical spontaneity and equilibrium criteria are found. For instance, if the whole system undergoes a transformation keeping constant volume and temperature, then the following general criterion is fulfilled.

$$dA \leq 0 \tag{2.31}$$

Now, if temperature is held constant and the volume V_j is allow to change during the process for each subsystem j forming the whole system, then Eq. (2.10) applied to each subsystem becomes

$$dA_j \leq -P_j dV_j \tag{2.32}$$

Combination of the two above expressions turns out to be

$$dA = \sum_{j=1}^{\sigma} dA_j \leq -\sum_{j=1}^{\sigma} P_j dV_j = -P_k dV_k - \sum_{j=1}^{\sigma} (1 - \delta_{jk}) P_j dV_j \leq 0 \tag{2.33}$$

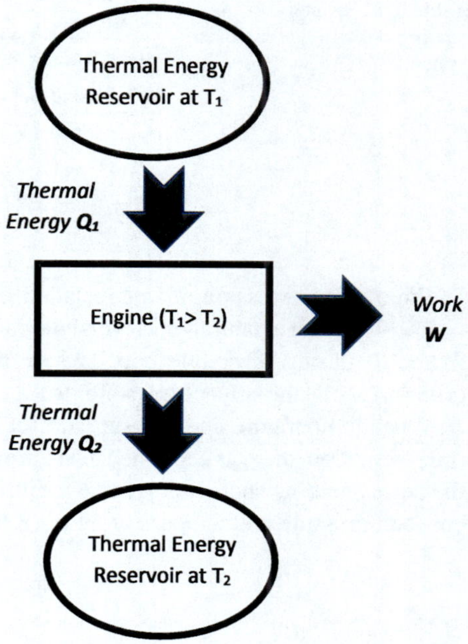

Fig. 2.1 Schematic representation of thermal machines

δ_{jk} is the kronecker delta and takes values of either one or zero according to

$$\delta_{jk} = \begin{cases} 1; \ j = k \\ 0; \ j \neq k \end{cases} \tag{2.34}$$

On the other hand, if the subsystem k suffers a volume expansion ($dV_k > 0$) at the expense of the remaining systems j, which therefore suffer a volume compression $\left(dV_j < 0\right)$ as illustrated in Fig. 2.1 again, then we have that

$$dV_k = -\sum_{j=1}^{\sigma}(1 - \delta_{jk})dV_j \tag{2.35}$$

Using this Eq. (2.33) transforms into

$$\sum_{j=1}^{\sigma}(1 - \delta_{jk})\left(P_k - P_j\right)dV_j \leq 0 \tag{2.36}$$

As all subsystems j are compressed, from the above expression, it is concluded that the following condition must be satisfied

$$P_k - P_j \geq 0 \tag{2.37}$$

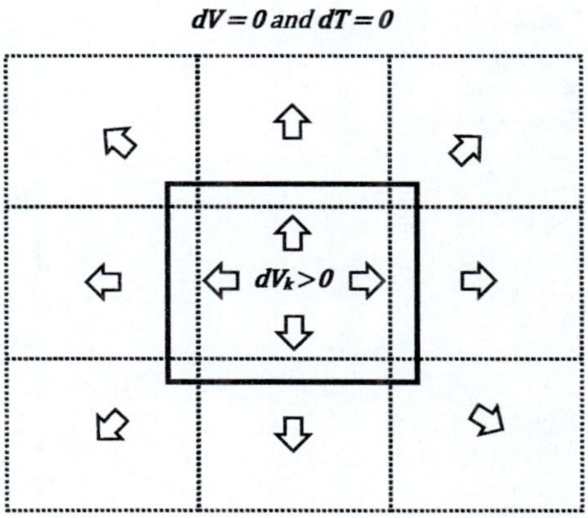

Fig. 2.2 Expansion-Compression of subsystems within a system that undergoes a change of state at constant temperature and volume

As seen from the above inequality, $P_k = P_j$ for reversible processes. This means that the pressure of all subsystems must be equal with each other at equilibrium conditions. For this reason, this equality is called the mechanical-equilibrium condition. On the other hand, $P_k - P_j > 0$ for irreversible processes, which indicates that the pressure of the subsystem k must be larger than pressures of the remaining subsystems j for the described expansion-compression transformation to occur. We may recognize this situation as natural, because the subsystem k grows while the remaining shrink. Thus, we have that $P_k - P_j > 0$ is the driving force for such a no-equilibrium transformation.

Alternatively, when the whole system is now isolated, but the forming subsystems are not isolated so that thermal energy is transferred from the subsystem k to the remaining subsystems j as illustrated in Fig. 2.2, then the entropy is now the suitable property of state to establish the spontaneity and equilibrium criteria. Thus, from criteria in Table 2.1 and Eq. (2.2), we arrive at

$$\sum_{j=1}^{\sigma}(1-\delta_{jk})\left(\frac{1}{T_j}-\frac{1}{T_k}\right)dQ_j \geq 0 \tag{2.38}$$

Because $dQ_j > 0$ for all subsystems different from subsystem k, the following inequality must be satisfy

$$\frac{1}{T_j}-\frac{1}{T_k} \geq 0 \tag{2.39}$$

or equivalently

$$T_k - T_j \geq 0 \tag{2.40}$$

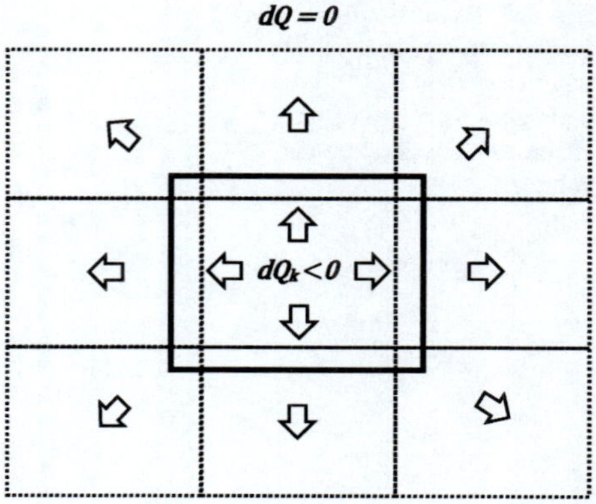

Fig. 2.3 Heat transfer exchange between subsystems within an isolated system

The above expression leads to $T_k = T_j$ for reversible processes, which is therefore the thermal-equilibrium criterion. On the other hand, $T_k - T_j > 0$ for irreversible processes, which is in turn the driving force for transformations where thermal energy is transferred. It is worthy noticing that, according to the described transformation, thermal energy transfers from systems at higher temperature to systems at lower temperature. A situation that we recognize as natural again.

Finally, as illustrated in Fig. 2.3, let the whole system be closed to the mass transfer with its surroundings and let temperature and pressure be constant for the whole system and its subsystems during the transformation. However, the mass transfer occurs from the subsystem k to the remaining subsystems j, that is, contrary to the whole system, the subsystems are open systems. Under these conditions, the suitable thermodynamic potential is G. Therefore, Eq. (2.30) is fulfilled for the whole system, while, for the forming subsystems, Eq. (2.26) reduces to

$$dG_j \le \sum_{i=1}^{C} \mu_{ij} dn_{ij} \tag{2.41}$$

Combination of Eqs. (2.30) and (2.41) results in

$$dG = \sum_{j=1}^{\sigma} dG_j \le \sum_{j=1}^{\sigma} \sum_{i}^{C} \mu_{ij} dn_{ij} = \mu_{ik} dn_{ik} - \sum_{j=1}^{\sigma} (1 - \delta_{jk}) \mu_{ij} dn_{ij} \le 0 \tag{2.42}$$

Because mass transfers from subsystem k to the others subsystems, the following mass balance must be satisfied

$$dn_{ik} = -\sum_{j=1}^{\sigma} (1 - \delta_{jk}) dn_{ij} \tag{2.43}$$

Substitution of this mass balance into Eq. (2.42) leads to

$$\sum_{j=1}^{\sigma}\sum_{i=1}^{C}(1-\delta_{jk})(\mu_{ij}-\mu_{ik})dn_{ij} \leq 0 \tag{2.44}$$

According to the hypothetical experiment, $dn_{ij} > 0$ for each subsystem j; therefore, for the above inequality to be satisfied unequivocally, it is concluded that

$$\mu_{ij} - \mu_{ik} \leq 0 \tag{2.45}$$

Similar to the previous hypothetical experiments, $\mu_{ij} = \mu_{ik}$ for reversible transformations and $\mu_{ik} - \mu_{ij} \geq 0$ for irreversible ones. The former corresponds to the chemical-equilibrium criterion, and the latter corresponds to the driving force for the mass transfer to occur between systems. Accordingly, this driving force indicates that a component i transfer from systems with high chemical potential to systems with low chemical potential. Indeed, unlike previous driving forces, we do not recognize $\mu_{ik}-\mu_{ij} > 0$ as the natural driving force for the mass transfer. μ_{ij} measures how fast energy changes because of mass exchanges, and we do not "feel" such variations. However, considering that, we recognize $T_k - T_j > 0$ and $P_k - P_j > 0$ as naturals, and that $\mu_{ik} - \mu_{ij} > 0$ is obtained from the same fundamental laws and from the same procedure, we should accept it as the natural driving force for the mass transfer to occur. This fact has been widely shown in specialized literature [3–5].

In summary, the criteria for the thermal-, mechanical and chemical-equilibrium between σ systems are

$$T_1 = T_2 = \cdots = T_\sigma \tag{2.46}$$

$$P_1 = P_2 = \cdots = P_\sigma \tag{2.47}$$

$$\mu_{i1} = \mu_{i2} = \cdots = \mu_{i\sigma};\ \forall i = 1, 2, \ldots, C \tag{2.48}$$

2.5 Additivity Rules and the Gibbs–Duhem Equation

As been shown before, the fundamental expressions of thermodynamics are differential equations; therefore, they must be integrated in order to be more useful in equilibrium calculations. The integration of such functions can be carried out by using the homogeneous function concept.

A function $f(x_1, x_2, \ldots, x_N)$ is called homogenous, if the following relation holds for any arbitrary number λ

$$f(\lambda x_1, \lambda x_2, \ldots, \lambda x_N) = \lambda^\eta f(x_1, x_2, \ldots, x_N) \tag{2.49}$$

The number η is the order of homogeneity. A homogeneous function can be integrated by means of the Euler's theorem, which states that if f is a homogeneous function of order η, then

$$\eta f(x_1, x_2, \ldots, x_N) = \sum_{i=1}^{N} x_i \left(\frac{\partial f}{\partial x_i}\right)_{x_j;i\neq j} \tag{2.50}$$

The Euler's theorem is important in thermodynamics because the fundamental equation is a homogeneous function of order one ($\eta = 1$), as can be seen from

$$\begin{aligned} d(\lambda U) &= Td(\lambda S) - Pd(\lambda V) + \sum_{i=1}^{C} \mu_i d(\lambda n_i) \\ &= \lambda\left(TdS - PdV + \sum_{i=1}^{C} \mu_i dn_i\right) = \lambda dU \end{aligned} \tag{2.51}$$

The constant number λ may be understood physically as a number that increases the mass of the system. For instance, $\lambda = 2$ means that the system now contains twice as much mass. Therefore, all variables depending on the mass of the system, which are called extensive variables (S, V, and U) are twice as large. Thus, the integration of the fundamental expression can be carried out by means of Eq. (2.50), the result is

$$U = ST - VP + \sum_{i=1}^{C} n_i \mu_i \tag{2.52}$$

By which and by using the corresponding definition, G turns out to be equal to

$$G = U - ST + VP = \sum_{i=1}^{C} n_i \mu_i \tag{2.53}$$

By dividing by the total number of moles n, we have that

$$\tilde{G} \equiv \frac{G}{n} = \sum_{i=1}^{C} x_i \mu_i \tag{2.54}$$

where the mole fraction is $x_i = n_i/n$. Notice that $\tilde{G}_i^0 = \mu_i^0$ for a pure component ($x_i = 1$), which is indicated by the superscript 0. $\tilde{G}$ is the molar Gibbs free energy, which is an intensive property of the system, that is, its value does not rely on the mass contained within the system. The molar property concept is valid for any extensive property of the system; thus, by diving by n, the total volume (V), the total internal energy (U) and the total entropy (S) transform into the molar volume $\tilde{V}$, the molar internal energy $\left(\tilde{U}\right)$ and the molar entropy $\left(\tilde{S}\right)$.

Returning to Eqs. (2.53) and (2.54) they are valid for any kind of system no matter its internal nature; hence, these equations can be written for a system composed of σ phases, such as those found in phase equilibrium calculations of heavy petroleum fluids. These phases may be conceived as those subsystems depicted in the previous section. Accordingly, we have that

$$G = \sum_{j=1}^{\sigma}\sum_{i=1}^{C} n_{ij}\mu_{ij} \tag{2.55}$$

or in terms of the mole fraction, $x_{ij} = n_{ij}/n_j$, for each phase j

$$\tilde{G}= \sum_{j=1}^{\sigma}\alpha_j \sum_{i=1}^{C} x_{ij}\mu_{ij} \tag{2.56}$$

where the phase fraction is $\alpha_j = n_j/n$. These equations are the fundamental expressions to be minimized during the calculation of both physical and chemical equilibriums when temperature and pressure are held constant. As will be shown in four and five chapters, the use of the Gibbs free energy in such calculations will consist in specifying temperature and pressure in order to determine volume and composition of all phases.

Similarly, the integrated form of A arises from Eq. (2.52) again, taking the form of

$$A = U - ST = -PV + \sum_{i=1}^{C} n_i\mu_i \tag{2.57}$$

With the corresponding molar Helmholtz free energy $\left(\tilde{A}\right)$ equal to

$$\tilde{A}= -P\,\tilde{V} + \sum_{i=1}^{C} x_i\mu_i \tag{2.58}$$

For a given reversible transformation where temperature and volume are set as independent variables, Eqs. (2.57) and (2.58) are the suitable expressions to be minimized for finding equilibrium conditions. If the system is formed by σ phases, then the corresponding equations are

$$A = -\sum_{j=1}^{\sigma}\left(P_j V_j + \sum_{i=1}^{C} n_{ij}\mu_{ij}\right) \tag{2.59}$$

$$\tilde{A} = -\sum_{j=1}^{\sigma}\alpha_j\left(P_j V_j + \sum_{i=1}^{C} x_{ij}\mu_{ij}\right) \tag{2.60}$$

In this case, equilibrium calculations will consist in specifying temperature and volume in order to determine pressure and composition of all involved phases. The integrated forms of the thermodynamic potentials are also called additivity rules.

As mentioned, equilibrium calculations involve specifying a set of experimental variables in order to determine the remaining ones. The number of experimental variables to be determined can be reduced because they are connect with each other through the so-called Gibbs–Duhem equation. This equation can be derived by taking the differential equation of the integrated form the internal energy given by Eq. (2.52), that is,

$$dU = SdT + TdS - VdP - PdV + \sum_{i=1}^{C} n_i d\mu_i + \sum_{i=1}^{C} \mu_i dn_i \tag{2.61}$$

Subtraction of the fundamental equation from this equation leads to the desired result

$$0 = SdT - VdP + \sum_{i=1}^{C} n_i d\mu_i \tag{2.62}$$

or in terms of molar properties

$$0 = \tilde{S}\, dT - \tilde{V}\, dP + \sum_{i=1}^{C} x_i d\mu_i \tag{2.63}$$

By knowing $C + 1$ variables, the one left can be calculated from the previous expressions. For instance, if temperature and chemical potential are known, then pressure can be calculated by integration.

2.6 Thermodynamic Potentials from the Ideal Gas Behavior

As can be observed from the previous sections, no assumptions were made about the internal nature of systems in deriving the fundamental expressions of thermodynamics. Therefore, they should be valid irrespective of such an internal nature.

For instance, the system may consist of molecules forming a human body, of human beings forming corporations and societies, or of galaxies forming our universe [6]. Hence, if the thermodynamic concepts and expressions can be applied to so very complex systems, then they can be applied to other complex systems such as heavy petroleum fluids. This general aspect implies a disadvantage, because, before the fundamental expressions can be applied, it is necessary to determine how the variables accounting for the system under study are related to each other. When systems are physical phases such as gases, liquids, and solids, this means that an equation of state must be suggested. In this context, an equation of state is any mathematical expression between pressure, temperature, volume, and composition of a given physical phase.

The simplest equation of state is the ideal gas law, which arises from the combination of the empirical laws of Boyle, and Gay-Lussac concerning the behavior of gases.

$$\frac{P\,\tilde{V}}{T} = R \tag{2.64}$$

The ideal-gas behavior states that the ratio $P\ \tilde{V}$ to T is equal to a constant value R for all gases: the universal gas constant. Any real gas approaches to the ideal-gas behavior when temperature goes to infinite and pressure goes to zero. Under these conditions, the molecules are so separated from each other that their combined volume is negligible in comparison with the total volume occupied by the gas, and besides, they are so scattered among them that they do not interact with each other. As inferred, this is an imaginary situation, because molecules always occupy a finite volume and interact with each other within real gases. However, because of its simplicity, ideal-gas behavior is suitable as starting point to estimate properties and to developed equations of state for any real fluid. For this purpose, parameters correcting the ideal-gas behavior are defined; for example, the compressibility factor Z

$$V = ZV^{ig} = \frac{ZnRT}{P} \tag{2.65}$$

The compressibility factor measures the deviation of the real volume of any fluid from the volume predicted by the ideal gas law; $Z = 1$ for an ideal gas, while $Z \neq 1$ for any real fluid. Therefore, a real fluid can be defined in terms of Z as the fluid for which the following limit is satisfied.

$$\lim_{P \to 0} \frac{Z - 1}{P} \neq 0 \tag{2.66}$$

On the contrary, an ideal can be defined as the fluid for which the ratio $(Z - 1)/P$ is equal to zero at any pressure.

With the ideal gas law as equation of state, changes of state for a gas behaving as an ideal one can be determined from the proper thermodynamic potential. Considering that an ideal gas behavior has been defined in terms of pressure according to Eq. (2.66), the Gibbs free energy is the proper thermodynamic potential to determine such changes. For transformations involving fluid phases, changes associated with the mass transfer between phases are the great significance; hence, these changes should be known as a function of pressure and temperature. For this purpose, we may start with Eq. (2.16), whose partial derivative with respect to the number of moles n_i of component i is

$$\frac{\partial}{\partial n_i}\left[\left(\frac{\partial G}{\partial P}\right)_T\right]_{T,P,n_j,j\neq i} = \left(\frac{\partial V}{\partial n_i}\right)_{T,P,n_j,j\neq i} \tag{2.67}$$

Because G is a function of state, we can change the order of differentiation, that is,

$$\frac{\partial}{\partial P}\left[\left(\frac{\partial G}{\partial n_i}\right)_{T,P,n_j,j\neq i}\right]_T = \left(\frac{\partial V}{\partial n_i}\right)_{T,P,n_j,j\neq i} \tag{2.68}$$

From which, the partial molar volume is defined as

$$\bar{V}_i \equiv \left(\frac{\partial V}{\partial n_i}\right)_{T,P,n_j,j\neq i} \tag{2.69}$$

In general, the partial derivative of any extensive property with respect to n_i is called partial molar property. Using $\bar{V}_i$ and the definition of chemical potential in terms of G according to Eqs. (2.28), (2.68) is expressed as

$$\left(\frac{\partial \mu_i}{\partial P}\right)_T = \bar{V}_i \tag{2.70}$$

This differential equation can be integrated as follows

$$\mu_i = \mu_i^0 + \int_{P^0}^{P_i} \bar{V}_i dP \tag{2.71}$$

where the lower limit corresponds to a Legendre transformation P^0, which is arbitrarily assigned. P^0 is the pressure from which the change of state is evaluated. The need for a reference pressure illustrates that absolute values of thermodynamic potentials are a matter of convenience, because energy and mass exchanges with the surroundings are the only magnitudes that can be measured during changes of state. On the other hand, the upper limit of integration is set equal to the partial pressure

($P_i = x_i P$) of component i within a mixture that obeys the ideal gas law. Finally, μ_i^0 is the chemical potential of pure component i at pressure P^0, whose value is arbitrarily assigned as in the case of P^0 itself. In other words, the integration is carried out from the thermodynamic state of a pure component at P^0 to the thermodynamic state of the same component in an ideal gas mixture, for which $\bar{V}_i$ is given by

$$\bar{V}_i = \tilde{V}_i = \frac{RT}{P} \tag{2.72}$$

This expression shows that $\bar{V}_i$ is equal to $\tilde{V}_i$ for an ideal gas. Thus, for a component i in an ideal gas mixture, the chemical potential as a function of pressure, temperature, and composition is

$$\mu_i^{ig}(T, P, \boldsymbol{x}) = \mu_i^0 + RT\, ln\left(\frac{P_i}{P^0}\right) \tag{2.73}$$

Substitution of the above expression into Eq. (2.54) leads to

$$\frac{\tilde{G}^{ig}}{RT} = ln\left(\frac{P}{P^0}\right) + \sum_{i=1}^{C} x_i\left[\frac{\mu_i^0}{RT} + ln(x_i)\right] \tag{2.74}$$

which is the molar Gibbs free energy of a mixture that behaves as an ideal gas. Because of its relative simplicity, Eq. (2.73) has been widely used as starting point for estimating the chemical potential in real fluids. This means that the ideal gas behavior can be used as fluid reference in order to estimate properties for real fluids.

The simplest way to adapt Eq. (2.73) to real fluids is by assuming that there is an analogous property to the partial pressure P_i so that the actual chemical potential can be predicted. The analogous property is called fugacity f_i of component i and is defined through

$$\mu_i(T, P, \boldsymbol{x}) = \mu_i^0 + RTln\left(\frac{f_i}{P^0}\right) \tag{2.75}$$

Although invented, the fugacity f_i may be conceived as a corrected pressure because of the deviations shown by real fluids with respect to the ideal-gas model.

Once P^0 and μ_i^0 have been established, μ_i is unreachable unless f_i can be calculated. For this purpose, it is advisable to define the residual chemical potential $\mu_i^R(T, P, \boldsymbol{x})$, which accounts for the difference between the chemical potential of any real fluid and the chemical potential of the same fluid behaving as an ideal gas; that is, subtraction of Eq. (2.73) from Eq. (2.75) gives

$$\frac{\mu_i^R(T, P, \boldsymbol{x})}{RT} \equiv \frac{\mu_i - \mu_i^{ig}}{RT} = ln(\phi_i) \tag{2.76}$$

ϕ_i is the so-called fugacity coefficient of component i and is defined as

$$\phi_i \equiv \frac{f_i}{x_i P} \tag{2.77}$$

As observed from Eq. (2.76), μ_i can be now determined as long as μ_i^{ig} and ϕ_i can be found. μ_i^{ig} is obtained from Eq. (2.73), but ϕ_i is still missing. To find ϕ_i, it is worthy reminding that any real fluid approaches to an ideal gas behavior when pressure goes to zero. Accordingly, from Eq. (2.76), it is concluded that

$$\lim_{P \to 0} \phi_i = 1 \tag{2.78}$$

Similar to the compressibility factor Z, ϕ_i is a measurement of the non-ideal behavior of any real fluid but in terms of energy and not in terms of volume as in the case of Z. The fugacity f_i^0 and the corresponding coefficient ϕ_i^0 of a pure component i are obtained by Eq. (2.77) from

$$\lim_{x_i \to 1} \phi_i = \phi_i^0 = \frac{f_i^0}{P} \tag{2.79}$$

To defined ϕ_i, we made use of the concept of residual property, which is valid for any function of state. For instance, the molar residual Gibbs free energy $\tilde{G}^R$ is defined as

$$\frac{\tilde{G}^R}{RT} \equiv \frac{\tilde{G}}{RT} - \frac{\tilde{G}^{ig}}{RT} \tag{2.80}$$

In other words, any residual property is the difference between the actual property and the same property but as an ideal gas. Actually, the definition given by Eq. (2.80) is useful to determine $\tilde{G}$ instead of $\tilde{G}^R$ as long as $\tilde{G}^R$ and $\tilde{G}^{ig}$ can be calculated. Equation (2.74) gives $\tilde{G}^{ig}$ and Eq. (2.53) allows $\tilde{G}^R$ to be determined because

$$\frac{\tilde{G}^R}{RT} = \sum_{i=1}^{C} x_i \mu_i^R \tag{2.81}$$

The result of substitution of Eq. (2.76) is

$$\frac{\tilde{G}^R}{RT} = ln(\phi) = \sum_{i=1}^{C} x_i ln(\phi_i) \tag{2.82}$$

Using Eqs. (2.74) and (2.82), the calculation of the actual molar Gibbs free energy $\tilde{G}$ is now possible as long as the reference state (μ_i^0 and P^0) is established and ϕ_i can be calculated for the real fluid under study. In conclusion, μ_i and $\tilde{G}$ can be calculated as long as ϕ_i can be found for the specific fluid under study.

For the purpose of finding ϕ_i, Eq. (2.70) is used again as starting point, but in this case an ideal gas is no longer assumed. Thus, a practical expression for ϕ_i can be derived from Eq. (2.70), but it is now combined with Eq. (2.75) to give

$$\frac{1}{RT}\left(\frac{\partial \mu_i}{\partial P}\right)_T = \left[\frac{\partial ln(f_i)}{\partial P}\right]_T = \frac{\bar{V}_i}{RT} \tag{2.83}$$

By introducing the fugacity coefficient definition according to Eq. (2.77) we have that

$$\left[\frac{\partial ln(x_i \phi_i P)}{\partial P}\right]_T = \frac{\bar{V}_i}{RT} \tag{2.84}$$

When the term $[\partial ln(V/RT)/\partial P]_T$ is added on both sides and the compressibility factor definition according to Eq. (2.65) is used, the above equation transforms into

$$\left[\frac{\partial ln(\phi_i)}{\partial P}\right]_T + \left[\frac{\partial ln(n_i Z)}{\partial P}\right]_T = \frac{\bar{V}_i}{RT} + \frac{1}{V}\left(\frac{\partial V}{\partial P}\right)_T \tag{2.85}$$

which can be integrated from ideal-gas conditions to arbitrary real-fluid conditions as follows:

$$\int_1^{\phi_i} d[ln(\phi_i)] + \int_1^{Z} d[ln(n_i Z)] = \frac{1}{RT}\int_0^{P}\left[\bar{V}_i + \frac{RT}{V}\left(\frac{\partial V}{\partial P}\right)_T\right]dP \tag{2.86}$$

Remembering that real fluids approach to ideal gas behavior when $P \rightarrow 0$, where by definition $\phi_i = 1$ and $Z = 1$. Finally, this expression takes the form of

$$ln(\phi_i) = \frac{1}{RT}\int_0^{P}\left[\bar{V}_i + \frac{RT}{V}\left(\frac{\partial V}{\partial P}\right)_T\right]dP - ln(Z) \tag{2.87}$$

This expression allows ϕ_i to be calculated as long as an equation of state or experimental data are available with pressure and temperature as independent variables. However, when the available information is function of volume and temperature, the Helmholtz free energy A and its derived expressions are more suitable than G. Hence, it is advisable to obtain expressions similar to those obtained previously, but in terms of volume, temperature, and composition. To this end, we proceed similarly, but we start now with Eq. (2.12), which is derived with respect to n_i, and by changing the order of differentiation, we find that

$$\frac{\partial}{\partial V}\left[\left(\frac{\partial A}{\partial n_i}\right)_{T,V,n_j,j\neq i}\right]_T = -\left(\frac{\partial P}{\partial n_i}\right)_{T,V,n_j,j\neq i} \tag{2.88}$$

The definition of μ_i as function of volume and pressure according to Eq. (2.28) transforms this partial derivative into

$$\left(\frac{\partial \mu_i}{\partial V}\right)_T = -\left(\frac{\partial P}{\partial n_i}\right)_{T,V,n_j,j\neq i} \tag{2.89}$$

which can be integrated from volume and chemical potential of pure component V_i^0 and μ_i^0 to arbitrary volume and chemical potential V and μ_i of the same component but in a mixture, that is,

$$\mu_i = \mu_i^0 - \int_{V_i^0}^{V}\left(\frac{\partial P}{\partial n_i}\right)_{T,V,n_j,j\neq i} dV \tag{2.90}$$

This expression is general; therefore, it remains valid for any fluid as long as the corresponding partial derivative can be found. For simplicity, in the case of Eq. (2.90), it is advisable to use the hypothetical ideal gas concept [2]. The idea behind this concept is that we can use the ideal gas law to represent the behavior of real fluids; that is, the limit from Eq. (2.66) is fulfilled always, even though the ideal gas law is used. We may recognize this situation as logical, because the ideal gas law arose from experimental observations at low pressure performed by Boyle and Gay-Lussac. From a mathematical point of view, the hypothetical ideal gas concept implies that $Z \neq 1$ even though the ideal gas law is used. Thus, for a hypothetical ideal gas, we have that

$$\left(\frac{\partial P}{\partial n_i}\right)_{T,V,n_j,j\neq i} = \frac{RT}{V} \tag{2.91}$$

with which, Eq. (2.90) is integrated, resulting in the expression of the chemical potential of a hypothetical ideal gas $\mu_i^{hg}(T, V, \boldsymbol{x})$ in terms of temperature, volume and composition

$$\mu_i^{hg}(T, V, \boldsymbol{x}) = \mu_i^0 - RTln\left(\frac{V}{V_i^0}\right) = \mu_i^0 + RTln\left(\frac{x_i RT}{P^0\,\tilde{V}}\right) \tag{2.92}$$

This expression allows the molar Helmholtz free energy of a hypothetical gas $\tilde{A}^{hg}$ to be determined by substitution into Eq. (2.58). The result is

$$\frac{\tilde{A}^{hg}}{RT} = ln\left(\frac{RT}{P^0\,\tilde{V}}\right) - \frac{P\,\tilde{V}}{RT} + \sum_{i=1}^{C} x_i\left[\frac{\mu_i^0}{RT} + ln(x_i)\right] \tag{2.93}$$

$\tilde{A}^{hg}$ can be expressed in terms of $\tilde{G}^{ig}$ by subtraction of Eq. (2.74) from Eq. (2.93).

$$\frac{\tilde{A}^{hg}}{RT} = \frac{\tilde{G}^{ig}}{RT} + ln\left(\frac{RT}{P\,\tilde{V}}\right) - \frac{P\,\tilde{V}}{RT} \tag{2.94}$$

or in terms of Z

$$\frac{\tilde{A}^{hg}}{RT} = \frac{\tilde{G}^{ig}}{RT} + ln(Z) - Z \tag{2.95}$$

Notice that Eqs. (2.94) and (2.95) connect a property of a real fluid $\tilde{A}^{hg}$ with a property of an ideal gas $\tilde{G}^{ig}$. A hypothetical ideal gas becomes an ideal one when $Z = 1$; therefore, from Eq. (2.95) we have that

$$\frac{\tilde{A}^{ig}}{RT} = \frac{\tilde{G}^{ig}}{RT} - 1 \tag{2.96}$$

Similarly, the difference between the chemical potentials comes from subtraction of Eq. (2.92) from Eq. (2.73)

$$\frac{\mu_i^{ig}(T, P, \boldsymbol{x}) - \mu_i^{hg}(T, V, \boldsymbol{x})}{RT} = ln\left(\frac{P\,\tilde{V}}{RT}\right) = ln(Z) \tag{2.97}$$

This equation can be written in terms of the residual chemical $\left(\mu_i^R = \mu_i - \mu_i^{hg}\right)$ by using Eq. (2.76)

$$\frac{\mu_i^R(T, V, \boldsymbol{x})}{RT} = ln(\phi_i) + ln(Z) \tag{2.98}$$

At this point, it is worthy verifying that chemical potentials are equal irrespective of the used independent variables for describing the transformation, as stated by Eq. (2.28). Thus, by combining Eqs. (2.92) and (2.98), we found that

$$\frac{\mu_i(T,V,\boldsymbol{x})-\mu_i^0}{RT}=ln\left(\frac{x_iRT}{P^0\tilde{V}}\right)+ln(\phi_i)+ln(Z)=ln\left(\frac{x_i\phi_iP}{P^0}\right)$$
$$=\frac{\mu_i(T,P,\boldsymbol{x})-\mu_i^0}{RT} \quad (2.99)$$

Once $\mu_i^R(T,V,\boldsymbol{x})$ is given, then the residual molar Helmholtz free energy $\tilde{A}^R$ can be calculated from Eq. (2.57)

$$\frac{\tilde{A}^R}{RT}=-\frac{P\tilde{V}^R}{RT}+\sum_{i=1}^{C}x_i\frac{\mu_i^R}{RT} \quad (2.100)$$

where $\tilde{A}^R$ and the residual molar volume $\tilde{V}^R$ are defined as

$$\frac{\tilde{A}^R}{RT}\equiv\frac{\tilde{A}}{RT}-\frac{\tilde{A}^{hg}}{RT} \quad (2.101)$$

$$\frac{\tilde{V}^R}{RT}\equiv\frac{\tilde{V}}{RT}-\frac{\tilde{V}^{ig}}{RT}=\frac{Z-1}{P} \quad (2.102)$$

Finally, substitution of Eqs. (2.98) and (2.102) into Eq. (2.100) gives

$$\frac{\tilde{A}^R}{RT}=1-Z+ln(Z)+\sum_{i=1}^{C}x_iln(\phi_i) \quad (2.103)$$

Which, together with Eq. (2.93), allows $\tilde{A}$ to be calculated from its definition given by Eq. (2.101). This implies again that the reference conditions and ϕ_i as a function of temperature, volume, and composition have to be known. In other words, an expression analogous to Eq. (2.87) is needed. For this purpose, we can combined Eq. (2.75) with Eq. (2.89)

$$\frac{1}{RT}\left(\frac{\partial\mu_i}{\partial V}\right)_T=\left[\frac{\partial ln(f_i)}{\partial V}\right]_T=-\frac{1}{RT}\left(\frac{\partial P}{\partial n_i}\right)_{T,V,n_j,j\neq i} \quad (2.104)$$

Notice that Eq. (2.89), which was used to introduce the fugacity, remains valid even in this case. In terms of the fugacity coefficient, the above equations takes the form of

$$\left[\frac{\partial ln(x_i\phi_iP)}{\partial V}\right]_T=-\frac{1}{RT}\left(\frac{\partial P}{\partial n_i}\right)_{T,V,n_j,j\neq i} \quad (2.105)$$

Integration from the ideal gas behavior ($V \rightarrow \infty$) to any arbitrary state allows the desired expression to be obtained. Thus, following similar steps in deriving Eq. (2.87), we arrive at

$$ln(\phi_i) = \frac{1}{RT}\int_{\infty}^{V}\left[\frac{RT}{V} - \left(\frac{\partial P}{\partial n_i}\right)_{T,V,n_j,j\neq i}\right]dV - ln(Z) \qquad (2.106)$$

Now, if information relating pressure to temperature, volume and composition is available, then the fugacity coefficient can be calculated. This situation occurs when the cubic equations of state or the statistical-associating-fluid theory equations of state are intended to be used for characterizing the phase behavior of hydrocarbons and their mixtures. The main features of these equations will be outlined in future chapters.

2.7 Thermodynamic Potentials from the Ideal Solution Behavior

In the previous section, the ideal gas concept was used as reference fluid in order to determine absolute values of the thermodynamic potentials through the fugacity and fugacity coefficient. However, the ideal gas is not the only reference fluid that can be used in order to determine such absolute values. Another reference fluid is the ideal solution, which is widely used when one intends to estimate properties of mixtures. An ideal solution concept assumed that molecular interactions between components in mixture are similar to those interactions at the pure state of each component. This means that molecules of a given component do not "feel" any difference in their interactions within an ideal solution regarding their interactions as pure component. From a physical point of view, a mixture approaches to the ideal-solution behavior when mixtures are dilute ($x_i \rightarrow 1$) or components forming mixtures are similar with each other from a chemical point of view, that is, their molecular interactions are similar. For instance, the so-called light and heavy petroleum are both mixtures formed by hydrocarbons, but the light one has more paraffinic and naphthenic components than aromatic ones, whereas the heavy one has more aromatic components than paraffinic and naphthenic ones. In an ideal mixture of both, one assumes that molecular interactions are not going to change with respect to those present in the original petroleum. Thus, if one combines 1 m^3 of light petroleum with 1 m^3 of heavy petroleum, 2 m^3 of mixture must be obtained. However, if the mixture does not behave as an ideal solution, the result is going to be different, as depicted in Fig. 2.4. Furthermore, the combination of two petroleum so different may result in phase changes, leading to the formation of organic solids such as asphaltenes and waxes [7, 8]. Except for the latter case, it is usual assumed that petroleum and its mixtures behaves like an ideal solutions. Finally, it is worthy pointing out that, unlike ideal gas, ideal solutions are

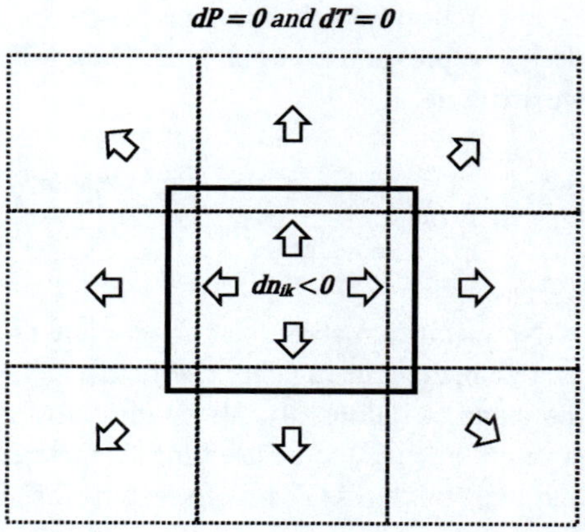

Fig. 2.4 Mass transfer exchange between subsystems within a system keeping constant temperature and pressure

formed by real phases and not by imaginary ones; consequently, gases, liquids and even solids can form ideal solutions.

Now we proceed to derive the expressions for thermodynamic potentials using the ideal solution model as reference fluid. To this end, we obtain the chemical potential of component i in an ideal solution of ideal gases from Eq. (2.73)

$$\frac{\mu_i^{is}(T, x)}{RT} = \frac{\mu_i^0}{RT} + ln(x_i) \tag{2.107}$$

where it is assumed that pressure do not change during the mixing processes, that is, $P = P_0$. Although derived from the ideal gas model, this equation is used as the mathematical definition of an ideal solution regardless of the type of fluids involved during the mixing process. It is assumed that it contains the simplest composition dependence as result of the increase of entropy due to the random mixing of components [9]. Indeed, μ_i^0 no longer refers to the chemical potential of a pure component as an ideal gas, but it should refers to the chemical potential of a pure component in its real fluid. Using the definition given by Eq. (2.107), the residual molar Gibbs free energy of an ideal solution $\tilde{G}^{is}$ can be obtained by means of Eq. (2.54) as follows:

$$\frac{\tilde{G}^{is}}{RT} = \sum_{i=1}^{C} x_i \left(\frac{\mu_i^0}{RT} \right) + \sum_{i=1}^{C} x_i ln(x_i) \tag{2.108}$$

This equation gives the absolute value of $\tilde{G}^{is}$, but it does not give the change of the Gibbs free energy during a mixing process. The mixing properties are understood as the difference between the property of the whole mixture and the average property from the pure components, that is,

$$\frac{\Delta \tilde{G}^{is}_{mix}}{RT} \equiv \frac{\tilde{G}^{is}}{RT} - \sum_{i=1}^{C} x_i \left(\frac{\tilde{G}^{0}_i}{RT} \right) = \sum_{i=1}^{C} x_i ln(x_i) < 0 \tag{2.109}$$

where $\tilde{G}^{0}_i = \mu_i^0$ by definition. Because $x_i < 1$, $\Delta \tilde{G}^{is}_{mix}$ is always negative. This means that ideal mixing processes occur in a spontaneous way always when pressure and temperature are kept constant. For instance, because all hydrocarbons form in general ideal solutions, they mix spontaneously.

When the ideal gas is used as reference fluid, absolute values of the thermodynamic properties are calculated from the corresponding residual properties. In the case of ideal solution, the analogous properties are the so-called excess properties, which are defined as the difference between the actual property of the mixture and the same property but as ideal solution. Thus, for the excess molar Gibbs free energy $\tilde{G}^{E}$, we have that

$$\frac{\tilde{G}^{E}}{RT} \equiv \frac{\tilde{G}}{RT} - \frac{\tilde{G}^{is}}{RT} = \sum_{i=1}^{C} x_i \left(\frac{\mu_i - \mu_i^0}{RT} \right) - \sum_{i=1}^{C} x_i ln(x_i) \tag{2.110}$$

Regarding the value of μ_i, Eq. (2.75) is used again, because the fugacity concept is not limited to gases; it is valid for any physical state such as liquid and solids. However, in this case, P^0 is changed by a reference fugacity of the pure component f_i^0, which accounts for the fact the reference fluid is no longer an ideal gas; it is the real physical state of the pure component at the mixing pressure and temperature. Therefore, Eq. (2.75) can be rewritten as

$$\mu_i(T, P, \boldsymbol{x}) = \mu_i^0 + RTln\left(\frac{f_i}{f_i^0} \right) \tag{2.111}$$

which can be substituted into Eq. (2.110) to give

$$\frac{\tilde{G}^{E}}{RT} = \sum_{i=1}^{C} x_i ln\left(\frac{f_i}{x_i f_i^0} \right) = \sum_{i=1}^{C} x_i ln(\gamma_i) \tag{2.112}$$

Therefore, the activity coefficient γ_i is defined as

$$\gamma_i \equiv \frac{f_i}{x_i f_i^0} \tag{2.113}$$

From Eq. (2.112), it is concluded that in the limit of dilute solution

$$\lim_{x_i \to 1} \gamma_i = 1 \tag{2.114}$$

Therefore, $\gamma_i = 1$ when an ideal solution is present. Otherwise $\gamma_i \neq 1$. Similarly to ϕ_i, γ_i is a measurement of non-ideality of the system, but in this case, the non-ideality is measured with respect to the ideal solution concept, and not with respect to the ideal gas law. This means that a given solution can be an ideal one, but at the same time, it can be a real fluid from the ideal gas point of view. In other words, all ideal gases form ideal solutions, but not all ideal solutions are formed by ideal gases. The general expression for ideal solutions is obtained from Eq. (2.113) by setting $\gamma_i = 1$

$$f_i^{is} = x_i f_i^0 \tag{2.115}$$

This equation is the Lewis-Randall rule, which is applied to each component of an ideal solution. Finally, the combination of Eqs. (2.108), (2.110), and (2.112) results in

$$\frac{\tilde{G}}{RT} = \sum_{i=1}^{C} x_i \left(\frac{\mu_i^0}{RT} \right) + \sum_{i=1}^{C} x_i ln(x_i \gamma_i) \tag{2.116}$$

This equation allows the absolute value of $\tilde{G}$ to be calculated by using the ideal solution model as reference fluid. Another important non-ideality measurement can be obtained from this equation: the so-called activity a_i, which can be defined as $a_i \equiv x_i \gamma_i$. As can be seen, $a_i = x_i$ when the component i is within an ideal solution ($\gamma_i = 1$). Otherwise $a_i \neq x_i$.

Although γ_i and ϕ_i measures the non-ideal behavior with respect to different referent fluids, they are related to each other because they share the same fugacity from their definitions. Thus, the relationship connecting both coefficients arises from the combination of Eq. (2.77) with Eq. (2.113), taking the form of

$$\gamma_i = \frac{\phi_i}{\phi_i^0} \tag{2.117}$$

An analogous relationship exists between $\tilde{G}^R$ and $\tilde{G}^E$, which is obtained when the above relationship is introduced into Eq. (2.82)

$$\frac{\tilde{G}^R}{RT} = \frac{\tilde{G}^E}{RT} + \sum_{i=1}^{C} x_i ln(\phi_i^0) \tag{2.118}$$

Equations (2.117) and (2.118) show that we do not have to deal with several references fluids in order to calculate thermodynamic properties. For instance, we can choose the ideal gas model as reference fluid and determine the mixing properties from these equations.

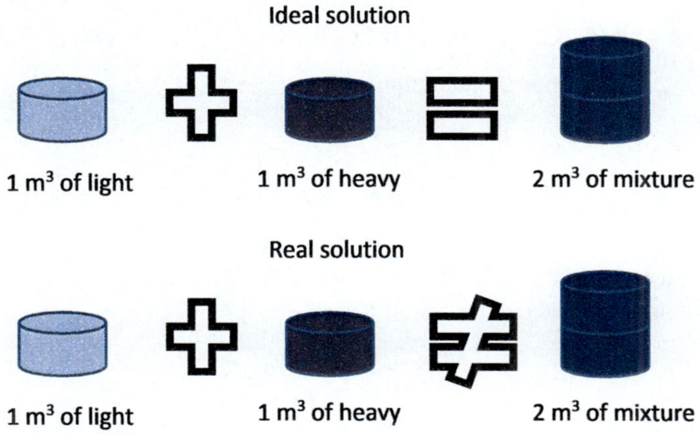

Fig. 2.5 Ideal solution versus real solution regarding mixture volume

Analogous to the derivations for the fugacity confident, when volume and temperature are held fixed, G is changed by A as the proper thermodynamic potential for describing the mixing process. In the case, by constant volume we mean

$$\tilde{V}^{is} = \sum_{i=1}^{C} x_i \tilde{V}_i^0 \tag{2.119}$$

which is the mathematical expression of the ideal mixing process illustrated in Fig. 2.5, but in terms of n_i. Equation (2.119) implies that Eq. (2.92) becomes

$$\frac{\mu_i^{is}(T, V, \boldsymbol{x})}{RT} = \frac{\mu_i^0}{RT} - ln\left(\frac{V^{is}}{V_i^0}\right) \tag{2.120}$$

Here again, the chemical potential is used to determine the corresponding thermodynamic potential. As can be observed from rewriting Eq. (2.58) for ideal solution

$$\frac{\tilde{A}^{is}}{RT} = -\frac{P\tilde{V}^{is}}{RT} + \sum_{i=1}^{C} x_i \frac{\mu_i^{is}}{RT} \tag{2.121}$$

Introducing Eq. (2.120) into this equation, the result is

$$\frac{\tilde{A}^{is}}{RT} = -\frac{P\tilde{V}^{is}}{RT} + \sum_{i=1}^{C} x_i \left(\frac{\mu_i^0}{RT}\right) - \sum_{i=1}^{C} x_i ln\left(\frac{V^{is}}{V_i^0}\right) \tag{2.122}$$

As in the case of $\Delta \overset{\sim is}{G}_{mix}$, the change of the Helmholtz free energy for an ideal mixing $\Delta \overset{\sim is}{A}_{mix}$ is obtained by definition as

$$\frac{\Delta \overset{\sim is}{A}_{mix}}{RT} \equiv \frac{\overset{\sim is}{A}}{RT} - \sum_{i=1}^{C} x_i \left(\frac{\overset{\sim 0}{A}_i}{RT} \right) \tag{2.123}$$

Substitution of Eqs. (2.119), (2.122), and (2.58) written for pure components into the above definition leads to

$$\frac{\Delta \overset{\sim is}{A}_{mix}}{RT} = \sum_{i=1}^{C} x_i ln\left(\frac{V_i^0}{V^{is}} \right) < 0 \tag{2.124}$$

Here again, $\Delta \overset{\sim is}{A}_{mix}$ is always negative due to the condition given by Eq. (2.119) meaning that the ideal mixing process is always spontaneous at constant volume and temperature as well.

Concerning the excess molar Helmholtz free energy $\overset{\sim E}{A}$, it is defined as the previous excess properties discussed before, that is,

$$\frac{\overset{\sim E}{A}}{RT} \equiv \frac{\tilde{A}}{RT} - \frac{\overset{\sim is}{A}}{RT} \tag{2.125}$$

By means of Eqs. (2.119) and (2.122), it can be written as

$$\frac{\overset{\sim E}{A}}{RT} = -\frac{P\overset{\sim E}{V}}{RT} + \sum_{i=1}^{C} x_i \left(\frac{\mu_i - \mu_i^0}{RT} \right) + \sum_{i=1}^{C} x_i ln\left(\frac{V^{is}}{V_i^0} \right) \tag{2.126}$$

where $\overset{\sim E}{V}$ stands for the excess molar volume

$$\frac{\overset{\sim E}{V}}{RT} \equiv \frac{\tilde{V}}{RT} - \frac{\overset{\sim is}{V}}{RT} = \frac{\tilde{V}}{RT} - \sum_{i=1}^{C} x_i \frac{\overset{\sim 0}{V}_i}{RT} \tag{2.127}$$

Using Eqs. (2.111), (2.126) can be rewritten in terms of the fugacities rather than chemical potentials in the following way

$$\frac{\overset{\sim E}{A}}{RT} = -\frac{P\overset{\sim E}{V}}{RT} + \sum_{i=1}^{C} x_i ln\left(\frac{f_i}{f_i^0} \right) + \sum_{i=1}^{C} x_i ln\left(\frac{V^{is}}{V_i^0} \right) \tag{2.128}$$

By adding and subtracting the term $\sum_{i=1}^{C} x_i ln(x_i)$, we found that

$$\frac{\tilde{A}^E}{RT} = -\frac{P\tilde{V}^E}{RT} + \sum_{i=1}^{C} x_i ln\left(\frac{f_i}{x_i f_i^0}\right) + \sum_{i=1}^{C} x_i ln\left(\frac{x_i V^{is}}{V_i^0}\right) \tag{2.129}$$

If the ratio $f_i / x_i f_i^0$ is replaced γ_i according to Eq. (2.113) then

$$\frac{\tilde{A}^E}{RT} = -\frac{P\tilde{V}^E}{RT} + \sum_{i=1}^{C} x_i ln(\gamma_i) + \sum_{i=1}^{C} x_i ln\left(\frac{x_i V^{is}}{V_i^0}\right) \tag{2.130}$$

or

$$\frac{\tilde{A}^E}{RT} = -\frac{P\tilde{V}^E}{RT} + \frac{\tilde{G}^E}{RT} + \sum_{i=1}^{C} x_i ln\left(\frac{\tilde{V}^{is}}{\tilde{V}_i^0}\right) \tag{2.131}$$

Because the ideal solution can be a real fluid with respect to the ideal gas law, then the above expression can be written in terms of Z by assuming that $\tilde{V}^E = 0$ as follows:

$$\frac{\tilde{A}^E}{RT} = \frac{\tilde{G}^E}{RT} + \sum_{i=1}^{C} x_i ln\left(\frac{Z}{Z_i}\right) \tag{2.132}$$

where the compressibility factor for a pure component i is

$$Z_i = \frac{P\tilde{V}_i^0}{RT} \tag{2.133}$$

Excess properties are not usual found in phase equilibrium calculations involving petroleum mixtures, because on the one hand, it is assumed that they form ideal solutions as already mentioned, and on the other hand, the cubic equations of state, which are the most widely used in such calculations, are derived from using the ideal gas model as reference fluid. Besides, the solution models, which uses ideal solution model as reference fluid, requires in general to know the exact composition and chemical structure of the components forming mixtures. From a practical and economic point of view, it is not possible to know in petroleum mixtures such information accurately. However, it will be shown and illustrated in futures chapters that chemical structures can be suggested for petroleum fractions, with which solution models are available for being used in solving equilibrium problems involving petroleum mixtures.

References

1. Castellan WG (1983) Physical chemistry, 3rd edn. Addison-Wesley, New York, USA
2. Michelsen ML, Mollerup J (2007) Thermodynamic models: fundamentals & computational aspects, 2nd edn. TIE-LINE, Holte, Denmark
3. Taylor R, Krishna R (1993) Multicomponent mass transfer. Wiley, New York, USA
4. Wesselingh JA, Krishna R (2000) Mass transfer in multicomponent mixtures. Delft Univerty Press, Delft, The Netherlands
5. Krishna R, Wesselingh JA (1997) The Maxwell-Stefan approach to mass transfer. Chem Eng Sci 52(6):861–911
6. Hershey D (2010) Entropy theory of aging systems: humans. Corporations and the Universe. Imperial College Press, London, UK
7. Gonzalez DL, Hirasaki GJ, Creek J, Chapman WG (2007) Modeling of asphaltene precipitation due to changes in composition using the perturbed chain statistical associating fluid theory equation of state. Energy Fuels 21(3):1231–1242
8. Rashid Z, Wilfred CD, Gnanasundaram N, Arunagiri A, Murugesan T (2018) A comprehensive review on the recent advances on the petroleum asphaltene aggregation. J Pet Sci Eng 2019(176):249–268
9. Smith JM, Van Ness HC, Abbott MM, Swihart MT (2018) Introduccion to chemical engineering thermodynamics, 8th edn. McGraw-Hill, New York, USA

Chapter 3
Cubic Equations of State

List of Abbreviations

CEoS	Cubic Equation of State
GCVOL	Group-Contribution-Volume
HR	Huron Vidal
MHV1	Modified Huron-Vidal of order one
MHV2	Modified Huron-Vidal of order two
NRTL	Non-Random-Two-Liquid
PSRK	Predictive SRK equation of state
UMR	Universal mixing rule
UNIFAC	Universal quasi-chemical functional group activity coefficient
VTPR	Volume translated Peng Robinson

Nomenclature

(Z_1)	Roots for vapor phase (largest)
(Z_2)	Roots for liquid phase (smallest)
a	Attractive forces between molecules
$\tilde{A}^E$	Helmholtz free energy molar excess
A, B	Dimensionless numbers (Eqs. (3.4–3.6))
a_i'	Partial derivatives (Eq. (3.33))
A_{mc}	CEoS parameters
a_{mn}	Interaction parameters for PSRK
b	Co-volume
B	Dimensionless co-volume
b_i'	Partial derivatives (Eq. (3.34))
b_{mn}	Interaction parameters for PSRK

B. Carreón-Calderón et al., *Thermophysical Properties of Heavy Petroleum Fluids*, Petroleum Engineering, https://doi.org/10.1007/978-3-030-58831-1_3

C	Components
c_1, c_2, c_3	Coefficients (Eq. (3.3))
c_{mn}	Interaction parameters for PSRK
D	Discriminant
E_{ij}	Binary interaction parameter Peneloux and Abdoul
F	Function
f_i	Fugacity
f_{il}	Occurrence fraction of functional group
k_0, k_1, k_2, k_4	Constants (Eq. (3.48))
k_{ij}	Binary interaction parameters
M	Molecular weight
Q	Numbers of the formula of Cardiano
q	Function (Eq. (3.73))
q_1	MHV1 parameter
q_i	Pure-component parameter
Q_k	Relative group area
r_i	Pure-component parameter
R_k	Relative group volume
S	Real root
s	Volume translation of component
S_j	Specific gravity
T	Real root
T_{bj}	Normal boiling temperature
U	Numbers of the formula of Cardiano
u, v	Parameters of Cubic Equations of State
x_i	Molar fraction
X_m	Group mole fraction
α	CEoS parameters
α_f	Empirical dimensionless function
$\alpha_{ij}, g_{ij}, g_{jj}$	Binary interaction parameters (Eq. (3.62))
β	CEoS parameters
γ_i	Activity coefficient
γ_i^{comb}	Combinatorial contribution
γ_i^{FH}	Flory-Huggins contribution
γ_i^{resi}	Residual contribution
γ_i^{SG}	Staverman-Guggenheim contribution
Γ_k	Activity coefficient corresponding to the functional group k
δ_i	Relationship beteewn a_i, b_i
ϵ	CEoS-dependent parameter
ϵ^*	Parameter (Eq. (3.84))
η	Packing fraction
η_{mc}	CEoS parameters
Θ_i	Molecular surface fraction
θ_m	Area fraction
ρ	Density

σ	Molecular diameter
υ_{ki}	Number of times (occurrence)
φ	CEoS-variable
Φ_i	Molecular volume fraction
ϕ_i	Fugacity coefficient
χ_{kl}, ψ_{kl}	Contributions for the binary interaction
Ψ_{mn}	Binary main group interaction parameters
ω	Acentric factor
Ω_a, Ω_b	CEoS-dependent coefficients

Superscripts

$\sim$	Molar
0	Pure component
∞	Infinite
C_{1i}, C_{2i}	Coefficients of function of the molecular weight and acentric factor
d_{jk}	Correlation parameters
E	Excess property
Exp	Experimental value
ext	Extrapolated
ig	Ideal gas
is	Ideal solution
l	Liquid
R	Real
R	Residual
$resi$	Residual (Eq. (3.108))
v	Vapor
ι_i, λ_j, χ_j	Parameters of GCVOL
$\hat{\rho}$	Mass density

In the previous Chapter, the thermodynamic potentials and the equilibrium criteria were obtained as a function of the fugacity and activity coefficients. In turn, the fugacity coefficient was expressed as a function of the two sets of measurable variables. The former consists of temperature, pressure and composition, while the latter consists of temperature, volume and composition. In either case, an equation of state is needed in order to obtain practical expressions. In this manuscript, two equations of state will be outlined, emphasizing applications to hydrocarbon mixtures in general and heavy petroleum fluids in particular.

The first equation of state that will be addressed is the so-called Cubic Equations of State (CEoS). These equations connect pressure (P), temperature (T) and molar volume ($\tilde{V}$) with each other of the fluid by means of the generalized form

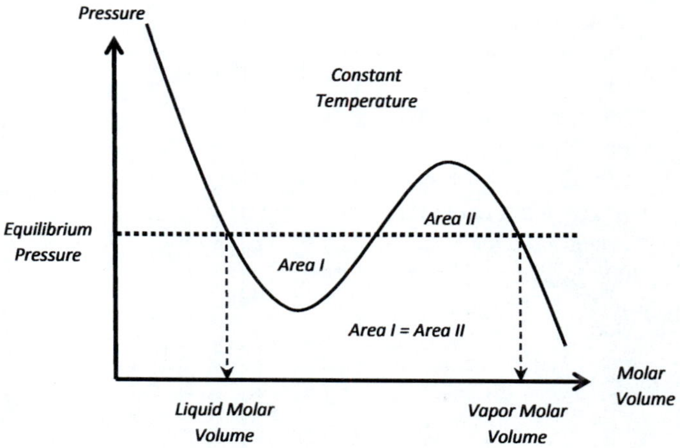

Fig. 3.1 Isothermal of Cubic Equations of State indicating the equilibrium pressure (Area I = Area II) and the corresponding vapor and liquid molar volumes $\left(\tilde{V}^l < \tilde{V}^v\right)$

$$P = \frac{RT}{\tilde{V} - b} - \frac{a}{\tilde{V}^2 + ub\tilde{V} + vb^2} \tag{3.1}$$

R is universal gas constant and is equal to 8.3144725 J/mol K in the international unit system. The parameters u and v give rise to several CEoS types. For instance, when $u = 0$ and $v = 0$, we obtain the famous Van der Walls equation of state. In contrast to the ideal gas law discussed in the Chap. 2, Van der Waals was the first to recognize the influence of molecular size and intermolecular forces on the behavior of fluids. In other words, his equation was the first equation describing the behavior of real fluids. The Van der Waals equation was the first CEoS as well. These are called cubic because they are cubic polynomials with respect to $\tilde{V}$, as illustrated in Fig. 3.1. As observed, the equilibrium or saturation pressure between liquid and vapor phases is the pressure for which both indicated areas are equal to each other. Because work is given by $W = \int Pd\tilde{V}$, these equality between areas may be understood as equality between works. Therefore, the equilibrium pressure corresponds to that pressure where the required works for each phase to exist are equal to each other. In Fig. 3.1, the liquid molar volume corresponds to the smallest root from CEoS, while the vapor molar volume corresponds to the largest root from CEoS. The middle root has no physical meaning. The distinction between liquid and vapor or gas is not always clear, as will be explained later.

When isothermals from CEoS are compared with vapor-liquid phase boundaries for pure components as depicted in Fig. 3.2, we observe that the respective behavior is well predicted from a qualitative point of view. Inside the vapor-liquid envelope, where vapor and liquid coexist at equilibrium, CEoS predict three real roots as those illustrated in Fig. 3.2. When pressure increases, both the vapor and liquid converge to the same conditions: the so-called critical point, whose temperature, pressure and

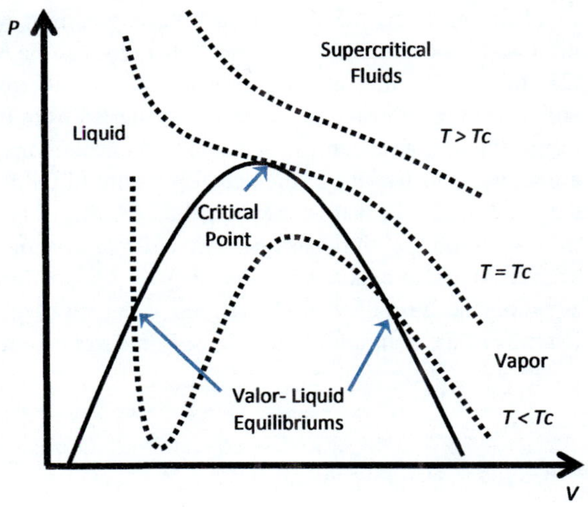

Fig. 3.2 Isothermals from Cubic Equations of State for a pure component on pressure-volume phases diagram

molar volume are indicated by T_c, P_c, and $\tilde{V}_c$, respectively. On the critical point, therefore, it is not possible to make a distinction between properties of the liquid and vapor phases at equilibrium. In this case, CEoS predict three real roots as well, but all of them are equal to each other. Outside of the vapor-liquid envelope, we find that one phase is present only. A liquid phase is present to the left and at temperatures below the critical one. A vapor phase is present to the right and temperatures below the critical one as well. Supercritical fluids are present above the critical temperature. For all these single phases, CEoS predict just one real root. It will be shown later, the way in which the roots are calculated from CEoS.

Returning to the general form given by Eq. (3.1), we can see that it is characterized by the parameters of a and b, which are specific for a given pure component. From the physical standpoint, the co-volume b accounts for the molecular volume; it is four-fold the effective volume of a hard sphere with molecular diameter σ, that is, $b = 4\left(\pi\sigma^3/6\right)$. On the other hand, the parameter a accounts for attractive forces between molecules. The values of a and b are determined by evaluating Eq. (3.1) on the critical point of a pure component, which is mathematically an inflexion point. This means that the first and second derivatives with respect to molar volume are equal to zero on the critical point, that is,

$$\left(\frac{\partial P}{\partial \tilde{V}}\right)_T = \left(\frac{\partial^2 P}{\partial \tilde{V}^2}\right)_T = 0 \tag{3.2}$$

Accordingly, three equations are present in order to determine three unknown variables: the parameters a, b, and the critical molar volume $\tilde{V}_c$ of a pure component. These unknowns are thus expressed in terms of the critical pressure P_c and temperature T_c of the same pure component, as shown below.

There are many CEoS that have been proposed [1] since J. D. van der Waals proposed his cubic equation of state in 1873, but the Soave-Redlich-Kwong (SRK) [2] and Peng-Robinson (PR) [3] equations are the most widely used in chemical and petroleum industry at present. As observed from Eq. (3.1), CEoS consist of two terms: the former accounts for repulsive contributions, while the latter accounts for attractive contributions to the actual pressure of the fluid system. All CEoS satisfy the ideal-gas limit, that is, they use the behavior of an ideal gas as a reference state in order to predict the phase behavior of real systems, which can be gases, liquids or even solids, as explained in the Chap. 2. As mention before, they are called cubic equations because they can be written as cubic polynomial in terms of either $\tilde{V}$, or in terms of the compressibility factor Z in a more concise form, that is,

$$Z^3 + c_1 Z^2 + c_2 Z + c_3 = 0 \tag{3.3}$$

where the coefficients are given by

$$c_1 = B(u-1) - 1 \tag{3.4}$$

$$c_2 = B^2(v-u) - uB + A \tag{3.5}$$

$$c_3 = -\left[v\left(B^2 + B^3\right) + AB\right] \tag{3.6}$$

This dimensionless polynomial is valid for any CEoS through the parameters u and v defined through Eq. (3.1). As explained Chap. 2, Z is a dimensionless number defined as the ratio of the actual molar volume to the ideal-gas molar volume, that is, $Z \equiv \tilde{V}/\tilde{V}^{ig}$; hence it can be equivalently defined as

$$Z = \frac{P\tilde{V}}{RT} \tag{3.7}$$

Thus, we have $Z = 1$ in the limit of ideal gas. The dimensionless numbers A and B from the coefficients given by Eqs. (3.4–3.6) depend on the CEoS-parameters a and b, or equivalently on P_c and T_c according to

$$A \equiv \frac{Pa}{(RT)^2} = \Omega_a \left(\frac{P}{P_c}\right)\left(\frac{T_c}{T}\right)^2 \alpha_f \tag{3.8}$$

$$\alpha_f = \left[1 + \left(p_0 + p_1\omega + p_2\omega^2 + p_3\omega^3\right)\left(1 - \sqrt{\frac{T}{Tc}}\right)\right]^2 \tag{3.9}$$

$$B \equiv \frac{Pb}{RT} = \Omega_b \left(\frac{P}{P_c}\right)\left(\frac{T_c}{T}\right) \tag{3.10}$$

Table 3.1 Parameters of cubic equations of state

Parameter	SRK[2]	PR[3]	
		ω ≤ 0.491	ω > 0.491
u	1	2	
v	0	−1	
x	$\frac{\sqrt[3]{2}-1}{3}$	$\frac{-1+\sqrt[3]{6\sqrt{2}+8}-\sqrt[3]{6\sqrt{2}-8}}{3}$	
Ω_a	$\frac{1}{27x}$	$\frac{8(5x+1)}{49-37x}$	
Ω_b	x	$\frac{x}{x+3}$	
p_0	0.480	0.37464	0.379642
p_1	1.574	1.54226	1.48503
p_2	−0.176	−0.26992	−0.164423
p_3	0	0	0.016666

The CEoS-dependent coefficients Ω_a and Ω_b are shown in Table 3.1 for both SRK and PR equations. Whereas, α_f is an empirical dimensionless function that depends on the acentric factor ω and temperature. As can be observed from Eq. (3.9), α_f becomes unity as temperature goes to the critical one; therefore, it may be considered as a correction factor of the parameter a with respect to its value on the critical point. An important characteristic of α_f that is often ignored is its thermodynamic consistency: α_f is said to be thermodynamically consistent when $\alpha_f \geq 0, d\alpha_f/dT \leq 0, d^2\alpha_f/dT^2 \geq 0$ and $d^3\alpha_f/dT^3 \leq 0$ all of them being continuous. These conditions guarantee that suitable tendencies are shown by thermodynamic properties calculated from CEoS; hence, the above inequalities establish what is called the consistency test for α_f [4]. This implies that not all α_f suggested in the open literature are consistent. Although, α_f as given by Eq. (3.9) is not consistent for all thermodynamic properties for the entire range of temperatures, it is thermodynamically consistent for temperatures lower than 1500 K. In other words, it can be claimed that it is consistent on the temperature interval of interest. For the special case of hydrogen and helium, it has been illustrated that the same α_f is unable to describe liquid and vapor phase simultaneously below 40-50 K. However, these components behave as regular components above 150 K; consequently, the α_f function given by Eq. (3.9) can be used above this temperature for such components [4]. The p_0, p_1, p_2, and p_3 values of the α_f are shown in Table 3.1 for both SRK and PR equations. The acentric factor and the critical temperature and pressure of relevant pure components are shown in Table 3.2.

As seen from Eq. (3.9), α_f depends on the so-called acentric factor ω, which is defined with respect to the saturation or equilibrium pressure P^{sat} of noble gases in vapor-liquid equilibriums. When $\log_{10}(P^{sat}/P_c)$ is plotted against T_c/T, noble gases lies on the same straight line taking the value of -1 at $T/T_c = 0.7$. In contrast, components different from nobles gases lie on different straight lines. The acentric factor is defined as the difference between the straight lines of a specific component and of noble gases, that is,

$$\omega \equiv -1 - \log_{10}\left(\frac{P^{sat}}{P_c}\right)_{\frac{T}{T_c}=0.7} \tag{3.11}$$

Because noble gases are almost spherical components, ω has been conceived as a number indicating how similar components are to a sphere, that is, $\omega = 0$ for noble gases and $\omega > 0$ for other components. However, this conception does not match for hydrogen because $\omega < 0$. In Table 3.2, ω of some gases and hydrocarbons are shown.

Although CEoS can predict phase behavior and volumetric properties of fluid systems in a relatively simple manner, they show several shortcomings. For instance, the critical compressibility factor Z_c is different for each pure component, for example, $Z_c = 0.286$ for methane and $Z_c = 0.184$ for eicosane. In contrast, CEoS predict a unique value of Z_c (0.333 for SRK and 0.307 for PR) regardless of component; consequently, it has been found that CEoS are inappropriate near the critical point in the interval of $0.95 < T/T_c < 1.0$ approximately [7]. Hence, several modifications have been proposed in order to overcome this and other deficiencies that will be illustrated later.

3.1 Analytic Solution of Cubic Equations

One of the most important advantages of CEoS is that the compressibility factor, and consequently, the molar volume can be determined analytically by using mathematical methods for solving equations of degree three [8]. To obtain the analytical solution by using the formula of Cardiano, we have to calculate the numbers $\boldsymbol{Q}$ and $\boldsymbol{U}$ as follows:

$$Q = \frac{3c_2 - c_1^2}{9} \tag{3.12}$$

$$U = \frac{9c_1c_2 - 27c_3 - 2c_1^3}{54} \tag{3.13}$$

Table 3.2 Critical properties of some Gases and Hydrocarbons [5, 6]

Component	T_c[K]	P_c[MPa]	$V_c 10^6 [\mathrm{m}^3/\mathrm{mol}]$	ω
Hydrogen	32.98	1.293	64.20	−0.217
Hydrogen sulfide	373.40	8.963	98.00	0.090
Nitrogen	126.20	3.398	90.10	0.037
Oxygen	154.58	5.043	73.37	0.022
Carbon monoxide	132.85	3.494	93.10	0.045
Carbon dioxide	304.12	7.374	94.07	0.225
Methane	190.56	4.599	98.60	0.011
Ethane	305.32	4.872	145.50	0.099
Propane	369.83	4.248	200.00	0.152
n-Butane	425.12	3.796	255.00	0.200
i-Butane	407.85	3.640	262.70	0.186
Pentane	469.70	3.370	311.00	0.252
i-Pentane	460.39	3.381	308.30	0.229
Neopentane	433.75	3.199	303.20	0.197
Cyclopentane	511.60	4.508	260.00	0.194
n-Hexane	507.60	3.025	368.00	0.300
2-methylpentane	497.50	3.010	366.70	0.278
3-methylpentane	504.40	3.120	366.70	0.273
2,2-dimethylbutane	488.70	3.080	359.10	0.233
2,3-dimethylbutane	499.90	3.130	357.60	0.248
Methylcyclopentane	532.79	3.784	319.00	0.227
Cyclohexane	553.50	4.073	308.00	0.211
Benzene	562.05	4.895	256.00	0.210
Toluene	591.75	4.108	316.00	0.264
n-Heptane	540.20	2.740	428.00	0.350
n-Octane	568.70	2.490	492.00	0.399
n-Decane	617.70	2.110	624.00	0.490
n-Undecane	639.00	1.980	689.00	0.537
n-Dodecane	658.00	1.820	754.00	0.576
n-Tetradecane	693.00	1.570	894.00	0.644
n-Hexadecane	723.00	1.400	1034.00	0.718
n-Octadecane	747.00	1.290	1189.00	0.8000
n-Eicosane	768.00	1.070	1190.00	0.865
Water	647.14	22.064	55.95	0.344

The coefficients c_1,c_2, and c_3 are those from the cubic polynomial given by Eq. (3.3). Using Q and U, the discriminant D is

$$D = Q^3 + U^2 \tag{3.14}$$

The discriminant establishes the nature of the roots of the cubic polynomial. If $D \geq 0$ then one root is real at least, and the other two roots are either real or conjugate complex numbers. If $D < 0$ all roots are real and different with each other. When $D > 0$ the only real root is calculated from

$$Z_1 = S + T - \frac{c_1}{3} \tag{3.15}$$

together with

$$S = \left(U + \sqrt{D}\right)^{1/3} \tag{3.16}$$

$$T = \left(U - \sqrt{D}\right)^{1/3} \tag{3.17}$$

Whereas the complex ones are determined from

$$Z_{2,3} = -\frac{(S+T)}{2} - \frac{c_1}{3} \mp \frac{\sqrt{3}(S-T)}{2} i \tag{3.18}$$

It is worth pointing out that even though the conjugate complex roots satisfies the cubic equation, they are meaningless for estimating thermophysical properties. In other words, when $D > 0$ the real root is the only one to be considered from a physical standpoint, which accounts for the presence of a single fluid phase, as explained in Fig. 3.2. If $D = 0$, $S = T$, with which Eq. (3.18) leads to $Z_2 = Z_3$, all roots are real and equal to each other. Finally, when $D < 0$ the roots are determined from

$$Z_{k+1} = 2\sqrt{-Q}\cos\left(\frac{\theta + 2\pi k}{3}\right) - \frac{c_1}{3}; k = 0, 1, 2 \tag{3.19}$$

$$\theta = \cos^{-1}\left(\frac{U}{\sqrt{-Q^3}}\right) \tag{3.20}$$

Once again, not all roots have physical meaning. In this case, the largest one corresponds to a vapor phase (Z_1), while the smallest one corresponds to a liquid phase (Z_2) on condition that they are positive numbers. This last case is depicted in Fig. 3.2.

3.2 Phase Identification and Extrapolation

During numerical simulations of processes involving petroleum fluids, iterative calculations are carried out at several levels. For instance, reservoir simulations, separations processes, or regression to experimental data needs that equation of states give properties efficiently. However, combinations of temperature, pressure or composition may occur in such iterative calculations so that phase properties are required where the phase does not exist according to the used equation of state. In other words, properties may be required at unrealistic conditions giving rise to converge problems. This problem can be addressed at either higher or lower level. As an example of converge strategies at a higher level, we have that the isothermal two-phase flash can be solved by searching for decreasing the total Gibbs energy in combination with tangent plane stability analysis without using some strategy at a lower lever [9]. On the other hand, CEoS being at the lowest level, it seems also appropriate to consider strategies at this level with the aim of determining properties on unfeasible regions for higher levels calculations to converge properly.

Using heuristic rules based on the variation of pressure with molar density($\tilde{\rho}$) at constant pressure and composition, Mathias et al. [11] suggested to extrapolate properties on unfeasible regions taking the mechanical instability envelope $\left[(\partial P/\partial\tilde{\rho})_T = 0\right]$ as starting point. These extrapolated properties, which follow suitable tendencies and continuities, are in a region of a tenth of the mechanical instability envelope. The most important characteristic of the suggested method is that it conceives properties, which do not exist actually; they are used as intermediate properties in higher-level iterative processes only. For the specific case of the SRK and PR equations, the respective procedure, which will be described in what follows, was developed by Bazúa-Rueda [10]. Because the mechanical instability envelope is taken as reference in order to establish unfeasible regions, it is necessary to determine the partial derivative with respect to molar density for any cubic equation of state as follows:

$$P'_\rho(\eta) \equiv \frac{1}{RT}\left(\frac{\partial P}{\partial \bar{\rho}}\right)_T = \frac{1}{(1-\eta)^2} - \left(\frac{A}{B}\right)\frac{\eta(2+u\eta)}{\left(1+u\eta+v\eta^2\right)^2} \tag{3.21}$$

The packing fraction η is defined as the ratio of the molecule volume to the actual fluid volume $\left(\eta \equiv B/Z = b/\tilde{V}\right)$. Then, using the CEoS parameters and the calculated Z, $P'_\rho(\eta)$ is evaluated. For either a liquid or a vapor phase, the criterion for a feasible region is $P'_\rho(\eta) > 0.1$. This means that if this criterion is satisfied according to Eq. (3.21), then it is not recommended to propose any extrapolation and

Table 3.3 Cubic equations of state parameters for phase identification and extrapolation [10]

	SRK	PR
A_{mc}	4.9339625	5.8773599
η_{mc}	0.2599211	0.2530766

we can proceed with calculations at a higher level. If $\eta > \eta_{mc}$ then Z corresponds to a liquid phase. Otherwise, it corresponds to a vapor phase $\eta < \eta_{mc}$ on condition that $A/B < A_{mc}$. In these inequalities, the subscript *mc* indicates conditions on the mechanical phase envelope; the respective values of η_{mc} and A_{mc} are shown in Table 3.3 for the PR and SRK equations. When $P'_\rho(\eta) \leq 0.1$, it is necessary to propose a pseudo compressibility factor, that is, to extrapolate Z. For which, Eq. (3.21) is solved iteratively to calculate η_1 as stated by

$$P'_\rho(\eta_1) - 0.1 = 0 \tag{3.22}$$

The above equation represents the condition on which the pseudo-compressibility factor will be extrapolated (a tenth of the mechanical instability envelope). With the knowledge of η_1, we proceed to calculate the function F_1 as follows

$$F_1 \equiv \eta_1 Z(\eta_1) = \frac{\eta_1}{1-\eta_1} - \left(\frac{A}{B}\right)\frac{\eta_1^2}{1+u\eta_1+v\eta_1^2} \tag{3.23}$$

If a vapor phase is wanted to be extrapolated then we have that

$$F_2 = \frac{\eta_{mc} - \eta_1 - 20F_1}{5(\eta_{mc}-\eta_1)^2 F_1^2} \tag{3.24}$$

$$F_3 = 2F_2\eta_1 + 0.1F_1^{-2} \tag{3.25}$$

$$\eta^{ext} = \frac{F_3 - \sqrt{F_3^2 - 4F_2\left(F_1^{-1} + F_3\eta_1 - F_2\eta_1^2 - B^{-1}\right)}}{2F_2} \tag{3.26}$$

On the other hand, if a liquid phase is wanted to be extrapolated then we have that

$$F_2 = F_1 - 0.1(\eta_1 - 0.7\eta_{mc})\ln(\eta_1 - 0.7\eta_{mc}) \tag{3.27}$$

$$\eta^{ext} = 0.7\eta_{mc} + e^{10(B-F_2)/(\eta_1-0.7\eta_{mc})} \tag{3.28}$$

Once, the extrapolated packing fraction η^{ext} has been determined, the respective compressibility factor is given by

$$Z^{ext} = \frac{B}{\eta^{ext}} \tag{3.29}$$

In addition to heuristic-based extrapolations, there are two more options of defining a hypothetical phase with its own set of physical properties from the mathematical point of view. One of them is to use the complex compressibility factor directly and to define complex-valued physical properties. Another option, which is easier to implement, is to make use of the modulus of the complex compressibility factor and to define real-valued physical properties [12]. In the latter case, this means that

$$Z^{ext} = \sqrt{\left[\frac{(S+T)}{2} + \frac{c_1}{3}\right]^2 + \left[\frac{\sqrt{3}(S-T)}{2}\right]^2} \quad (3.30)$$

S and T are determined from Eqs. (3.16) to (3.17), respectively.

3.3 Fugacity Coefficient

As discussed in Chap. 2, the idea behind the compressibility factor is to correct the molar volume predicted by the ideal-gas law. Similarly, the fugacity coefficient ϕ_i is a quantity that indicates deviations of a given system (gas, liquid or solid) with respect to an ideal-gas behavior, but from an energetic standpoint. Accordingly, we showed that the thermodynamic potentials, and consequently the chemical potentials, can be expressed in terms of ϕ_i. The needed expressions for determining from any CEoS were derived in the same Chap. 2. Because CEoS are explicit functions of pressure, it is advisable to use the ϕ_i expression as function of volume

$$\ln(\phi_i) = \frac{1}{RT}\int_{\infty}^{V}\left[\frac{RT}{V} - \left(\frac{\partial P}{\partial n_i}\right)_{T,V,n_j,j\neq i}\right]dV - \ln(Z) \quad (3.31)$$

The partial derivative $(\partial P/\partial n_i)_{T,V,n_j,j\neq i}$ is obtained from the general form of CEoS given by Eq. (3.1)

$$\left(\frac{\partial P}{\partial n_i}\right)_{T,V,n_j} = \frac{RT}{V-nb} + \frac{nb_i' RT}{(V-nb)^2} - \frac{a_i'}{V^2 + u(nbV) + v(nb)^2} + \frac{ab_i' n^2(uV + 2vnb)}{\left[V^2 + u(nbV) + v(nb)^2\right]^2} \quad (3.32)$$

where a_i' and b_i' are partial derivatives defined as:

$$a_i' \equiv \left[\frac{\partial\left(an^2\right)}{\partial n_i}\right]_{T,V} \quad (3.33)$$

$$b_i' \equiv \left[\frac{\partial(bn)}{\partial n_i}\right]_{T,V} \quad (3.34)$$

Substituting Eq. (3.32) into Eq. (3.31) and performing the corresponding integration, we arrive at

$$\ln(\phi_i) = -\ln(Z - B) + \frac{b'_i}{b}(Z - 1) + \frac{A}{B(\alpha + \beta)}\left(\frac{b'_i}{b} - \frac{a'_i}{na}\right)\ln\left(\frac{Z + B\beta}{Z - B\alpha}\right) \tag{3.35}$$

where

$$\alpha = \frac{\sqrt{u^2 - 4v} - u}{2} \tag{3.36}$$

$$\beta = \frac{\sqrt{u^2 - 4v} + u}{2} \tag{3.37}$$

It is worth highlight that this expression is valid for any CEoS through α and β parameters. Besides, it remains valid for any mixing rule on the condition that the partial derivatives a'_i and b'_i can be established. For a pure component, we have that the terms involving these partial derivatives in Eq. (3.35) are

$$\frac{a'_i}{na} = 2 \tag{3.38}$$

$$\frac{b'_i}{b} = 1 \tag{3.39}$$

with which the fugacity coefficient for a pure component ϕ_i^0 is

$$\ln(\phi_i^0) = -\ln(Z_i - B_i) + (Z_i - 1) - \frac{A_i}{B_i(\alpha + \beta)}\ln\left(\frac{Z_i + B_i\beta}{Z_i - B_i\alpha}\right) \tag{3.40}$$

In terms of the properties of the whole mixture, this equation becomes

$$\ln(\phi) = -\ln(Z - B) + (Z - 1) - \frac{A}{B(\alpha + \beta)}\ln\left(\frac{Z + B\beta}{Z - B\alpha}\right) \tag{3.41}$$

3.4 Mixing Rules

It addition to their relative simplicity for pure components, CEoS can be applied to mixtures by the so-called mixing rules, in which pure-component properties and interaction parameters between pairs of either components or functional groups are used in order to estimate the required mixture parameters. All mixing rules discussed here are based on the one-fluid approximation, which assumes that the form of CEoS remains unchanged for a mixture; the mixture effects such as composition are taking into account only through the parameters a and b. Although this could be evident and the only way to address extension to mixtures, pure-component cubic equation of state transforms into an equation of fifth degree for multicomponent mixtures when ternary interactions are taken into account [7]. This robust theoretical basis, however, leads to more parameters to be determined and none of the ternary interaction parameters can be estimated based on chemical formulas and molecular structures, as can be done for binary parameters, limiting its possible application to petroleum fractions.

Most of the mixing rules based on the one-fluid approximation are been developed for the parameter a because it affects stronger prediction capabilities, some of them will be explained later. For the co-volume b, the simplest mixing rule is a linear function of the mole fraction x_i of component i.

$$b = \sum_{i=1}^{C} x_i b_i \tag{3.42}$$

where C is the total number of components in the mixture. x_i is the mole fraction of component i, while b_i is its corresponding co-volume as pure component, which depends on the acentric factor, the critical pressure and temperature according to Eq. (3.10). Using the dimensionless co-volume B defined by the same Eq. (3.10) the above equation becomes

$$B = \sum_{i=1}^{C} x_i B_i \tag{3.43}$$

As shown, the estimation of fugacity coefficients in mixtures requires the ratio of partial derivative $b_i^{'}$ to b. When using the above mixing rule, this ratio is equal to

$$\frac{b_i'}{b} = \frac{b_i}{b} = \frac{B_i}{B} \tag{3.44}$$

In the case of the attractive energy parameter a, the simplest mixing rule is a quadratic function of the mole fraction in the form of

$$a = \sum_{i=1}^{C} \sum_{j=1}^{C} x_i x_j \left(1 - k_{ij}\right) \sqrt{a_i a_j} \tag{3.45}$$

a_i and a_j correspond to the attractive parameter of pure components i and j, respectively. Similar to b_i, these parameters are depend on temperature, the acentric factor, and the critical pressure and temperature according to Eq. (3.8) k_{ij} stand for the so-called binary interaction parameters, which are symmetric $\left(k_{ij} = k_{ji}\right)$ and equal to zero for a pure component $\left(k_{ii} = k_{jj} = 0\right)$. This means that, for C-component mixture, there are only $C(C-1)/2$ independent binary interaction parameters. Using the dimensionless attractive parameter A defined by Eq. (3.8) the above mixing rule becomes

$$A = \sum_{i=1}^{C}\sum_{j=1}^{C} x_i x_j (1 - k_{ij})\sqrt{A_i A_j} \tag{3.46}$$

From which, the ratio of the partial derivative a_i' to na required for the calculation of the fugacity coefficients is

$$\frac{a_i'}{na} = \frac{2}{A}\sum_{j=1}^{C} x_j (1 - k_{ij})\sqrt{A_i A_j} \tag{3.47}$$

The above mixing rules, as expressed by Eqs. (3.42) and (3.45) are called classical mixing rules. They have been successfully used in describing phase behavior of nearly ideal mixtures as those formed by nonpolar components such as hydrocarbons. They are therefore inadequate for complex chemically mixtures as those formed by polar or asymmetric components. As discussed in the Chap. 2, by ideal mixtures, we mean ideal solutions, which must not be confused with ideal gases. The successful of CEoS relies on the empirical parameters k_{ij}, which accounts for molecular interactions between unlike components. Hence, values different from zero are connected to the extent of nonideality of the mixture. Because of their empirical nature, they are usually determined by regression schemes to experimental data from phase equilibriums. The k_{ij} values so obtained depends on the used cubic equation of state and the available data bank. Therefore, they should not be used with other cubic equation of state different from that used in the regression process. Although there are available tabulations for the PR and SRK equations, these are obtained from experimental data of mixtures where forming components are known. This implies that adjustments of k_{ij} are usually required in the presence of petroleum fractions, in which the exact amount and nature of such components is unknown. Instead of adjusting each one of $C(C-1)/2$ values of k_{ij}, one can use generalized expressions. For instance, the following empirical expression can be used [13]

$$k_{ij} = \left(\frac{k_0}{M_j} + k_1\right) + \left(\frac{k_2}{M_j} + k_3\right)T \tag{3.48}$$

with which the k_0, k_1, k_2, and k_3 constants are the only ones to be determined regardless of the total number of components C. This expression has been successfully applied to asymmetric mixtures of hydrocarbons, including those in the presence of heavy petroleum fluids. Besides, the use of Eq. (3.48) is not limited to CEoS; it has been used with the so-called simplified Perturbed-Chain Statistical Associating Fluid Theory (sPC-SAFT) equation of state, which will be discussed in the next Chapter.

As seen from Eq. (3.48), the molecular weight M_j is used as parameter in order to take into account the chemical nature of component j. Other pure-component properties can be used as well. For instance, we can use the critical volumes of the involved components as follows:

$$k_{ij} = 1 - \left[k_0 \frac{\left(V_{ci} V_{cj}\right)^{1/6}}{V_{ci}^{1/3} + V_{cj}^{1/3}} \right]^{k_1} \tag{3.49}$$

Similarly, k_0 and k_1 are the only adjustable parameters for the whole mixture. The above expression was suggested in estimating the second virial coefficients from the Tsonopoulos correlation [14]. However, even though it was not suggested to be used with the classical mixing rules described above, good results have been found when heavy petroleum fractions are present by setting $k_0 = 2$ and $k_1 = 1$ [15]. For mixtures of solvents and heavy petroleum fractions, it has been used a similar expression, but in terms of critical temperatures, that is [16, 17],

$$k_{ij} = 1 - \left[k_0 \frac{\left(T_{ci} T_{cj}\right)^{1/2}}{T_{ci} + T_{cj}} \right]^{k_1} \tag{3.50}$$

Setting $k_0 = 2$, the only adjustable parameters is k_1. For prediction purposes, where no adjustments are carried out, one can set $k_1 = Z_{cij}$ [18], where

$$Z_{cij} = \frac{Z_{ci} + Z_{cj}}{2} \tag{3.51}$$

Z_{ci} stands for the compressibility factor on the critical point of pure component i. From Figs. 3.3, 3.4 and 3.5, it is shown the comparison between experimental data and calculated values using the classical mixing rules with k_{ij} from Eqs. (3.50) and (3.51).

As may be inferred from the above discussion, it is common practice in the petroleum industry to make use of adjustments of model parameters to diminish deviations between calculated values and experimental data. Hence, several adjustment procedures have been suggested not only for k_{ij} but also for the different parameters entering CEoS such as the critical properties of the petroleum fractions [22]. However, there is no guarantee that the parameters so obtained remain valid for other thermophysical properties or for other temperatures, pressures and compositions not considered in the adjustment process [4]. This becomes more critical in the presence of heavy petroleum fractions, because their physicochemical nature may change drastically far away from the adjustment conditions. For instance, they may change from Newtonian fluid to non-Newtonian fluid as temperature decreases, as will be shown in the second part of this manuscript; vapor-liquid equilibrium may turn into vapor-liquid-liquid equilibrium as composition of light components increases

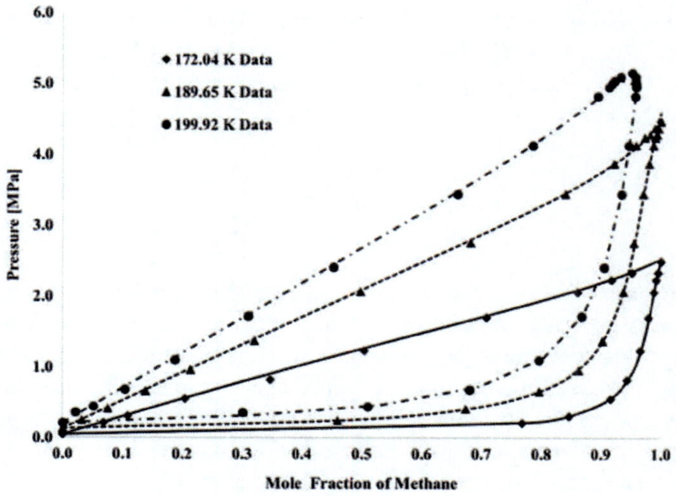

Fig. 3.3 Vapor-Liquid Equilibrium for Methane-Ethane Mixtures using PR-CEoS and Eq. (3.50) setting $k_0 = 2$ and $k_1 = Z_{cij}$. Data from Wichterle and Kobayashi [19]

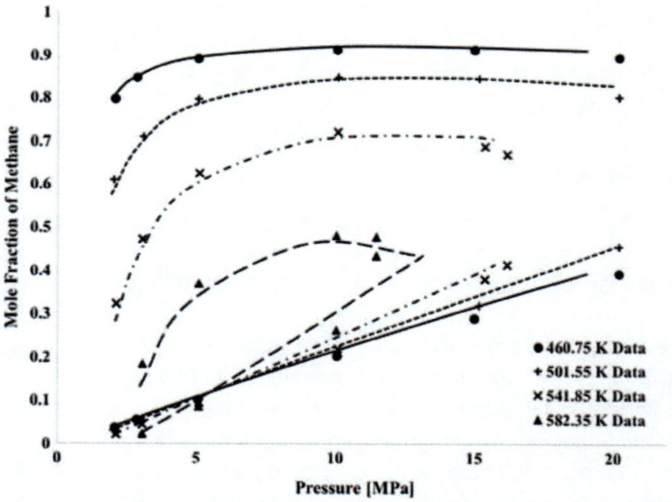

Fig. 3.4 Vapor-Liquid Equilibrium for Methane-m-Xylene Mixtures using PR-CEoS and Eq. (3.50) setting $k_0 = 2$ and $k_1 = Z_{cij}$. Data from Simnick et al. [20]

[23], etc. This is the reason why in this book emphasis is placed upon methods with strong fundamental basis (predictive ones), avoiding as long as possible specific adjustments of the models parameters to experimental data.

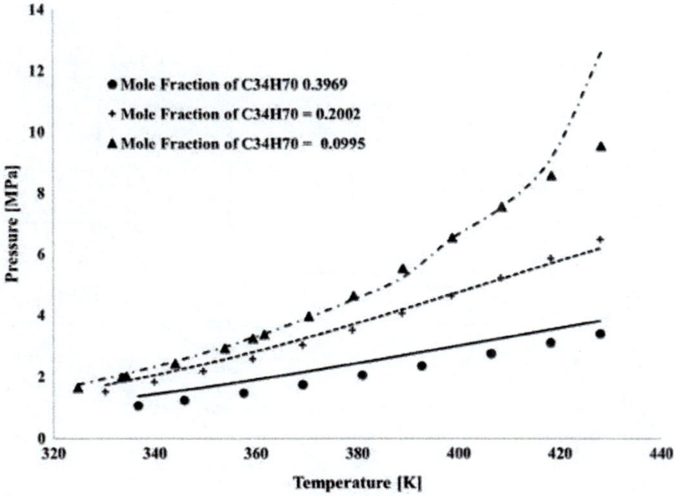

Fig. 3.5 Vapor-Liquid Equilibrium for Propane-Tetratriacontane Mixtures using PR-CEoS and Eq. (3.50) setting $k_0 = 2$ and $k_1 = Z_{cij}$. Data from Peters et al. [21]

3.4.1 Mixing Rules Based on Excess Gibbs Energy Models

Although the classical mixing rule is the most widely used in petroleum industry, it shows several drawbacks. One of them is that unsatisfactory results are obtained for non-ideal mixtures, as those formed by highly polar components, for example water, or by components very different in size (asymmetric systems), or in their properties in general, for example mixtures of supercritical gases and heavy fractions. Hence, the classical mixing rule as well as the PR and SRK equations of state are limited to mixtures formed by non-polar, nearly polar or symmetric components. The second major drawback of the classical mixing rule is that the binary interaction parameters relies generally on temperature, and consequently, the PR and SRK parameters as well. To overcome these drawbacks empirical or semi-empirical mixing rules have been suggested by incorporating temperature, composition or even pure-component properties in several ways. These expressions have been developed for estimating either binary interaction parameters or CEoS parameters directly [1, 24].

On the other hand, parallel to CEoS, thermodynamic models based on the solution theory have been developed [25], which have been successful in describing phase equilibriums of those systems where CEoS have not, that is, of non-ideal mixtures where temperature effects are important. However, unlike CEoS, pressure is usually not taken into account in these thermodynamic models, because these models have sought to characterize molecular interactions from the liquid-phase behavior, where pressure effects can be ignored in the range from low to moderate pressures. Thus, with the aim of combining advantages from both approaches, solution theory models

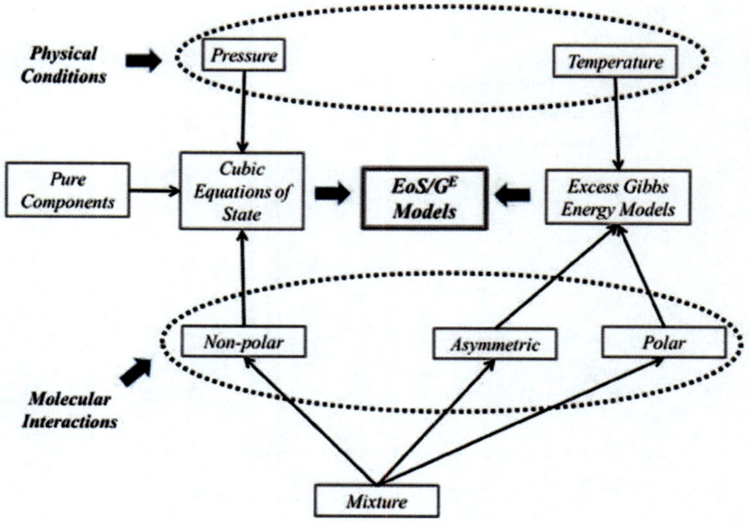

Fig. 3.6 Combination of cubic equations of state and excess Gibbs energy models into EoS/G^E models

have been incorporated into CEoS, giving rise to new mixing rules, which are symbolized in general as EoS/G^E. The abbreviation EoS invokes cubic equations of state, while G^E invokes the excess Gibbs free energy from solution theory models. When comparisons have been carried out, it has been found that EoS/G^E-type mixing rules give in general better results than other mixing rules [26]. In summary, the introduction of solution models into CEoS searches for including intermolecular interactions properly, as illustrated in Fig. 3.6. Despite this fact, this type of mixing rules is not routinely used in the presence of petroleum fractions, because thermodynamic models based on the solution theory generally require knowing the components forming the mixture, which are unknown in such fractions by definition. As will be shown, this inconvenient can be overcome by assigning either pure components or chemical structures to the petroleum fractions.

3.4.2 Huron-Vidal

The first EoS/G^E-type mixing rule that was successfully applied to industrial process was obtained by Huron and Vidal (the Huron-Vidal mixing rule) [27]. As any EoS/G^E-type mixing rule, the aim is to combine solution-theory models with CEoS. To carry out such a combination, it is important to remember that solution-theory models use the ideal-solution model as reference fluid in order to estimate properties of actual mixtures, whereas CEoS search for estimating properties of actual fluids by using the ideal-gas model as reference fluid. In turn, the use of the ideal-solution and ideal-gas concepts leads to excess and residual properties respectively. In other words, EoS/G^E-type mixing rules require that excess and residual properties can be related to each other. An expression relating both type of properties was obtained in the Chap. 2:

$$\frac{\tilde{G}^E}{RT} = \frac{\tilde{G}^R}{RT} - \sum_{i=1}^{C} x_i \ln\left(\phi_i^0\right) \tag{3.52}$$

As explained then, $\tilde{G}^E$ and $\tilde{G}^R$ stand for the total molar excess and residual Gibbs free energy of the mixture. It is worth reminding that any excess property is conceived as the difference between the property of the actual mixture and the property of the ideal solution. Therefore, if a given mixture behaves as an ideal solution, then $\tilde{G}^E = 0$. Excess properties have no meaning for pure components only for mixtures, whereas residual properties are defined for both pure components and mixtures.

On the other hand, the residual Gibbs free energy of the mixture as a whole can be expressed in terms of fugacity coefficients through $\tilde{G}^R/RT = ln(\phi)$, with which Eq. (3.52) takes the form of

$$\frac{\tilde{G}^E}{RT} = \ln(\phi) - \sum_{i=1}^{C} x_i \ln\left(\phi_i^0\right) \tag{3.53}$$

Equations (3.40) and (3.41) can used for determining ϕ_i^0 and ϕ, respectively. Thus, (3.53) is rewritten as

$$\begin{aligned}\frac{\tilde{G}^E}{RT} = {} & Z - \ln(Z - B) - \frac{A}{B(\alpha + \beta)} \ln\left(\frac{Z + B\beta}{Z - B\alpha}\right) \\ & - \sum_{i=1}^{C} x_i \left[Z_i - \ln(Z_i - B_i) \right. \\ & \left. - \frac{A_i}{B_i(\alpha + \beta)} \ln\left(\frac{Z_i + B_i\beta}{Z_i - B_i\alpha}\right)\right]\end{aligned} \tag{3.54}$$

This equation requires one pressure at which the solution model accounting for $\tilde{G}^E$ and the used CEoS can be equated. In other words, a reference pressure must be defined, for which the term $Z - B$ is expressed in terms of $\tilde{V}$ and b using the general form of a cubic equation of state given by Eq. (3.1) as follows:

$$Z - B = \frac{P\left(\tilde{V} - b\right)}{RT} = 1 - \frac{a\left(\tilde{V} - b\right)}{\left(\tilde{V}^2 + ub\tilde{V} + vb^2\right)RT} \tag{3.55}$$

The form of this equation is the same for either the mixture as one or a given pure component *i*. The above equation allows us to show that $\lim_{P\to\infty}(Z-B)=1$ or $\lim_{P\to\infty}(Z_i-B_i)=1$, because if $P\to\infty$ then $\tilde{V}\to b$ and $\tilde{V}_i^0\to b_i$; that is, the volume is equal to the volume occupied by the molecules as pressure goes to infinite. Besides,

$$Z-\sum_{i=1}^{C}x_iZ_i=Z-B+B-\sum_{i=1}^{C}x_i(Z_i-B_i+B_i)=B-\sum_{i=1}^{C}x_iB_i=0 \quad (3.56)$$

where Eq. (3.43) has been used. Therefore, we conclude that Eq. (3.54) reduces to

$$\frac{\tilde{G}_\infty^E}{RT}=-\frac{1}{(\alpha+\beta)}\ln\left(\frac{1+\beta}{1-\alpha}\right)\left(\sum_{i=1}^{C}x_i\frac{A_i}{B_i}-\frac{A}{B}\right)=\epsilon\left(\sum_{i=1}^{C}x_i\frac{A_i}{B_i}-\frac{A}{B}\right) \quad (3.57)$$

This expression is the so-called Huron-Vidal mixing rule [27], in which the subscript ∞ indicates the infinite pressure condition. The CEoS-dependent parameter ϵ is defined as

$$\epsilon\equiv\frac{1}{\alpha+\beta}\ln\left(\frac{1+\beta}{1-\alpha}\right) \quad (3.58)$$

Equation (3.57) together with Eq. (3.43) allows the *A* and *B* mixture parameters to be calculated, which in turn are used in Eq. (3.3) for determining the compressibility factor Z. The Huron-Vidal (HV) mixing rule can be expressed in terms of activity coefficients γ_i^∞ using the fundamental thermodynamic expression derived in Chap. 2:

$$\frac{\tilde{G}_\infty^E}{RT}=\sum_{i=1}^{C}x_i\ln\left(\gamma_i^\infty\right) \quad (3.59)$$

Hence, HV mixing rule is

$$\frac{A}{B}=\sum_{i=1}^{C}x_i\left[\frac{A_i}{B_i}-\frac{\ln\left(\gamma_i^\infty\right)}{\epsilon}\right] \quad (3.60)$$

According to Eq. (3.35), in addition to the *A* and *B* mixture parameters, it is needed the partial derivative a_i' given by Eq. (3.33) for ϕ_i to be determined. In the case of HV mixing rule, it is found that

$$\frac{a_i'}{na}=\frac{B}{A}\left[\frac{A_i}{B_i}-\frac{\ln\left(\gamma_i^\infty\right)}{\epsilon}\right]+\frac{B_i}{B} \quad (3.61)$$

At this point, it is worth recalling that the activity coefficient γ_i of a component i indicates deviations with respect to an ideal solution. If $\gamma_i = 1$ then component i behaves as if it would be in an ideal solution, whereas $\gamma_i \neq 1$ indicates a non-ideal behavior. If $\gamma_i > 1$. then a positive deviation from ideality is present. On the contrary, if $\gamma_i < 1$, then a negative deviation from ideality is present.

As observed from Eqs. (3.60) to (3.61), a solution model is required for γ_i^{∞} to be calculated. Huron and Vidal [27] modified the Non-Random-Two-Liquid (NRTL) model [28] by introducing the b parameter from CEoS as follows:

$$\ln(\gamma_i^{\infty}) = \frac{\sum_{j=1}^{C} x_j \tau_{ji} b_i e^{-\alpha_{ij}\tau_{ij}}}{\sum_{k=1}^{C} x_k b_k e^{-\alpha_{ki}\tau_{ki}}} + \sum_{j=1}^{C} \frac{x_j b_j e^{-\alpha_{ji}\tau_{ji}}}{\sum_{k=1}^{C} x_k b_k e^{-\alpha_{kj}\tau_{kj}}} \left(\tau_{ij} - \frac{\sum_{k=1}^{C} x_k \tau_{kj} b_k e^{-\alpha_{kj}\tau_{kj}}}{\sum_{k=1}^{C} x_k b_k e^{-\alpha_{kj}\tau_{kj}}} \right) \tag{3.62}$$

$$\tau_{ij} = \frac{g_{ij} - g_{jj}}{RT} \tag{3.63}$$

Notice that pressure does not appear as independent variable, because solution models have been developed at conditions from low to moderate pressure. This is the reason why solution models have not been attractive for petroleum processes where high pressures are present. However, in combination with mixing rules such as HV mixing rule, they become an important tool in solving phase equilibrium problems of petroleum fluids at high pressure.

Returning to Eq. (3.62), the binary interaction parameters α_{ij}, g_{ij} and g_{jj} must be known for each pair of components in the mixture. Similar to k_{ij} from either Eqs. (3.45) or (3.46), these binary parameters are in general determined by regression to experimental data. However, it can be shown that the classical Van der Waals mixing rule is retrieved from Eqs. (3.62) and (3.63) by means of the following choice [22]

$$\alpha_{ij} = 0 \tag{3.64}$$

$$\frac{g_{jj}}{RT} = -\epsilon \frac{A_j}{B_j} \tag{3.65}$$

$$\frac{g_{ij}}{RT} = -2\epsilon \frac{\sqrt{A_i A_j}}{B_i + B_j} (1 - k_{ij}) \tag{3.66}$$

Accordingly, when the binary interaction parameter k_{ij} is suitable for a given pair of components, no additional parameters are needed in order to use HV mixing rule. Otherwise, α_{ij}, g_{ij} and g_{jj} must be determined. For instance, for systems formed by hydrocarbons and water, only parameters between hydrocarbons and water should be estimated independently [29], because the respective values between pairs of hydrocarbons come from k_{ij} in accordance with Eqs. (3.64) to (3.65).

3.4.3 Modified Huron—Vidal Mixing Rule

Although HV mixing rule establishes a connection between solution models and CEoS, the respective activity coefficients from excess Gibb energy models must be reevaluated, but at high pressure this time. This means that the widely available data banks of activity coefficient models and their corresponding parameters obtained at low pressures are not accessible to be used into HV mixing rule, because the results from both sets of data type differ considerably [9].

To overcome this issue, combination of excess Gibbs energy and CEoS is sought again, but now the pressure reference should be changed from $P \to \infty$ to $P \to 0$. To accomplish this objective, we make use of Eq. (3.41), which can be written as

$$\ln(\phi B) = -\ln\left(\frac{Z}{B} - 1\right) + (Z - 1) - \frac{A}{B(\alpha + \beta)}\ln\left(\frac{Z + B\beta}{Z - B\alpha}\right) \tag{3.67}$$

or

$$\ln(\phi B) = -\ln\left(\frac{\tilde{V}}{b} - 1\right) + \left(\frac{P\tilde{V}}{RT} - 1\right) - \frac{a}{bRT(\alpha + \beta)}\ln\left(\frac{\tilde{V} + b\beta}{\tilde{V} - b\alpha}\right) \tag{3.68}$$

In the limit of zero pressure, this equation reduces to

$$\ln(\phi B) = -\ln\left(\frac{\tilde{V}}{b} - 1\right) - 1 - \frac{a}{bRT(\alpha + \beta)}\ln\left(\frac{\tilde{V} + b\beta}{\tilde{V} - b\alpha}\right) \tag{3.69}$$

If the packing factor η is defined as $b/\tilde{V}$ and φ as a/bRT, then we arrive at

$$\ln(\phi B) = q(\eta, \varphi) \tag{3.70}$$

with

$$q(\eta, \varphi) \equiv -\ln\left(\frac{1}{\eta} - 1\right) - 1 - \frac{\varphi}{(\alpha + \beta)}\ln\left(\frac{1 + \beta\eta}{1 - \alpha\eta}\right) \tag{3.71}$$

The η value can be obtained analytically using

$$\frac{1}{\eta} = \frac{\varphi - u - \sqrt{\varphi^2 - 2\varphi(u + 2) + u^2 - 4v}}{2} \tag{3.72}$$

which has not a real solution for components with small values of φ for example helium. This expression comes from the general form of a CEoS given by Eq. (3.1) evaluated in the limit of $P \to 0$, where η corresponds to the positive root of the quadratic expression so obtained. Equation (3.72) allows η to be removed as independent variable from Eq. (3.71), transforming $q(\eta_0, \varphi)$ into a function of the form $q(\varphi)$.

For a pure component, the above procedure is repeated but Eq. (3.40) is used as starting point rather than Eq. (3.41), with which the result is now

$$\ln\left(\phi_i^0 B_i\right) = q(\eta_i, \varphi_i) \tag{3.73}$$

where $\eta_i = b_i/\tilde{V}_i^0$. On the other hand, Eq. (3.53) can be written as

$$\frac{\tilde{G}^E}{RT} = \ln(\phi B) - \ln(B) - \sum_{i=1}^{C} x_i\left[\ln\left(\phi_i^0 B_i\right) - \ln(B_i)\right] \tag{3.74}$$

Equations (3.70) and (3.73) simplify this equation to

$$\frac{\tilde{G}^E}{RT} = q(\varphi) - \sum_{i=1}^{C} x_i\left[q_i(\varphi_i) - \ln\left(\frac{B_i}{B}\right)\right] \tag{3.75}$$

It is important to notice that $\varphi = a/bRT = A/B$.Therefore, on condition that Eq. (3.72) has a solution, Eq. (3.75), together with Eq. (3.71), defines a mixing rule implicit for φ, which requires an iterative process in order to calculate φ. However, a simplified version can be obtained, if $q(\varphi)$ is replaced by linear approximation of the form [30]:

$$q(\varphi) = q_0 + q_1\varphi \tag{3.76}$$

which transforms Eq. (3.75) into

$$\frac{\tilde{G}^E}{RT} = q_1\left(\frac{A}{B} - \sum_{i=1}^{C} x_i \frac{A_i}{B_i}\right) + \sum_{i=1}^{C} x_i \ln\left(\frac{B_i}{B}\right) \tag{3.77}$$

This mixing rule is the so-called modified Huron-Vidal of order one (MHV1), because of its similarity with the HV mixing rule. If the fundamental expression derived in the Chap. 2 is used

$$\frac{\tilde{G}^E}{RT} = \sum_{i=1}^{C} x_i \ln(\gamma_i) \tag{3.78}$$

Then Eq. (3.77) takes the form of

$$\frac{A}{B} = \sum_{i=1}^{C} x_i \left[\frac{A_i}{B_i} + \frac{1}{q_1} \ln\left(\frac{\gamma_i B}{B_i} \right) \right] \tag{3.79}$$

From this form of the MHV1 mixing rule, the partial derivative entering into the generalized expression for ϕ_i given by Eq. (3.35) is

$$\frac{a_i'}{na} = \frac{B_i}{B} + \frac{B}{q_1 A} \left[\ln\left(\frac{\gamma_i^0 B}{B_i} \right) + \frac{B_i - B}{B} \right] + \frac{BA_i}{AB_i} \tag{3.80}$$

As can be observed, for MHV1 to be useful in calculations, the q_1 parameter must be established. There have been several suggestions in order to determine q_1. For instance, one of them implies the use of pure-component data in order to calculate η_i^0 from Eq. (3.72), which, in turn, is used in Eq. (3.71) for calculating $q\left(\eta_i^0, \varphi_i\right)$. By fitting the results to a linear equation in the interval of $9 < \varphi < 14$, Michelsen [30] found that q_1 takes values of -0.593 and -0.530 for the SRK and PR equations of state respectively. By taking the data bank of pure components provided by Poling et al. [5] and by evaluating Eqs. (3.71) and (3.72) at the normal boiling temperature ($P = 0.101325$ MPa), one obtains the resulting linear approximations that are shown in Figs. 3.7 and 3.8. As observed, q_1 takes values a little different from those obtained by Michelsen; q_1 is now equal to -0.578 for the SRK equation of state and it is equal to -0.519 for the PR equation of state. Besides, the considered intervals are wider ($8 < \varphi < 15$ for SRK and $9 < \varphi < 16$ and PR), covering components from hydrogen to selenium. It is worth pointing out that the heaviest components in the used data bank are those with twenty carbon atoms, which are in the interval $\varphi < 9$ approximately. As can be noticed from the same figures a quadratic approximation

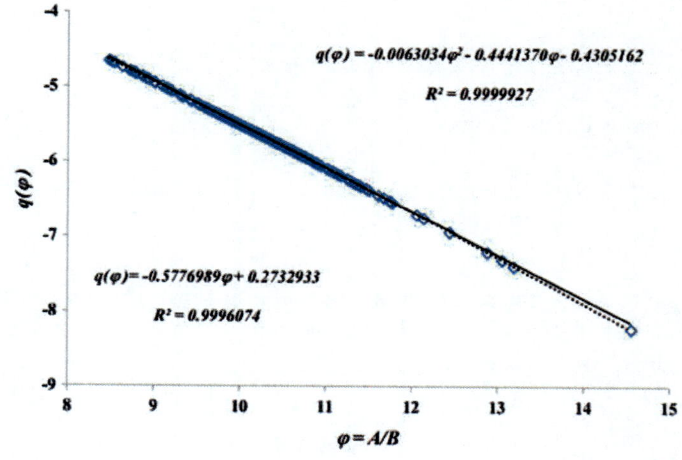

Fig. 3.7 Determination of the MHV1 parameter q_1 for the SRK equation

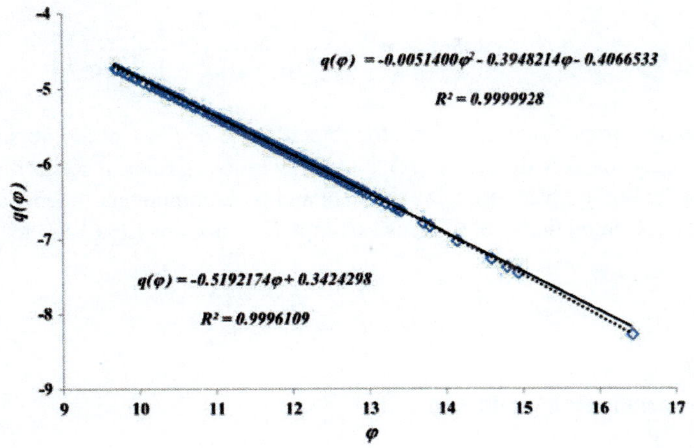

Fig. 3.8 Determination of the MHV1 parameter q_1 for the PR equation

can be employed as well. The use of this approximation leads however to another mixing rule the so-called modified Huron-Vidal of order two (MHV2) [31].

There is another manner of deriving HV and MHV mixing rules without making use of either the zero-pressure or infinite-pressure limit, allowing us to make use of solution models regardless of pressure condition from which they are derived. The derivation starts with

$$\frac{\tilde{A}^E}{RT} = \frac{\tilde{G}^E}{RT} + \ln(Z) - \sum_{i=1}^{C} x_i \ln(Z_i) \tag{3.81}$$

which was obtained in Chap. 2. The substitution of Eq. (3.54), transforms the above equation into

$$\begin{aligned}\frac{\tilde{A}^E}{RT} &= -\ln\left(1 - \frac{B}{Z}\right) - \frac{A}{B(\alpha+\beta)}\ln\left(\frac{Z + B\beta}{Z - B\alpha}\right) \\ &\quad + \sum_{i=1}^{C} x_i\left[\ln\left(1 - \frac{B_i}{Z_i}\right) + \frac{A_i}{B_i(\alpha+\beta)}\ln\left(\frac{Z_i + B_i\beta}{Z_i - B_i\alpha}\right)\right]\end{aligned} \tag{3.82}$$

where the excess compressibility factor $\left(Z^E = Z - \sum_{i=1}^{C} x_i Z_i\right)$ is assumed to be negligible. In terms of the packing factor η, it becomes

$$\frac{\tilde{A}^E}{RT} = -\ln(1-\eta) - \frac{A}{B(\alpha+\beta)}\ln\left(\frac{1+\eta\beta}{1-\eta\alpha}\right)$$

$$+\sum_{i=1}^{C} x_i \left[\ln(1-\eta_i) + \frac{A_i}{B_i(\alpha+\beta)} \ln\left(\frac{1+\eta_i\beta}{1-\eta_i\alpha}\right) \right] \quad (3.83)$$

Instead of extrapolating Eq. (3.83) to either $P \to 0$ or $P \to \infty$, to obtain a more concise expression as in the case of HV and MHV1 mixing rules, it seems reasonable to assume for liquid phases that η is constant and pure-component independent, that is, it is now assumed that $\eta = \eta_i$. This assumption simplifies Eq. (3.83) to

$$\frac{\tilde{A}^E}{RT} = \epsilon^* \left(\sum_{i=1}^{C} x_i \frac{A_i}{B_i} - \frac{A}{B} \right) \quad (3.84)$$

where, the parameter ϵ^* is defined as

$$\epsilon^* = \frac{1}{(\alpha+\beta)} \ln\left(\frac{1+\eta\beta}{1-\eta\alpha}\right) \quad (3.85)$$

which is similar to similar to the ϵ parameter given by Eq. (3.58). In the limit as pressure goes to infinity $\eta = 1$ and therefore $\epsilon = \epsilon^*$. This result allows us to retrieve HV mixing rule from Eq. (3.84).

On the other hand, from $\eta = \eta_i$ it follows that $Z_i/Z = B_i/B$. The substitution of this results and Eq. (3.84) into the relationship given by Eq. (3.81) leads to MHV1 mixing rule as stated by Eq. (3.77), where $q_1 = -\epsilon^*$. It is worth underlying that extrapolation to either zero pressure or infinite pressure are not required in deriving Eq. (3.84). Therefore, any solution model can be used into such mixing rule on the condition that a constant value of η can be established properly. Considering that the most widely used solution models are those obtained under a low-pressure condition, it is advisable then to find a value of η satisfying such condition so that they can be incorporated into the above mixing rule. In the context of the so-called Predictive SRK equation of state (PSRK) [32, 33], experimental data of liquid molar volumes of pure components at the normal boiling temperature were used to estimate an average value of 1.1^{-1} for η, which, through, Eq. (3.85) leads to $q_1 = -0.6466272$. The corresponding value of η for the PR equation is found from

$$\left(\frac{\eta^{PR}}{\eta^{SRK}}\right) = \left(\frac{\Omega_b^{PR}}{\Omega_b^{SRK}}\right) \quad (3.86)$$

The substitution of $\eta^{SRK} = 1.1^{-1}$, Ω_b^{PR}, and Ω_b^{SRK} from Table 3.4 in the above expression results in $\eta^{PR} = 1.22505^{-1}$, which leads to $q_1 = -0.5308486$. It is worth noticing that the calculated η^{PR} compares well with the estimated value of $\eta^{PR} = 1.22498^{-1}$ in the context of the volume translated Peng Robinson (VTPR) equation of state [34]. This average value was estimated in the same way as η^{SRK}; that is, it was estimated from experimental data of pure-component liquid volumes

Table 3.4 Values of the parameter q_1 for HV-type mixing rules using a reference pressure equal to 0.101325 MPa

	MHV1 [30]	PSRK [32, 33]	VTPR [34]	This work
SRK	−0.593	−0.647	N/A	−0.578
PR	−0.530	N/A	−0.531	−0.519

at the normal boiling point. The summary of the proposed values for q_1 from the different approaches is shown in Table 3.4.

3.4.4 UNIFAC Solution Model

The universal quasi-chemical functional group activity coefficient (UNIFAC) is a solution model for the predication of activity coefficients [35]. The UNIFAC solution model is based on the group contribution concept. The main idea is that, whereas hundreds or thousands of components form mixtures, a few tens of functional groups are present in the same mixtures. For instance, a mixture composed of five normal alkanes can be expressed as a mixture of just two functional groups, as illustrated in Fig. 3.9. From this Figure, we observe that the n-heptane molecule consists of two $-CH_3$ and five $-CH_2-$ functional groups, whereas the propane molecule consists of two $-CH_3$ and just one $-CH_2-$ functional groups.

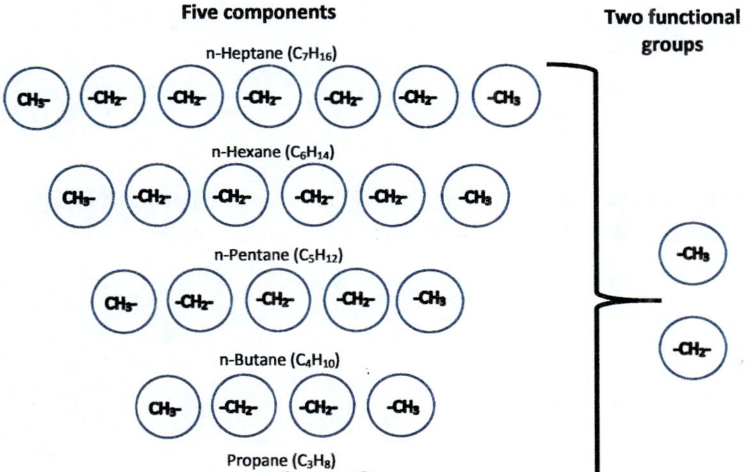

Fig. 3.9 Concept of functional groups contribution

The activity coefficient γ_i of component i in the mixture results from two separate contributions: the combinatorial contribution symbolized by γ_i^{comb}, and the residual contribution symbolized by γ_i^{resi}, that is,

$$\ln(\gamma_i) = \ln\left(\gamma_i^{comb}\right) + \ln\left(\gamma_i^{resi}\right) \tag{3.87}$$

The former accounts for differences in size and shape of component molecules, while the latter accounts for the corresponding energy interactions. In turn, γ_i^{comb} consists of the Flory-Huggins contribution γ_i^{FH}, which accounts for effects due to components very different in size, and the Staverman-Guggenheim contribution γ_i^{SG}, which accounts for effects due to differences in shape. These contributions are determined from the following equations:

$$\ln\left(\gamma_i^{comb}\right) = \ln\left(\gamma_i^{FH}\right) + \ln\left(\gamma_i^{SG}\right) \tag{3.88}$$

$$\ln\left(\gamma_i^{FH}\right) = 1 + \ln\left(\frac{\Phi_i}{x_i}\right) - \frac{\Phi_i}{x_i} \tag{3.89}$$

$$\ln\left(\gamma_i^{SG}\right) = -5q_i\left[1 + \ln\left(\frac{\Phi_i}{\Theta_i}\right) - \frac{\Phi_i}{\Theta_i}\right] \tag{3.90}$$

where Θ_i is the molecular surface fraction and Φ_i is the molecular volume fraction. These fractions are in turn a function of the van der Waals molecular surface q_i and molecular volume r_i in the form of

$$\Theta_i = \frac{x_i q_i}{\sum_{j=1}^{C} x_j q_j} \tag{3.91}$$

$$\Phi_i = \frac{x_i r_i}{\sum_{j=1}^{C} x_j r_j} \tag{3.92}$$

The pure-component parameters q_i and r_i are calculated as sums in terms of the relative group volume R_k, and relative group area Q_k as follows:

$$r_i = \sum_{k=1}^{G} \upsilon_{ki} R_k \tag{3.93}$$

$$q_i = \sum_{k=1}^{G} \upsilon_{ki} Q_k \tag{3.94}$$

υ_{ki} is stands for the number of times (occurrence) that group k appears in molecule i and G for the types of functional groups present in the molecule. The parameters R_k and Q_k are input data in the UNIFAC solution model. For groups relevant to petroleum fluids, R_k and Q_k values are given in Table 3.5. As observed from this

Table 3.5 R_k and Q_k parameters for PSRK cubic equation of state [36]

Main group		Sub-group		R_k	Q_k
1	CH_2	1	CH_3	0.9011	0.8480
		2	CH_2	0.6744	0.5400
		3	CH	0.4469	0.2280
		4	C	0.2195	0.0000
2	C = C	5	$CH_2 = CH$	1.3454	1.1760
		6	CH = CH	1.1167	0.8670
		7	$CH_2 = C$	1.1173	0.9880
		8	CH = C	0.8886	0.6760
		70	C = C	0.6605	0.4850
		109	$CH_2 = CH_2$	1.3564	1.3098
3	ACH	9	ACH	0.5313	0.4000
		10	AC	0.3652	0.1200
4	$ACCH_2$	11	$ACCH_3$	1.2663	0.9680
		12	$ACCH_2$	1.0396	0.6600
		13	ACCH	0.8121	0.3480
7	H_2O	16	H2O	0.9200	1.4000
56	CO_2	117	CO_2	1.3000	0.9820
57	CH_4	118	CH_4	1.1292	1.1240
60	N_2	115	N_2	0.8560	0.9300
61	H_2S	114	H_2S	1.2350	1.2020
62	H_2	113	H_2	0.4160	0.5710

table, the UNIFAC solution model defines two sets of functional groups: main groups and subgroups. A main functional group consists of several subgroups. For instance, the main functional groups CH_2 includes four subgroups: CH_3, CH_2, CH, and C each with different parameters R_k, and Q_k.

On the other hand, the residual contribution γ_i^{resi} is given by

$$\ln\left(\gamma_i^{resi}\right) = \sum_{k=1}^{G} \upsilon_{ki}[\ln(\Gamma_k) - \ln(\Gamma_{ki})] \tag{3.95}$$

Γ_k is activity coefficient corresponding to the functional group k in the mixture, while Γ_{ki} is the activity coefficient for the same group k but in the pure component i. For either the mixture or the pure component, these activity coefficients are calculated from the same equation:

$$\ln(\Gamma_k) = Q_k\left[1 - \ln\left(\sum_{m=1}^{G} \theta_m \Psi_{mk}\right) - \sum_{m=1}^{G} \frac{\theta_m \Psi_{km}}{\sum_{n=1}^{G} \theta_n \Psi_{nm}}\right] \tag{3.96}$$

θ is the area fraction and X_m is the group mole fraction of the functional group m. For the mixture these fractions are

$$\theta_m = \frac{Q_m X_m}{\sum_{n=1}^{G} Q_n X_n} \tag{3.97}$$

$$X_m = \frac{\sum_{k=1}^{C} x_k \upsilon_{mk}}{\sum_{n=1}^{G} \sum_{j=1}^{C} x_j \upsilon_{nj}} \tag{3.98}$$

For the pure component, the only difference lies in the way that the group mole fraction is calculated. The sum is only over the functional groups present in the pure component, that is,

$$X_{mi} = \frac{\upsilon_{mi}}{\sum_{n=1}^{G} \upsilon_{ni}} \tag{3.99}$$

Finally, the Ψ_{mn} is related to the binary group interaction parameters a_{mn} through

$$\Psi_{mn} = \exp\left(-\frac{a_{mn}}{T}\right) \tag{3.100}$$

a_{mn} is the other input datum in the UNIFAC solution model. The corresponding values are given in Table 3.6. As observed from this table, the functional groups m and n correspond to main groups only, that is, interaction parameters are not defined for subgroups. Besides, it can be observed from this table that, for each pair of main groups m and n, there are two binary group interaction parameters a_{mn} and a_{nm}, that is, they are not symmetric ($a_{mn} \neq a_{nm}$). By definition, the interaction parameters between two identical main groups is equal to unity ($\Psi_{nn} = 1$).

Because of the low-pressure condition under which UNIFAC parameters are determined, pairs of components containing gases are absent. Hence, the use of the modified HV-type mixing rules described previously requires determining the corresponding MHV-UNIFAC parameters. As the exact form of MHV mixing rule depends on the selected CEoS and on the q_1 value, these parameters are also MVH-mixing-rule dependent. For instance, PSRK equation results from selecting the SRK equation and liquid volumes at the normal boiling temperature as reference state for establishing the q_1 value. In this case, original UNIFAC parameters are kept and the missing ones are fitted by assuming temperature-dependent parameters according to Eq. (3.102). The resulting parameters are shown from Tables 3.5, 3.6, 3.7 and 3.8. Although the parameters so obtained should not be used with other combinations of CEoS and q_1 value, it will be shown that, in the case of petroleum fluids in the presence of heavy fractions, comparable results are obtained in predicting vapor-liquid equilibriums regardless of the considered combination. For instance, vapor-liquid equilibriums are shown from Figs. 3.10, 3.11 and 3.12, where calculated values are obtained by using UNIFAC parameters from Tables 3.5, 3.6, 3.7 and 3.8, but with the PR equation of state and $q_1 = -0.519$.

Table 3.6 a_{mn} Interaction parameters for PSRK cubic equation of state [36]

Main group	1	2	3	4	7	56	57	60	61	62
1	0.0	86.02	61.13	76.5	1318	919.8	−39.101	271.12	893.01	613.3
2	−35.36	0.0	38.81	74.15	270.6	−52.107	38.602	−231.09	0.0	585.0
3	−11.12	3.446	0.0	167.0	903.8	219.25	131.51	309.59	49.132	734.87
4	−69.7	−113.6	−146.8	0.0	5695	296.88	37.876	−303.76	23.158	320.0
7	300.0	496.1	362.3	377.6	0.0	−1163.5	−1573.2	−3123.4	724.93	−1531.6
56	−38.672	148.57	−29.4	249.32	1720.6	0.0	73.563	−580.82	78.98	838.06
57	68.141	18.780	−65.123	5.5762	−1149.1	196.16	0.0	64.108	511.99	253.92
60	3.7506	667.10	58.687	406.32	1459.6	694.28	11.865	0.0	862.84	77.701
61	742.31	0.0	−98.107	40.011	−8.1881	114.96	278.10	648.2	0.0	628.0
62	315.96	399.44	16.884	126.44	5564.1	3048.9	128.55	247.42	137.0	0.0

Table 3.7 b_{mn} interaction parameters for PSRK cubic equation of state [36]

Main group	1	2	3	4	7	56	57	60	61	62
1	0.0	0.0	0.0	0.0	0.0	−3.9132	0.8459	−0.9903	−3.1342	−2.5418
2	0.0	0.0	0.0	0.0	0.0	1.5473	0.0	5.8221	0.0	−0.8727
3	0.0	0.0	0.0	0.0	0.0	0.0	0.0	0.0	0.0	0.0
4	0.0	0.0	0.0	0.0	0.0	−0.2073	0.0	2.1966	0.0	0.0
7	0.0	0.0	0.0	0.0	0.0	5.4765	11.993	20.683	−0.0136	12.174
56	0.8615	−1.1151	0.0	−0.9249	−4.3437	0.0	0.0	3.6997	0.0	−1.0158
57	−0.7386	0.0	0.0	0.0	5.8604	0.0	0.0	0.0	−1.1761	0.0
60	0.3229	−8.0999	0.0	−1.5230	−7.6407	−3.0173	0.0	0.0	−2.1569	0.0
61	−5.7074	0.0	0.0	0.0	−0.2077	0.0	−0.23	−0.3007	0.0	0.0
62	−0.4563	−0.5806	0.0	0.0	−27.41	−10.247	0.0	0.0	0.0	0.0

Table 3.8 c_{mn} Interaction parameters for PSRK cubic equation of state [36]

Main group	1	2	3	4	7	56	57	60	61	62
1	0.0	0.0	0.0	0.0	0.0	0.004631	0.0	0.003190	0.001302	0.006638
2	0.0	0.0	0.0	0.0	0.0	0.0	0.0	−0.01655	0.0	0.0
3	0.0	0.0	0.0	0.0	0.0	0.0	0.0	0.0	0.0	0.0
4	0.0	0.0	0.0	0.0	0.0	0.0	0.0	0.0	0.0	0.0
7	0.0	0.0	0.0	0.0	0.0	−0.002603	−0.01225	−0.01956	0.001058	−0.006931
56	−0.001791	0.0	0.0	0.0	0.00131	0.0	0.0	0.0	0.0	0.0
57	0.0	0.0	0.0	0.0	−0.005122	0.0	0.0	0.0	0.0	0.0
60	−0.002011	0.02256	0.0	0.0	0.00945	0.0	0.0	0.0	0.0	0.0
61	0.01265	0.0	0.0	0.0	0.0001303	0.0	0.0	0.0	0.0	0.0
62	−0.00156	0.0	0.0	0.0	0.07115	0.0	0.0	0.0	0.0	0.0

Generally, MHV-type mixing rules have shown to be successful models in predicting vapor-liquid equilibria of complex mixtures as those find in the petroleum industry. However, several shortcomings persist. Among them, we have that liquid volumes are not well predicted, though this is a deficiency from CEoS irrespective of the used mixing rule. Besides, unsatisfactory results are obtained in predicting vapor-liquid equilibria for asymmetric mixture, encouraging modifications to the mixing rules already proposed, as will be seen in the next section. Finally, deviations between calculated and experimental excess enthalpies and activity coefficients at infinite dilution are significant, which may be a consequence of UNIFAC solution model itself, rather than a deficiency of the MHV mixing rule, as explained in what follows.

The UNIFAC solution model has been successfully applied to phase equilibrium problems; however, it shows shortcomings concerning activity coefficients at infinite solution particularly for asymmetric systems, and the temperature dependence of excess properties. Thus, to obtain a more robust model, the so-called Modified UNIFAC (Dortmund) solution model was developed by modifying empirically the combinatorial contribution through Eq. (3.84) as follows [37, 38]:

$$\Phi_i' = \frac{x_i r_i^{3/4}}{\sum_{j=1}^{C} x_j r_j^{3/4}} \tag{3.101}$$

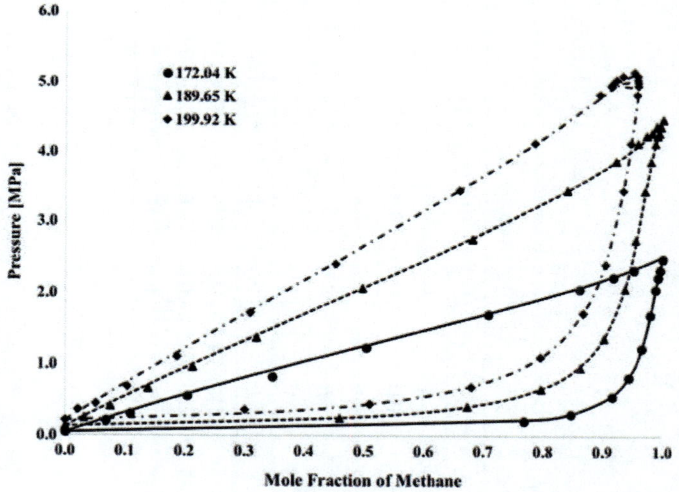

Fig. 3.10 Vapor-Liquid equilibrium for methane-ethane mixtures using PR-CEoS, MHV1 mixing rule ($q_1 = -0.519$), and UNIFAC solution model using parameters from Tables 3.5, 3.6, 3.7 and 3.8. Data from Wichterle and Kobayashi [19]

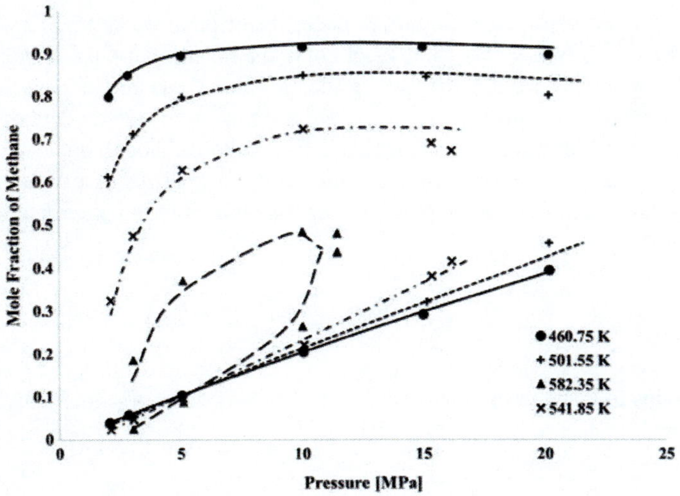

Fig. 3.11 Vapor-Liquid equilibrium for methane-m-Xylene mixtures using PR-*CEoS*, MHV1 mixing rule ($q_1 = -0.519$), and UNIFAC solution model using parameters from Tables 3.5, 3.6, 3.7 and 3.8. Data from Simnick et al. [20]

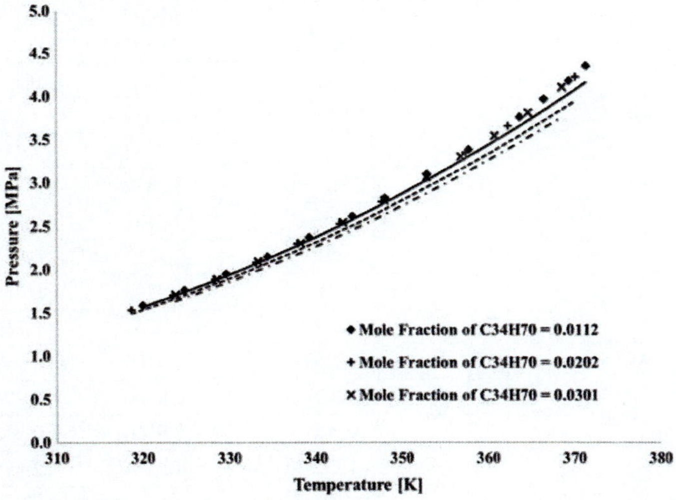

Fig. 3.12 Vapor-Liquid equilibrium for Propane-Tetratriacontane mixtures using PR-CEoS, MHV1 mixing rule ($q_1 = -0.519$), and UNIFAC solution model using parameters from Tables 3.5, 3.6, 3.7 and 3.8. Data from Peters et al. [21]

This molecular volume fraction Φ_i' replaces the original Φ_i in the Flory-Huggins contribution γ_i^{FH} given by Eq. (3.89), but not in the Staverman-Guggenheim contribution γ_i^{SG} given by Eq. (3.90), where the original Φ_i is retained. Besides, the required interaction parameters Ψ_{mn} were adjusted to experimental data of not only vapor-liquid equilibria as in the original UNIFAC solution model, but also of liquid-liquid equilibria, solid-liquid equilibria, and excess properties in a wider temperature interval. In addition, the temperature-dependent interaction parameters are now determined by

$$\Psi_{mn} = \exp\left(-\frac{a_{mn} + b_{mn}T + c_{mn}T^2}{T}\right) \tag{3.102}$$

For groups relevant to petroleum fluids, parameters of the Modified UNIFAC are shown from Tables 3.9, 3.10, 3.11 and 3.12.

Table 3.9 R_k and Q_k parameters for the Modified UNIFAC [38]

Main group		Sub-group		R_k	Q_k
1	CH_2	1	CH_3	0.6325	1.0608
		2	CH_2	0.6325	0.7081
		3	CH	0.6325	0.3554
		4	C	0.6325	0
2	C = C	5	$CH_2 = CH$	1.2832	1.6016
		6	CH = CH	1.2832	1.2849
		7	$CH_2 = C$	1.2832	1.2489
		8	CH = C	1.2832	0.8962
		70	C = C	1.2832	0.4582
3	ACH	9	ACH	0.3763	0.4321
		10	AC	0.3763	0.2113
4	$ACCH_2$	11	$ACCH_3$	0.9100	0.9490
		12	$ACCH_2$	0.9100	0.7962
		13	ACCH	0.9100	0.3769
7	H_2O	16	H_2O	1.7334	2.4561

Table 3.10 The Modified UNIFAC interaction parameters a_{mn} [38]

Main group	1	2	3	4	7
1	0.0	189.66	114.20	7.3390	1391.3
2	−95.418	0.0	174.10	117.3	778.3
3	16.070	−157.2	0.0	139.20	792.0
4	47.200	−113.10	−45.33	0.0	1050.2
7	−17.253	−1301.0	332.3	24.144	0.0

Table 3.11 The modified UNIFAC interaction parameters b_{mn} [38]

Main group	1	2	3	4	7
1	0.0	−0.2723	0.09330	−0.4538	−3.6156
2	0.06171	0.0	−0.5886	−0.8552	0.1482
3	−0.2998	0.6166	0.0	−0.6500	−1.7260
4	0.3575	1.1720	0.4223	0.0	−1.9939
7	0.8389	4.0720	1.1580	1.6504	0.0

Table 3.12 The modified UNIFAC interaction parameters c_{mn}[38]

Main group	1	2	3	4	7
1	0.0	0.0	0.0	0.0	0.001144
2	0.0	0.0	0.0	0.0	0.0
3	0.0	0.0	0.0	0.0	0.0
4	0.0	0.0	0.0	0.0	0.0
7	0.0009021	0.0	0.0	0.0	0.0

3.4.5 Universal Mixing Rule

The main purpose of incorporating solution models into CEoS has been to increase the range of applicability of such equations to chemically more complex mixtures, combining thus the advantages from both types of thermodynamic models. As shown previously, of special interest are those schemes incorporating models based on functional groups such as UNIFAC, because the resulting approaches are predictive. That is, neither specific correlations nor parameters are required to be known for a given mixing. In principle, this seems an attractive characteristic of such schemes for being applied to petroleum fractions, on condition that the respective functional groups can be determined. In the specific case of MHV-type mixing rules in combination with UNIFAC solution model, however, there are still some drawbacks that have been discussed in the open literature [34, 39]. Two of them are particularly important when heavy petroleum fractions are present: the deviations shown in predicting liquid volumes and phase equilibriums of asymmetric mixtures, for example, supercritical gases (CH_4, N_2, H_2, CO_2) in heavy petroleum fluids.

It has been illustrated that the combinatorial contributions from cubic equation of state $\left(\tilde{G}^{comb}/RT = \sum_{i=1}^{C} x_i \ln(B_i/B)\right)$ and from UNIFAC solution model $\left(\tilde{G}^{comb}/RT = \sum_{i=1}^{C} x_i \ln\left(\gamma_i^{FH}\right)\right)$ are responsible for deviations shown by MHV1-type mixing rules in predicting phase equilibriums of asymmetric mixtures [40]. It has been found that they differ from each other for very asymmetric mixtures. The larger the size difference between components in the mixture, the larger the difference between the combinatorial contributions from both thermodynamic models. The postulate is that they should be equal in an independent way for the entire range of conditions and mixtures and not in combination with the residual contribution. However, when solution models are introduced into MHV mixing rules, it is assumed that the excess Gibbs energies from both models, including combinatorial and residual contributions, are equal to each other. To overcome this issue, it has been suggested to discard both combinatorial contributions, which gives rise to, for example, VTPR equation of state [34, 41, 42], but at the expense of recalculating UNIFAC interaction parameters. On the other hand, to avoid UNIFAC parameters to be estimated again, the Flory-Huggins (FG) contribution from UNIFAC solution model has been suggested as the only one to be discarded from the mixing rule. That is, the contribution due to differences in shape are kept. The result is the so-called universal mixing rule (UMR) [43, 44], which takes the form of

$$\frac{A}{B} = \sum_{i=1}^{C} x_i \left[\frac{A_i}{B_i} + \frac{1}{q_1} \ln\left(\gamma_i^{SG} \gamma_i^{resi}\right) \right] \tag{3.103}$$

Besides, the mixing rule for the co-volume is changed by

$$B = \sum_{i=1}^{C} \sum_{j=1}^{C} x_i x_j B_{ij} \tag{3.104}$$

$$B_{ij} = \left(\frac{\sqrt{B_i} + \sqrt{B_j}}{2} \right)^2 \tag{3.105}$$

As in the case of VTPR, the aim of these changes is to cancel out the differences between the combinatorial contributions from both thermodynamic models, but in this case using the existing interaction parameters of UNIFAC solution model. It is important to point out that UMR combined with the PR equation has produced goods results not only for phase equilibriums of asymmetric mixtures, but also for several types of phase equilibrium problems such as liquid-liquid, vapor-liquid-liquid, and solid-gas as well as in estimating heats of mixing. Finally, the needed terms for calculating fugacity coefficients according to Eq. (3.35) are

$$\frac{a_i'}{na} = \frac{B}{A} \left[\frac{\ln\left(\gamma_i^{SG} \gamma_i^{resi}\right)}{q_1} + \frac{A_i}{B_i} \right] + \frac{b_i'}{b} \tag{3.106}$$

$$\frac{b_i'}{b} = \frac{2}{B} \sum_{j=1}^{C} x_j B_{ij} - 1 \tag{3.107}$$

3.4.6 Predictive-PR78 and VTPR Equations

The assumption of constant packing factor allows MHV mixing rules to be derived from a unified framework. However, as mentioned before, the resulting mixing rules still shows some shortcomings such as those found in phase equilibria of asymmetric systems. Searching for a more generalized MHV mixing rule, where such shortcomings vanish, one may further assume no difference between the molecular surface fraction Θ_i and the molecular volume fraction Φ_i ($\Theta_i = \Phi_i$). In other words, molecules similar in shape are assumed, which leads to the respective contribution to disappear from the solution model. On the other hand, considering that the molecular volume r_i from the UNIFAC solution model has similar meaning to the co-volume b_i from CEoS (both provide information about the molecule volume), one can assume that $r_i \approx b_i$. These additional assumptions transform Eq. (3.84) into

$$\frac{\tilde{G}^{resi}}{RT} = \epsilon^* \left(\sum_{i=1}^{C} x_i \frac{A_i}{B_i} - \frac{A}{B} \right) \tag{3.108}$$

This mixing rule is similar to the HV mixing rule, but the infinite-pressure limit is not required in its derivation and the residual contribution to the excess Gibbs energy is the only present. As explained previously, the fact that the combinatorial contributions are not considered avoids in theory the vapor-liquid equilibrium problems for asymmetric systems. Considering that the infinite-pressure condition is not required in deriving the above mixing rule, any solution model developed under the low-pressure condition can be used; however, in such cases, the corresponding parameters are determined including all contributions. Accordingly, they must be recalculated considering the residual contribution only. When the PR equation is selected in combination with the residual contribution from the UNIFAC solution model, the resulting parameters lead to the so-called volume translated Peng-Robinson equation of state (VTPR) [34, 41, 42].

As has been explained, any solution model can be incorporated into the above mixing rule providing the corresponding parameters are recalculated. For instance, another option is to select a Van Laar-type solution model as suggested by Peneloux [45] and Abdoul [46]:

$$\frac{\tilde{G}^{resi}}{RT} = \epsilon^* \frac{\sum_{i=1}^{C} \sum_{j=1}^{C} x_i x_j b_i b_j E_{ij}}{\sum_{i=1}^{C} x_i b_i} \tag{3.109}$$

E_{ij} is the binary interaction parameter, which, like UNIFAC solution model, can be expressed in terms of functional groups by means of

$$E_{ij} = -\frac{1}{2} \sum_{k=1}^{G} \sum_{l=1}^{G} (f_{ik} - f_{jk})(f_{il} - f_{jl}) \chi_{kl} \left(\frac{298.15}{T} \right)^{\left(\frac{\psi_{kl}}{\chi_{kl}} - 1 \right)} \tag{3.110}$$

where χ_{kl} and ψ_{kl} are contributions for the binary interaction between functional groups k and l. f_{il} stands for the occurrence fraction of functional group l in the component i, that is,

$$f_{il} = \frac{\upsilon_{il}}{\sum_k^G \upsilon_{ik}} \tag{3.111}$$

An important characteristic of the Van Laar-type solution model as stated by Eq. (3.109) is its connection with the classical van der Waals mixing rule, because b_i is the co-volume from CEoS. The connection between the solution model given by Eq. (3.109) and the mixing rule given by Eq. (3.108) results from introducing the classical mixing rules given by Eq. (3.45) into Eq. (3.108), with which the following results is reached [47]

$$\frac{\tilde{G}^{resi}}{RT} = \frac{\epsilon^*}{2} \frac{\sum_{i=1}^C \sum_{j=1}^C x_i x_j b_i b_j \left[2\delta_i \delta_j k_{ij} + \left(\delta_i - \delta_j\right)^2\right]}{\sum_{i=1}^C x_i b_i} \tag{3.112}$$

with

$$\delta_i = \frac{\sqrt{a_i}}{b_i} \tag{3.113}$$

Comparison of Eqs. (3.109) and (3.112) leads to

$$k_{ij} = \frac{E_{ij} - \left(\delta_i - \delta_j\right)^2}{2\delta_i \delta_j} \tag{3.114}$$

That is, the classical binary interaction parameters k_{ij} can be expressed as a function of temperature and chemical structures of pure components by means of the functional group contribution concept as stated by Eq. (3.111). The PR equation of state, together with Eq. (3.114), is the so-called predictive Peng-Robinson equation of state (PPR78) [48, 49], whose binary interaction parameter χ_{kl} and ψ_{kl} entering Eq. (3.110) are shown in Tables 3.13 and 3.14. To avoid these parameters to be recalculated when the SRK equation is intended to be used in combination with Eq. (3.114), an analogous expression can be obtained by assuming that the residual excess Gibbs energy from Eq. (3.108) holds regardless of the used cubic equation of state, that is, assuming that $\tilde{G}_{PR}^{resi} = \tilde{G}_{SRK}^{resi}$ together with $\left(\frac{B_i^{PR}}{B_i^{SRK}}\right) = \left(\frac{\Omega_b^{PR}}{\Omega_b^{SRK}}\right)$, we obtain [47]

$$k_{ij}^{SRK} = \frac{2\xi k_{ij}^{PR} \delta_i^{PR} \delta_j^{PR} + \xi\left(\delta_i^{PR} - \delta_j^{PR}\right)^2 - \left(\delta_i^{SRK} - \delta_j^{SRK}\right)^2}{2\delta_i^{SRK} \delta_j^{SRK}} \tag{3.115}$$

together with

Table 3.13 Group interaction parameters ($\chi_{lm} = \chi_{ml}$) [MPa] for the PR equation of state [48, 49]

	CH_3	CH_2	CH	C	CH_4	C_2H_6	ACH	AC	AC_f	$CH2_c$	CH_c or C_c	CO_2	N_2	H_2S	H_2	H_2O
CH_3	0															
CH_2	74.81	0														
CH	261.5	51.47	0													
C	396.7	88.53	−305.7	0												
CH_4	32.94	36.72	145.2	263.9	0											
C_2H_6	8.579	31.23	174.3	333.2	13.04	0										
ACH	90.25	29.78	103.3	158.9	67.26	41.18	0									
AC	62.80	3.775	6.177	79.61	139.3	−3.088	−13.38	0								
AC_f	62.80	3.775	6.177	79.61	139.3	−3.088	−13.38	0	0							
CH_{2c}	40.38	12.78	101.9	177.1	36.37	8.579	29.17	34.31	34.31	0						
CH_c or C_c	98.48	−54.90	−226.5	17.84	40.15	10.29	−26.42	−105.7	−105.7	−50.10	0					
CO_2	164.0	136.9	184.3	287.9	137.3	135.5	102.6	110.1	267.3	130.1	91.28	0				
N_2	52.74	82.28	365.4	263.9	37.90	61.59	185.2	284	718.1	179.5	100.9	98.42	0			
H_2S	158.4	134.6	193.9	305.1	181.2	157.2	21.96	1.029	1.029	120.8	−16.13	134.9	319.5	0		
H_2	202.8	132.5	415.2	226.5	156.1	137.6	284.8	377.5	549.0	232.0	−314.0	265.9	65.20	145.8	0	
H_2O	3557.0	4324.0	971.4	0	2265.0	2333.0	2268.0	543.5	1340.0	4211.0	244.0	559.3	2574.0	603.9	830.8	0

f: fused aromatic rings; c: cyclic

Table 3.14 Group interaction parameters ($\psi_{lm} = \psi_{ml}$) [MPa] for the PR equation of state [48, 49]

	CH_3	CH_2	CH	C	CH_4	C_2H_6	ACH	AC	ACf	CH_{2c}	CHc or Cc	CO_2	N_2	H_2S	H_2	H_2O
CH_3	0															
CH_2	165.7	0														
CH	388.8	79.61	0													
C	804.3	315.0	−250.8	0												
CH_4	−35.0	108.4	301.6	531.5	0											
C_2H_6	−29.51	84.76	352.1	203.8	6.863	0										
ACH	146.1	58.17	191.8	613.2	167.5	50.79	0									
AC	41.86	144.8	−33.97	−326.0	464.3	13.04	20.25	0								
ACf	41.86	144.8	−33.97	−326.0	464.3	13.04	20.25	0	0							
CH_2c	95.9	28.37	−90.93	601.9	26.42	76.86	69.32	95.39	95.39	0						
CHc or Cc	231.6	−319.5	−51.47	−109.5	255.3	−52.84	−789.2	−286.5	−286.5	−891.1	0					
CO_2	269.0	254.6	762.1	346.2	194.2	239.5	161.3	637.6	444.4	225.8	82.01	0				
N_2	87.19	202.8	521.9	772.6	37.20	84.92	490.6	1892.0	1892.0	546.6	249.8	221.4	0			
H_2S	241.2	138.3	307.8	−143.1	288.9	217.1	13.04	−8.579	−8.579	163.0	−147.6	201.4	550.1	0		
H_2	317.4	147.2	726.4	1812.0	92.99	150.0	175.0	1201	1476.0	167.5	−225.8	268.3	70.10	823.5	0	
H_2O	11195.0	12126.0	567.6	0	4722.0	5147.0	6218.0	411.8	−65.88	13031.0	−60.39	277.9	5490.0	599.1	−137.9	0

f: fused aromatic rings; c: cyclic

$$\xi = \left(\frac{\epsilon^*_{PR}}{\epsilon^*_{SRK}}\right)\left(\frac{\Omega_b^{PR}}{\Omega_b^{SRK}}\right) \quad (3.116)$$

Alternatively, one can use the same group interaction parameters χ_{kl} and ψ_{kl} from Tables 3.13 to 3.14 for the PR equation it order to estimate the corresponding parameters for the SRK equation. This is carried out using the relationship between binary interaction parameter from both CEoS; that is, $E_{ij}^{SRK} = \xi E_{ij}^{PR}$ where $\xi = 0.8073411$ [47, 50]. It is important to underline that this value of ξ is obtained by assuming that $\eta = 1$, that is, taking the infinite- pressure limit ($P \to \infty$), with which we have that $\epsilon = \epsilon^*$.

From Figs. 3.13 to 3.15, the performance of the classical (Eq. (3.49) y PPR78), MHV1, and UMR mixing rules are compared with each other. As observed, vapor-liquid equilibriums are well predicted by MHV1 and UMR mixing rules in combination with the UNIFAC solution model On the other hand, the classical mixing rule overestimates pressures when k_{ij} are estimated from Eq. (3.49), while the calculated values from PPR78 agree well with experiments. For the most asymmetric mixture (propane-eicosane mixture), the worst predictions are obtained from the classical mixing rule, while the best predictions are obtained from UMR mixing rule and PPR78. These results justify the elimination of effects due to components very different in size.

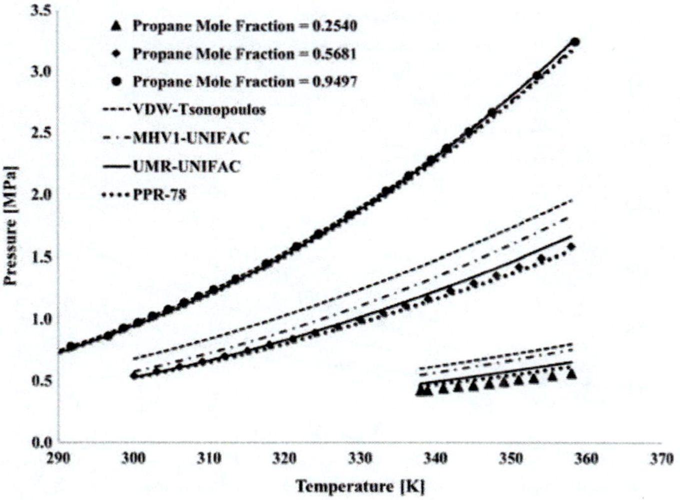

Fig. 3.13 Vapor-Liquid equilibrium of Propane-Eicosane mixtures using PR-CEoS with MHV1, UMR ($q_1 = -0.519$), and the classical mixing rule (PPR78, and Eq. (3.50)). UNIFAC solution model using parameters from Tables 3.5, 3.6, 3.7 and 3.8. Data from Gregorowtez et al. [51]

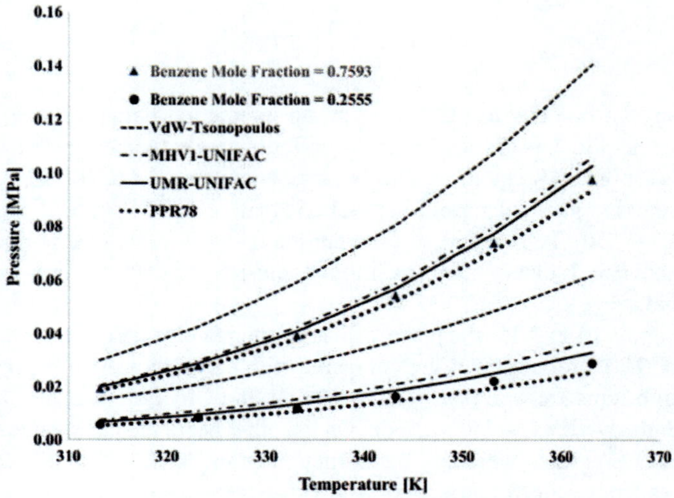

Fig. 3.14 Vapor-Liquid equilibrium of Benzene-Eicosane mixtures using PR-CEoS with MHV1, UMR ($q_1 = -0.519$), and the classical mixing rule (PPR78, and Eq. (3.50)). UNIFAC solution model using parameters from Tables 3.5, 3.6, 3.7 and 3.8. Data from Liu [52]

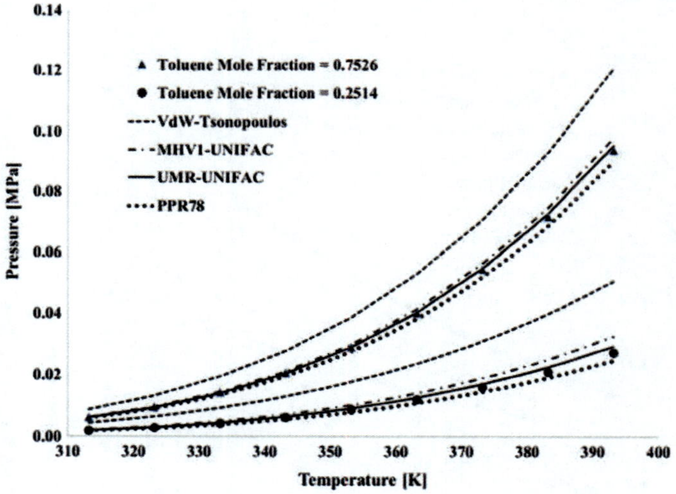

Fig. 3.15 Vapor-Liquid equilibrium of Toluene-Eicosane mixtures using PR-CEoS with MHV1, UMR ($q_1 = -0.519$), and the classical mixing rule (PPR78, and Eq. (3.50)). UNIFAC solution model using parameters from Tables 3.5, 3.6, 3.7 and 3.8. Data from Liu [52]

3.5 Volume Translation

As shown in previous sections, the use of excess Gibbs energy models allows the PR and SRK equations to be applied to non-ideal mixtures, being thus suitable to predict fluid phase behavior and volumetric properties for a wide variety of fluid systems. However, one shortcoming remains; the liquid volume is not predicted properly. The simplest way to overcome this deficiency is by volume translations [53], where the predicted liquid molar volume is translated or corrected by means of

$$\tilde{V} = \tilde{V}^{CEoS} + \sum_{i=1}^{C} x_i s_i \tag{3.117}$$

The parameter s_i is the volume translation of component i, $\tilde{V}$ is the actual volume of the liquid mixture, while $\tilde{V}^{CEoS}$ corresponds to the volume calculated from some cubic equation of state. The above equation indicates that a systematic difference between experimental and calculated molar volumes is found from these type of equations of state. Although this way of correcting $\tilde{V}^{CEoS}$ improves significantly predictions, an additional correction term is necessary near the critical point [54].

The most outstanding characteristic of s_i is that the phase equilibrium calculations remain unchanged. In other words, the use of a given mixing rule either classical or non-classical is independent of the way that s_i is established. Another advantage of the volume translation is that s_i can calculated so that experimental and calculated values of the molar liquid volume coincide with each other; that is,

$$s_i = \tilde{V}_i^{Exp} - \tilde{V}_i^{CEoS} = \frac{M_i^{Exp}}{\hat{\rho}_i^{Exp}} - \tilde{V}_i^{CEoS} \tag{3.118}$$

The superscript *Exp* indicates an experimental value, while M is the molecular weight and $\hat{\rho}$ the mass density. The above expression is particularly convenient for pure components, whose properties are easier to estimate. Considering that M and $\hat{\rho}$ are known values for petroleum fractions at standard conditions (0.101325 MPa and 288.15 K), Eq. (3.118) ensures that the calculated volume coincides with the experimental value at least at these conditions. However, when an experimental value is not available, estimations of s_i are needed. For the SRK equation, the following correlation can be used [53]:

$$s_i = 0.40768 \frac{RT_{ci}}{P_{ci}} (0.00385 + 0.08775\omega_i) \tag{3.119}$$

This expression was obtained by regression to saturated liquid volumes of the first ten n-alkanes at reduce temperature T_r ($T_r \equiv T/T_{ci}$) equal to 0.7. On the other hand, for the PR equation, we can use [22]

Table 3.15 Parameters for estimating the volume translation s_i [cm^3/mol]

Parameters	SRK		PR	
	C_{1i}	C_{2i}	C_{1i}	C_{2i}
d_{1k}	0.2300	−0.3471	−4.1034	−0.3489
d_{2k}	46.843	−29.748	31.723	−28.547
d_{3k}	0.0571	0.0644	0.0531	0.0687
d_{4k}	23161.0	−347.04	188.68	−817.73
d_{5k}	0.0003	0.001	0.0057	0.0007
d_{6k}	267.4	−88.547	20.196	−65.067
d_{7k}	0.0053	0.0048	0.0003	0.0076

$$s_i = 0.50033\frac{RT_{ci}}{P_{ci}}(-0.03087 + 0.08775\omega_i) \tag{3.120}$$

On the other hand, s_i has been suggested as a function of temperature and other properties different from critical ones. Baled et al. [55] compared several options available in the open literature including the PC-SAFT equation of state, especially at high pressure and temperature. For the interval pressure from 7 to 276 MPa and the interval temperature from 278 to 533 K, they developed a correlation in terms of temperature, molecular weight and acentric factor in the form of

$$s_i = -(C_{1i} + C_{2i}T_{ri}) \tag{3.121}$$

The coefficients C_{1i} and C_{2i} are a function of the molecular weight and acentric factor according to Eq. (3.122) and the correlation parameters d_{jk} are shown in Table 3.15. Equation (3.121) is particularly attractive for extreme conditions, obtaining results comparable to those obtained from other approaches at low pressure.

$$C_{ki} = d_{1k} + \frac{d_{2k}}{e^{M_i\omega_i d_{3k}}} + \frac{d_{4k}}{e^{M_i\omega_i d_{5k}}} + \frac{d_{6k}}{e^{M_i\omega_i d_{7k}}} \tag{3.122}$$

When chemical structures are available for petroleum fractions, s_i has been proposed to be calculated in terms of the pure-component molar volume and the functional groups as follows [56, 57]:

$$s_i = \sum_{j=1}^{G} \upsilon_{ij}\Delta v_j - V_i^{CEoS} \tag{3.123}$$

where υ_{ij} stands for the occurrence of the functional group j in molecule i and G for the types of functional groups considered. Δv_j is the group volume increment corresponding to the functional group j, which depends on temperature according to

$$\Delta v_j = \iota_j + \lambda_j T + \chi_j T^2 \tag{3.124}$$

Table 3.16 Parameters of the GCVOL method for estimating liquid densities

GCVOL group	$\iota_i 10^5$ [m^3/mol]	$\lambda_j 10^8$[m^3/mol K]	$\chi_j 10^{10}$ [m^3/mol K^2]
$-CH_3$	1.643	5.562	0.000
$-CH_2-$ (chain)	1.202	1.410	0.000
>CH− (chain)	0.730	−2.606	0.000
>C < (chain)	8.780	−61.990	8.822
ACH	0.993	1.741	0.000
$ACCH_3$	2.471	2.111	0.000
$ACCH_2$	1.684	−0.464	0.000
ACCH	5.439	−29.080	3.301
ACC	3.937	−27.210	2.492
$ACNH_2$	1.300	1.211	0.000
$-CH_2=$	3.269	−6.014	1.628
>CH=	−0.165	9.342	−1.439
>C=	−1.093	6.241	−1.413
$-CH_2-$ (cyclic)	1.565	0.599	0.000
>CH− (cyclic)	−5.295	29.560	−3.138
>C < (cyclic)	11.580	−76.700	12.260

where ι_i, λ_j, and χ_j are parameters of the Group-Contribution-Volume (GCVOL) method [58], which are shown in Table 3.16. This method is used in the prediction of liquid densities of complex compounds from the melting temperature to the normal boiling temperature. It is worth noticing that comparison of Eq. (3.118) with Eq. (3.123) leads to

$$\tilde{V}_i = \frac{M_i}{\hat{\rho}_i} = \sum_{j=1}^{G} \upsilon_{ij} \Delta v_j \tag{3.125}$$

which is the equation suggested in the GCVOL method for estimating the liquid density of pure components in terms of functional groups.

3.6 CEoS Parameters for Heavy Petroleum Fractions

From the previous sections, it is observed that CEoS and the corresponding mixing rules described previously, critical properties and chemical structures of the pure components are required in either estimating properties of a pure component itself or their mixtures. Accordingly, to be able to use them in the presence of petroleum fractions, it is necessary to develop methods that allows such properties to be estimated for mixtures whose forming components are unknown, and therefore, their critical properties, chemical structures, and composition. The determination of all needed information in order to use CEoS and their respective mixing rules for phase behavior calculations is what we call the thermodynamic characterization of petroleum fractions. As will be seen, such determination is often performed from a limited or incomplete experimental data.

3.6.1 Correlations for Estimating Critical Properties and Acentric Factor

Because components forming petroleum fractions are unknown, many sets of correlations have been developed for estimating their critical properties and acentric factor, which are in terms of other properties that are experimentally accessible, such as the normal boiling temperature T_{bj}, the specific gravity S_j, and the molecular weight M_j. Such correlations are based on properties of pure components with M_j smaller than 350 and with T_{bj} lower than 700 K approximately, because critical properties and even normal boiling temperatures are experimentally unattainable for bigger and more complex components due to thermal cracking, as can be seen from Figs. 3.16 and 3.17. For instance, M_j may be as large as 2500 or even more for asphaltene fractions and they chemically decompose before something like a critical point can be reached. In other words, properties entering CEoS are not well-defined properties. Therefore, the use of CEoS implies that properties are extrapolated estimations when heavy components or petroleum fractions are involved. This extrapolation has led to suggest several strategies in order to experiments and calculations coincide with each other. In general, one may follow two options to carry out a regression process to experimental data: adjust the parameters entering into CEoS such as binary interaction parameters or critical properties, or search for a combination of CEoS

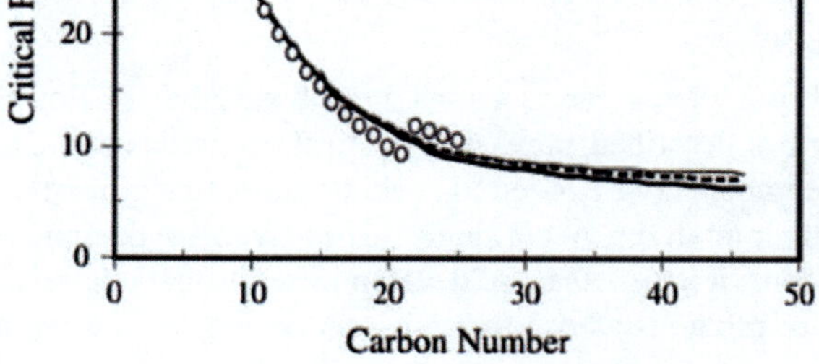

Fig. 3.16 Estimation of various correlations for calculation of critical pressure of n-alkylcyclopentanes [59]

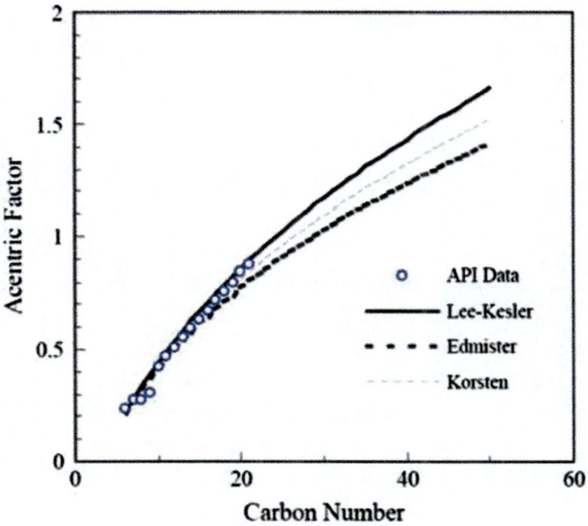

Fig. 3.17 Evaluation of various correlations for calculation of acentric factor of n-alkylcyclopentanes [59]

and mixing rule so that the available data can be predicted properly. Both will be illustrated in the next Chapter.

Comparisons between several sets of correlations for estimating properties entering CEoS are available in the open literature, where applications to light and medium petroleum fluids are illustrated [22, 60, 61]. However, few assessments have been performed to establish what sets of correlations are suitable for petroleum fluids in the presence of heavy fractions. One of these assessments involved solvent-heavy fraction mixtures [16, 62], where it is concluded that the set of correlations proposed by Lee and Kesler provides the best results [63]. In this set of correlations, T_{cj} is determined from

$$T_{cj} = 189.8 + 450.6S_j + \left(0.4244 + 0.1174S_j\right)T_{bj} + \left(0.1441 - 1.0069S_j\right)\frac{10^5}{T_{bj}} \tag{3.126}$$

while for P_{cj} the correlation is

$$\ln\left(P_{cj}\right) = 3.3864 - \frac{0.0566}{S_j} - \left(0.43639 + \frac{4.1216}{S_j} + \frac{0.21343}{S_j^2}\right)\frac{T_{bj}}{10^3} + \left(0.47579 + \frac{1.182}{S_j} + \frac{0.15302}{S_j^2}\right)\frac{T_{bj}^2}{10^6} - \left(2.4505 + \frac{9.9099}{S_j^2}\right)\frac{T_{bj}^3}{10^{10}} \tag{3.127}$$

When ω_j is intended to be estimated, two cases are considered. If the reduced temperature evaluated at the normal boiling temperature $\left(T_{rb} \equiv T_{bj}/T_{cj}\right)$ is smaller than 0.8 ($T_{rb} < 0.8$) then

$$\omega_j = \frac{\ln(P_{rb}) - 5.92714 + 6.09648/T_{rb} + 1.28862\ln(T_{rb}) - 0.1693471T_{rb}^6}{15.2518 - 15.6875/T_{rb} - 13.4721\ln(T_{rb}) + 0.43577T_{rb}^6} \tag{3.128}$$

P_{rb} is the reduced pressure evaluated at the normal boiling pressure ($P_b = 0.101325MPa$) and is defined as $P_{rb} \equiv P_b/P_{cj}$. On the other hand, if $T_{rb} \geq 0.8$ then

$$\omega_j = -7.904 + 0.1352K_W - 0.007465K_W^2 + 8.359T_{rb} + \frac{1.408 - 0.01063K_W}{T_{rb}} \tag{3.129}$$

where K_W is the so-called Watson characterization factor, which is defined as

$$K_W \equiv \frac{\sqrt[3]{1.8T_{bj}}}{S_j} \tag{3.130}$$

This factor may be seen as a measure of the chemical nature of the petroleum fraction: it varies from 12.5 to 13.5 for paraffinic fractions, while it varies from 11.0 to 12.5 for naphthenic fractions; and it varies from 8.5 to 11.0 for aromatic fractions [60]. Finally, the corresponding correlation for M_j is given by

$$M_j = -12272.6 + 9486.4S_j + \left(8.3741 - 5.99177S_j\right)T_{bj} + \left(1 - 0.77084S_j - 0.02058S_j^2\right)\left(\frac{0.7465}{T_{bj}} - \frac{222.466}{T_{bj}^2}\right)10^7 + \left(1 - 0.80882S_j + 0.02226S_j^2\right)\left(\frac{0.3228}{T_{bj}^3} - \frac{17.335}{T_{bj}^4}\right)10^{12} \tag{3.131}$$

When the available experimental data are M_j and S_j, one must find T_{bj} by solving Eq. (3.131) iteratively. Instead, one can use the following correlation that is explicit for T_{bj} [60]:

$$T_{bj} = 1071.28 - 9.4167x10^4 M_j^{-0.03522} S_j^{3.266} e^{-4.922x10^{-3}M_j - 4.7685S_j + 3.462x10^{-3}M_jS_j} \tag{3.132}$$

3.6.2 *Estimation of Parameters from Chemical Pseudostructures*

As most of forming components of petroleum fractions are unknown, it is not possible to establish accurately the corresponding physicochemical properties such as critical properties and chemical structures. As shown previously, critical properties can be estimated using empirical correlations in terms of other bulk properties, but only a few similar methods have been developed for estimating chemical structures. This implies that methods such as UNIFAC, GCVOL and any method requiring chemical structures cannot be used routinely in the presence of petroleum fractions. To deal with this problem, an option is to make use of molecular reconstruction techniques, which allows generate numerically mixtures of defined components that represent physicochemical properties of petroleum fractions properly [64–69]. In other words, these methods transform petroleum fractions into mixtures with well-identified components. Two are the main disadvantages of these type of techniques: the considerable amount of required experimental information as input data and the number of components in the mixtures (from tens to thousands for each petroleum fraction). A more practical method is to make use of the concept of the chemical pseudostructure.

The concept of chemical pseudostructures is encouraged by the concept of pseudo-property used in estimating critical pressure and temperature for petroleum fractions. To estimate such properties, it is assumed that petroleum fractions do not behave as mixtures but like pure components. This means that they can be treated as pure components from a mathematical standpoint as well. This is the reason why critical properties and acentric factors of pure components are taken as starting point to build the procedure described previously. Accordingly, it is also possible to treat a petroleum fraction as pure component not only from a physical standpoint, but also from a chemical standpoint by determining one chemical structure (pseudostructure) for such fraction. Therefore, the chemical pseudostructure of a petroleum fraction consists of a set of either atoms or functional groups so that the chemical reactivity of the petroleum fraction can be characterized as it behaves like a pure component. In Fig. 3.18, the chemical-pseudostructure concept is illustrated. Taking the chemical structure of ethyl benzene as reference, one might suggest a similar pseudostructure for the shown hypothetical petroleum fraction. It is important to underline that only one pseudostructure is suggested because the aim is to treat the petroleum fraction as single pure component. In contrast to actual molecular structure of the ethyl benzene (5 ACH, 1 AC 1 $-CH_2-$ and 1 $-CH_3$), the molecular pseudostructure of the hypothetical petroleum fraction can be described by non-integer occurrences of the functional groups [70]. In fact, the non-integer occurrences (4.75 ACH, 0.75 AC 0.75 $-CH_2-$ and 0.75 $-CH_3$) do not exist physically; there is no 0.75 carbon or hydrogen atoms. Therefore, this molecule does not exist either; the functional groups are arranged in such a fashion only for illustrative purposes. This means that none of the experimental compositional analysis will return chemical structures like the pseudostructure shown in Fig. 3.18. In other words, this pseudostructure is merely

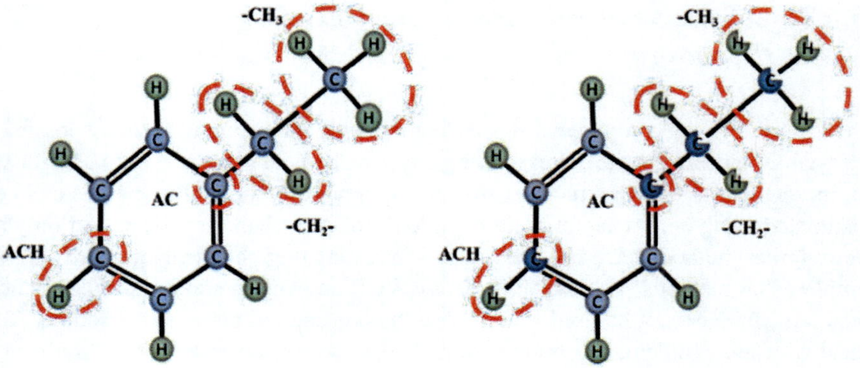

Fig. 3.18 Comparison between a pure-component molecular structure and a petroleum-fraction molecular pseudostructure in terms of functional groups. Ethyl benzene: 5 ACH, 1 AC 1 $-CH_2-$ and 1 $-CH_3$. Petroleum fraction: 4.75 ACH, 0.75 AC 0.75 $-CH_2-$ and 0.75 $-CH_3$

a mathematical representation accounting for the reactivity of the whole petroleum fraction, subject to some physicochemical constraints, as will be shown next.

PNA Content

For the purpose of applying the PPR78 mixing rules to petroleum fractions, Xu et al. [71] developed a group contribution method to estimate T_c, P_c and ω by considering only three chemical families: paraffinic (P), naphthenic (N) and aromatic (A). The following equations were suggested:

$$T_{cj} = T_a \ln\left(\sum_{i=1}^{3} \upsilon_{ij} t_{ci}\right) \tag{3.133}$$

$$P_{cj} = P_a + \left(P_b + \sum_{i=1}^{3} \upsilon_{ij} p_{ci}\right)^{-2} \tag{3.134}$$

$$\omega_j = \omega_a \left(\omega_c + \sum_{i=1}^{3} \upsilon_{ij} w_i\right)^{\frac{1}{\omega_b}} \tag{3.135}$$

In the above equations, υ_{ij} is the occurrence of the functional group i in component j. The universal constants of the above equations, as well as the group contributions, were determined by regression to experimental data of T_c, P_c and ω, together with their respective chemical structures of pure components. Accordingly, we have that t_{ci}, p_{ci}, and w_i account for contributions to T_c, P_c and ω, because of the presence of functional group i. The used data bank in regressions was composed of hydrocarbons ranged from 6 to 24 carbon atoms; the obtained values are shown in Tables 3.17 and 3.18.

Table 3.17 Model constants for the group contribution method given by Eqs. (3.133) to (3.135)

T_a[K]	238.2735
P_a[MPa]	0.335196
P_b[MPa]$^{-0.5}$	0.370214
ω_a	0.597969
ω_b	0.451450
ω_c	1.393320

Table 3.18 Group contributions to estimate T_c, P_c and ω from Eqs. (3.133) to (3.135)

Functional group	t_{ci}	p_{ci}	w_i
Paraffinic (P)	1.304497	0.036416	0.093981
Naphthenic (N)	1.601452	0.027210	0.066826
Aromatic (A)	1.828457	0.018064	0.083756

At this point, equations from (3.133) to (3.135) allows T_c, P_c and ω to be determined when the chemical structure of a pure component is known. However, in the case of a petroleum fraction, an inversion procedure is required because the chemical structure is unknown. The inversion procedure begins with T_c, P_c and ω of the petroleum fraction under study as input data. Using these physical properties, occurrences υ_{Pj},υ_{Nj}, and υ_{Aj} of the three functional groups (paraffinic, naphthenic and aromatic) are determined by solving equations from (3.133) to (3.135) iteratively. It is assumed that υ_{Pj},υ_{Nj}, and υ_{Aj} correspond to groups CH_2, $CH_{2,c}$, and ACH from Tables 3.13 and 3.14, respectively. This means that each petroleum fraction will consist of three functional groups only: one paraffinic, one naphthenic, and one aromatic. Once this procedure has been carried out for each petroleum fraction forming the petroleum fluid, these calculated values of υ_{Pj},υ_{Nj}, and υ_{Aj} can be used into the PPR78 mixing rule in order to estimate binary interaction parameters for each pair of petroleum fractions, becoming the petroleum fluid into a mixture of known components (pseudocomponents) from a mathematical point of view. The major drawback of this method is the difficulty in knowing experimental T_c, P_c and ω for petroleum fractions, especially for heavy ones. Hence, other methods must be used to estimate such properties from available experimental properties such as M_j, S_j and T_{bj}. For instance, Nasrifar et al. [50] used the correlations suggested by Twu [72] in order to estimate T_c, P_c, and ω. Despite this drawback, the procedure described above illustrates that procedures developed for estimating properties in terms of chemical structures can be used in principle for the thermodynamic characterization of petroleum fluids through the concept of the chemical pseudostructure.

Minimization of the Gibbs Free Energy of Mixing

Instead of calculating occurrences υ_{ij} by solving the system of equations from (3.133) to (3.135), matching critical properties and acentric factor that are not experimental data. We may search for calculating υ_{ij} by satisfying properties that are available experimentally for petroleum fractions such as M_j, S_j and T_{bj}. This implies that an adjustment process to experimental data determines υ_{ij}. In either case, the resulting set of nonlinear equations may be satisfied by more than one solution from a mathematical standpoint. To seek a mathematical solution with physical meaning in addition to satisfying the available experimental data, one could make use of the equilibrium criteria discussed in Chap. 2. It was illustrated that the fundamental relationships of thermodynamics and the corresponding spontaneity and equilibrium criteria remain valid regardless of the internal nature of the considered system. Thus, if the system is a petroleum fraction, and it is assumed to be composed of functional groups instead of identifiable components, we could ask for the type and the occurrence of functional groups so that the petroleum fraction is thermodynamically stable, that is, for which the petroleum fraction is at thermodynamic equilibrium. Because experimental data are measured at standard conditions (288.71 K and 0.101325 MPa), the Gibbs free energy is the proper thermodynamic potential to establish such equilibrium state. Therefore, we may search for occurrences υ_{ij} so that the Gibbs free energy of the petroleum fraction becomes minimum. The result is a minimization process where the experimental data are mathematical constraints that must be satisfied. Carreón-Calderón et al. proposed this characterization method [56], which has been applied to the thermodynamic characterization of petroleum fractions from gas-condensate [73], light [74] and heavy petroleum fluids [15, 75, 76]. As all group contribution methods, the idea is that even though a petroleum fraction may contain from tens to thousands of components, it contains only a few functional groups, reducing thus the number of degrees of freedom. The objective function to be minimized is obtained from the molar Gibbs free energy for a pure component derived in Chap. 2:

$$\frac{\tilde{G}_j^0}{RT} = \ln\left(\phi_j^0\right) + \frac{\mu_j^0}{RT} + \ln\left(\frac{P_j}{P^0}\right) \tag{3.136}$$

Bearing in mind that $\tilde{G}_j^0$ is minimum for a given pressure and temperature when a system is in equilibrium, the ideal-gas contribution (the second and third terms on the right-hand side) can be discarded, because it depends on these variables only, and consequently, it will remain constant during any minimization process. Thus, we have that the objective function to be minimized reduces to

$$\frac{\tilde{G}_j^0}{RT} = \ln\left(\phi_j^0\right) \tag{3.137}$$

Because temperature and pressure remain fixed, the only variable responsible for the energy of the petroleum fraction j is its chemical nature; that is, its chemical pseudostructure that consists of functional groups. Therefore, the functional groups that form the chemical pseudostructure of a given petroleum fraction j are found from

$$\min_{\upsilon_{1j},\upsilon_{2j},\ldots\upsilon_{1G}} \ln\left(\phi_j^0\right) \tag{3.138}$$

The subscript G stands for the types of functional groups that are likely present in the petroleum fraction. According to what has been shown previously, ϕ_j^0 depends on ω_j, P_{cj} and T_{cj} of the pure component j, and not on υ_{ij} of the functional groups. Hence, relationships connecting these properties with υ_{ij} are required. The needed relationships are given by group-contribution methods such as the Joback-Reid [77] or the Marreno-Gani methods [78]. Although the latter consists of contributions at three levels to estimate properties, only the first one is enough here. These methods allow the objective function to be rewritten as

$$\min_{\upsilon_{1j},\upsilon_{2j},\ldots\upsilon_{1G}} \ln\left\{\phi_j^0\left[T_{cj}(\upsilon_{1j},\upsilon_{2j},\ldots\upsilon_{1G}), P_{cj}(\upsilon_{1j},\upsilon_{2j},\ldots\upsilon_{1G}), \omega_j(\upsilon_{1j},\upsilon_{2j},\ldots\upsilon_{1G})\right]\right\} \tag{1.139}$$

or merely

$$\min_{\upsilon_{1j},\upsilon_{2j},\ldots\upsilon_{1G}} \ln\left[\phi_j^0(\upsilon_{1j},\upsilon_{2j},\ldots\upsilon_{1G})\right] \tag{3.140}$$

In Table 3.19, expressions for determining critical properties and the normal boiling temperature T_{bj} in terms of occurrences υ_{ij} are shown for the two considered group-contribution methods. t_{ci}, p_{ci}, v_{ci}, and t_{bi} are contributions to T_{cj}, P_{cj}, V_{cj}, and T_{bj} because of the functional group i. The corresponding values are shown Table 3.20 for the used functional groups in the thermodynamic characterization of petroleum fractions. As can be seen from Table 3.19, in the Joback-Reid method, an experimental value of T_{bj} is necessary in order to calculate T_{cj}. If such a value is not available, then the expression suggested in the Marrero-Gani can be used, because the expression in the Joback-Reid method does not give good predictions. Another option is to make use of some correlation in terms of bulk properties, for example Eq. (3.132).

Table 3.19 Critical properties and normal boiling temperature T_{bj} in terms of occurrences υ_{ij} of functional groups

Property	Group-contribution method	
	Joback-Reid [77]	Marrero-Gani [78]
T_{cj} [K]	$\frac{T_{bj}}{0.584+0.965\sum_{i=1}^{G}\upsilon_{ij}t_{ci}-\left(\sum_{i=1}^{G}\upsilon_{ij}t_{ci}\right)^2}$	$231.239ln\left(\sum_{i=1}^{G}\upsilon_{ij}t_{ci}\right)$
P_{cj} [MPa]	$\frac{0.1}{\left(0.113+0.0032n_A-\sum_{i=1}^{G}\upsilon_{ij}p_{ci}\right)^2}$	$\frac{0.1}{\left(0.108998+\sum_{i=1}^{G}\upsilon_{ij}p_{ci}\right)^2}+0.59827$
V_{cj} [m^3/mol]	$\frac{17.5+\sum_{i=1}^{G}\upsilon_{ij}v_{ci}}{10^6}$	$\frac{7.95+\sum_{i=1}^{G}\upsilon_{ij}v_{ci}}{10^6}$
T_{bj}[K]	$198.2+\sum_{i=1}^{G}\upsilon_{ij}t_{bi}$	$222.543ln\left(\sum_{i=1}^{G}\upsilon_{ij}t_{bi}\right)$

Table 3.20 Contributions to T_{cj}, P_{cj}, V_{cj}, and T_{bj} for functional groups needed during the minimization process of the fugacity coefficient

Functional group			Joback-Reid [77]				Marrero-Gani [78]			
	M_i	RHC_i	t_{ci}	p_{ci}	v_{ci}	t_{bi}	t_{ci}	p_{ci}	v_{ci}	t_{bi}
$-CH_3$	15.0344	3.0	0.0141	−0.0012	65.00	23.58	1.7506	0.018615	68.35	0.8491
$-CH_2-$ (chain)	14.0265	2.0	0.0189	0.0000	56.00	22.88	1.3327	0.013547	56.28	0.7141
>CH− (chain)	13.0186	1.0	0.0164	0.0020	41.00	21.74	0.5960	0.007259	37.50	0.2925
>C < (chain)	12.0107	0.0	0.0067	0.0043	27.00	18.25	0.0306	0.001219	16.01	−0.0671
$-CH_2-$ (cyclic)	14.0265	2.0	0.0100	0.0025	48.00	27.15	1.8815	0.009884	49.24	0.8234
>CH- (cyclic)	13.0186	1.0	0.0122	0.0004	38.00	21.78	1.1020	0.007596	44.95	0.5946
>C < (cyclic)	12.0107	0.0	0.0042	0.0061	27.00	21.32	-0.2399	0.003268	33.32	0.0386
AC	12.0107	0.0	0.0143	0.0008	32	31.01	5.4979	0.003564	35.71	1.7324
ACH	13.0186	1.0	0.0082	0.0011	41.00	26.73	2.0337	0.007260	42.39	0.8365
$ACCH_3$	28.0530	1.5	0.0284	−0.0004	97.00	54.59	3.4611	0.020907	97.33	1.5633

Once T_{cj}, P_{cj}, and T_{bj} have been established, the missing property entering into CEoS is ω_j, which can be estimated from [5]

$$\omega_j = -\frac{\ln\left(P_{cj}/0.101325\right) + f_j^{(0)}}{f_j^{(1)}} \tag{3.141}$$

with

$$f_j^{(0)} = \frac{-5.97616\tau + 1.29874\tau^{1.5} - 0.60394\tau^{2.5} - 1.06841\tau^5}{T_{rb}} \tag{3.142}$$

$$f_j^{(1)} = \frac{-5.03365\tau + 1.11505\tau^{1.5} - 5.41217\tau^{2.5} - 7.46628\tau^5}{T_{rb}} \tag{3.143}$$

$f_j^{(0)}$ and $f_j^{(1)}$ are functions of T_{bj} and T_{cj} through the parameters $\tau = 1 - T_{rb}$, where T_{rb} is the reduced temperature evaluated at T_{bj} $\left(T_{rb} \equiv T_{bj}/T_{cj}\right)$.

Thus, either the Joback-Reid or Marrero-Gani method allows ϕ_j^0 to be minimized as a function of υ_{ij}. The available experimental properties of the petroleum fraction are written as linear constraints of such a minimization process. The corresponding constraint for M_j is

$$\sum_{i=1}^{G} \upsilon_{ij} M_i - M_j = 0 \tag{3.144}$$

while for the mass density $\tilde{\rho}_j$ is

$$\sum_{i=1}^{G} \upsilon_{ij} \Delta v_i - \frac{M_j}{\hat{\rho}_j} = 0 \tag{3.145}$$

Δv_i is taken from Table 3.16. It is worth remembering that the specific gravity S_j is related to $\hat{\rho}_j$ by means of

$$S_j = \frac{\hat{\rho}_j}{\hat{\rho}_{H_2O@288.71\ \mathrm{K}}} \tag{3.146}$$

where $\hat{\rho}_{H_2O} = 999.02\mathrm{kg/m^3}$ at standard conditions. Typically, the API_j is known instead of S_j for a given petroleum fraction, but they are related to each other by

$$API_j = \frac{141.5}{S_j} - 131.5 \tag{3.147}$$

In addition to M_j and S_j, the calculation of υ_{ij} may require to satisfy the molar ratio of atomic hydrogen to carbon of the petroleum fraction (HCR_j), especially for petroleum fractions with high content of aromatic components. This ratio in the form of linear constraints is

$$\sum_{i=1}^{G} \upsilon_{ij}\left[HCR_i - HCR_j\right] = 0 \tag{3.148}$$

The first constraint states that the sum of the product of all occurrences υ_{ij} of the functional groups i and its respective molecular weight M_i must be equal to the molecular weight of the petroleum fraction M_j. This constraint may be understood

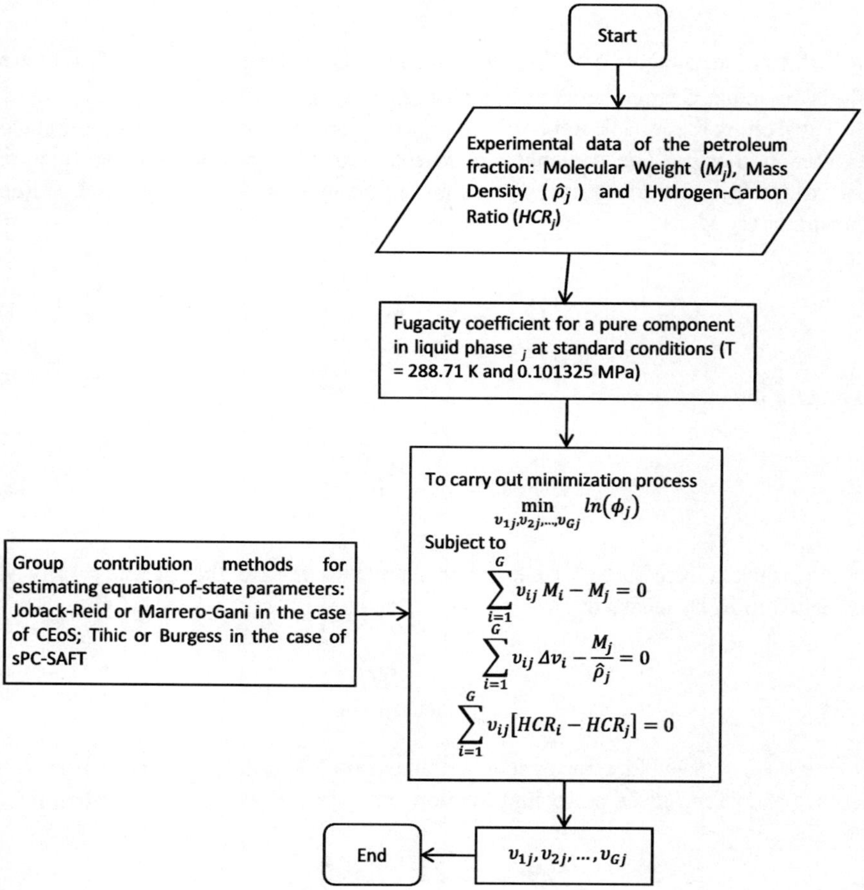

Fig. 3.19 Determination of chemical pseudostructures of petroleum fractions

as a mass balance for the petroleum fraction. Similarly, the second constraint states that the sum of the product of all υ_{ij} and its respective volume increment Δv_i must be equal to the ratio of the molecular weight to the mass liquid density $\hat{\rho}_j$, that is, to the molar volume $\tilde{V}_j$. This second constraint may be seen as a volume balance. Finally, the third constraint states that the sum of the product of all υ_{ij} and its respective HCR_i must be equal to the HCR_j value corresponding to the petroleum fraction j. This last constraint may be understood as a mass balance as well, but for the carbon and hydrogen atoms. Indeed, other constraints accounting for other experimental properties may be included in the minimization process, for example, T_{bj}. However, the minimization problem may become more difficult to converge, especially when nonlinear constraints are included. Besides, it has been found that either two or three constraints are enough for most of the characterization problems. In Fig. 3.19, the determination of chemical pseudostructures of petroleum fractions is schematically illustrated. Notice that in this figure, the use of the so-called PC-SAFT equation of state is also included, because the same general algorithm is valid in this case. The PC-SAFT equation of state equation of state and the respective group-contribution methods will be addressed in Chap. 4

Finally, from Figs. 3.20, 3.21, 3.22 and 3.23, it is illustrated the comparison between the results obtained from the Lee-Kesler correlations and the minimization of the Gibbs free energy. In these figures, the normal boiling temperature, critical

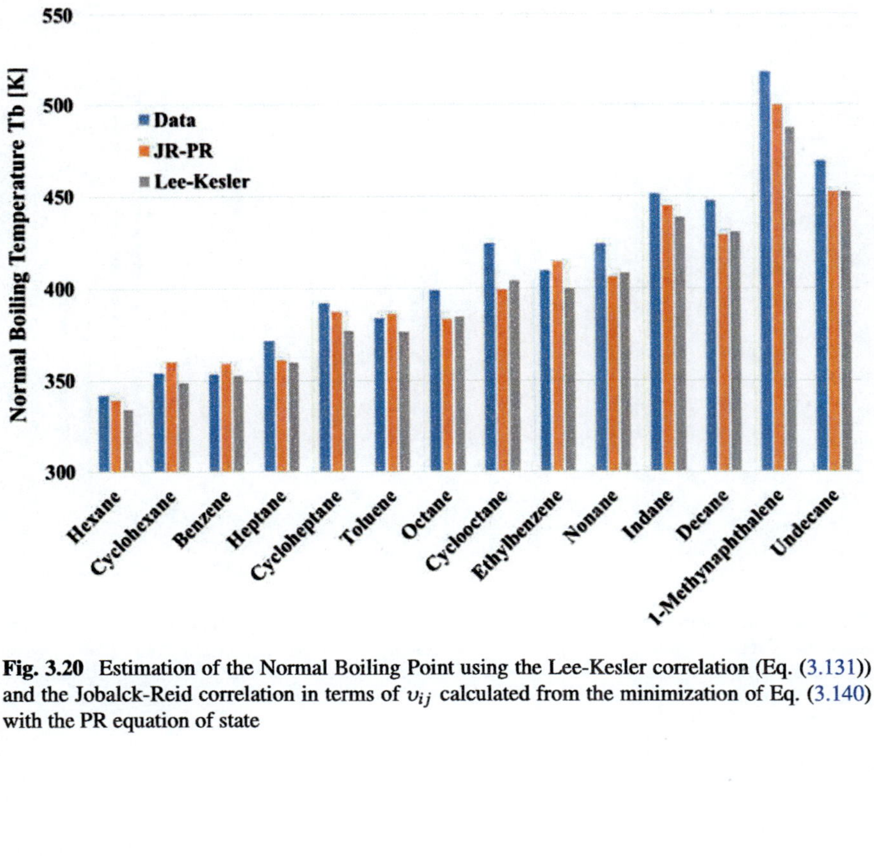

Fig. 3.20 Estimation of the Normal Boiling Point using the Lee-Kesler correlation (Eq. (3.131)) and the Jobalck-Reid correlation in terms of υ_{ij} calculated from the minimization of Eq. (3.140) with the PR equation of state

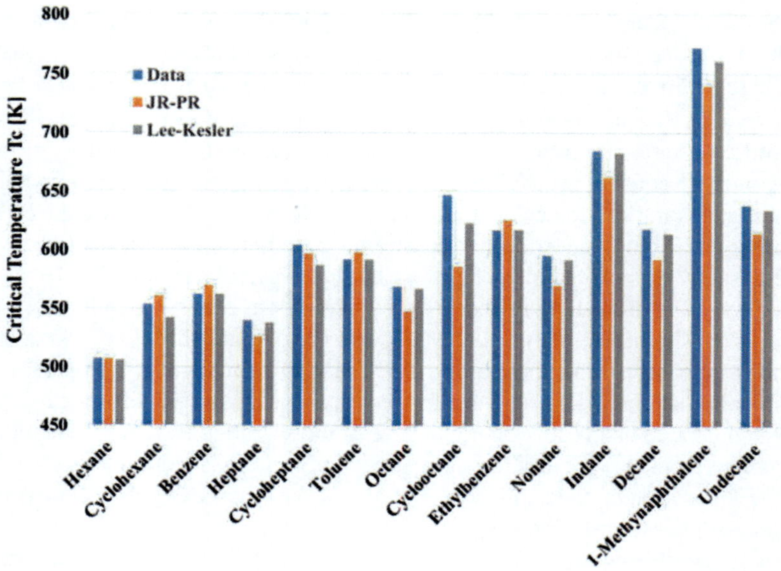

Fig. 3.21 Estimation of the Critical Temperature using the Lee-Kesler correlation (Eq. (3.131)) and the Jobalck-Reid correlation in terms of υ_{ij} calculated from the minimization of Eq. (3.140) with the PR equation of state

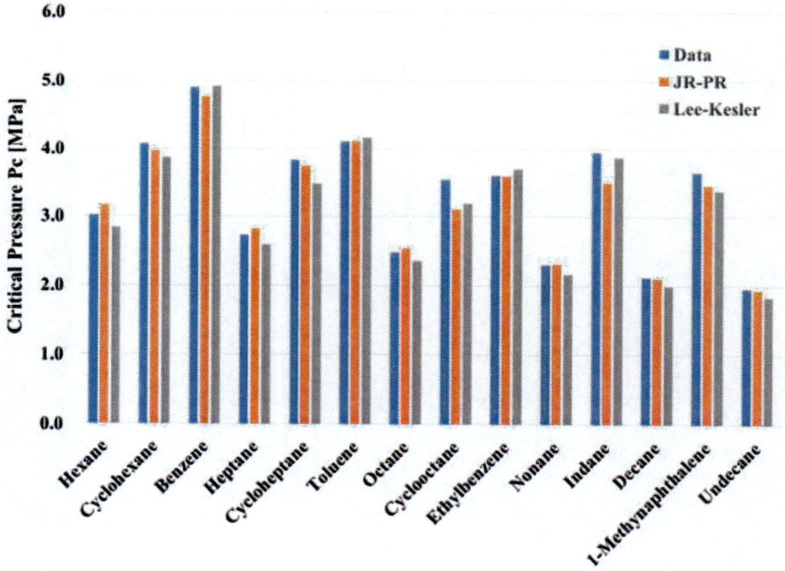

Fig. 3.22 Estimation of the Critical Pressure using the Lee-Kesler correlation (Eq. (3.131)) and the Jobalck-Reid correlation in terms of υ_{ij} calculated from the minimization of Eq. (3.140) with the PR equation of state

properties and acentric factor are calculated for pure components, whose molecular weight, specific gravity and ratio hydrogen to carbon are exactly known. As observed, both methods provide good estimations for the considered pure components. The Lee-Kesler correlations are easier to implement than the minimization process; therefore, it should be preferred to characterize petroleum fractions as first option. However, because of the complexity of heavy petroleum fractions, the classical mixing rule might not be enough for estimation purposes, being necessary to make use of more complex mixing rules, as those discussed before. In this case, it would be needed the UNIFAC solution model, and consequently the knowledge of chemical pseudostructures. For pure components, Fig. 3.24 shows pseudostructures obtained from the minimization process of the Gibbs free energy.

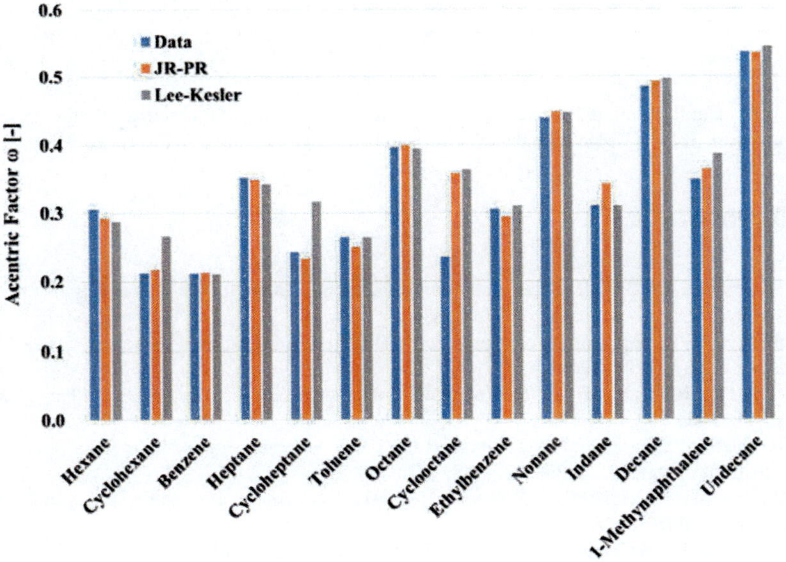

Fig. 3.23 Estimation of the Acentric Factor using the Lee-Kesler correlation and the Jobalck-Reid correlation in terms of υ_{ij} calculated from the minimization of Eq. (3.140) with the PR equation of state

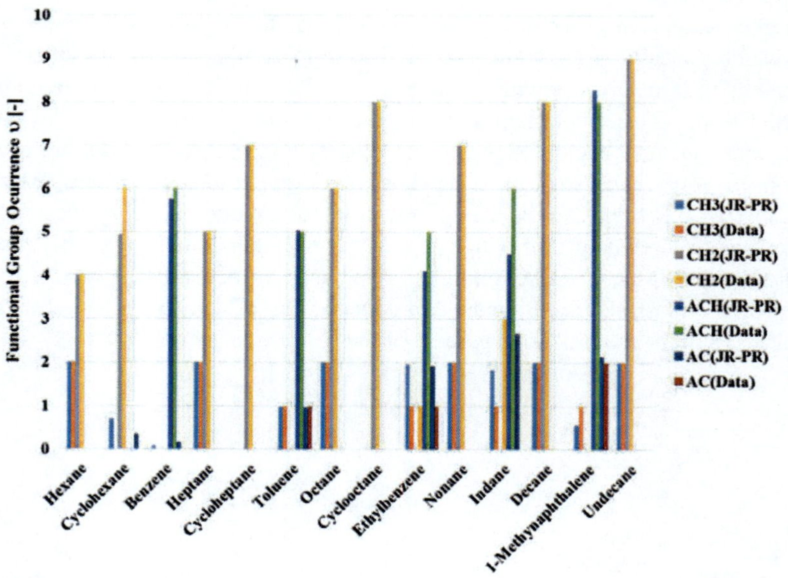

Fig. 3.24 Comparison of actual and calculated occurrences of functional groups from the minimization of Eq. (3.140) with the PR equation of state and the Joback-Reid correlation

References

1. Valderrama JO (2003) The state of the cubic equations of state. Ind Eng Chem Res 42(8):1603–1618
2. Soave G (1972) Equilibrium constants from a modified Redlich-Kwong equation of state. Chem Eng Sci 27(6):1197–1203
3. Peng DY, Robinson DB (1976) A new two-constant equation of state. Ind Eng Chem Fundam 15(1):59–64
4. Le Guennec Y, Lasala S, Privat R, Jaubert JN (2016) A consistency test for α-functions of cubic equations of state. Fluid Phase Equilib 427:513–538
5. Poiling BE, Prausnitz JM, O'Conell JP (2001) The properties of gases and liquids, fifth. McGraw-Hill, New York, USA
6. Yaws CL (1999) Chemical properties handbook, New York, USA
7. Abovsky V (2014) Cubic equation of state: limit of expectations. Fluid Phase Equilib 376:141–153
8. Bronshtein IN, Semendyayev KA, Musiol G, Muehlig H (2004) Handbook of mathematics. Springer, Berlin
9. Michelsen ML, Mollerup J (2007) Thermodynamic models: fundamentals & computational aspects, second. TIE-LINE, Holte, Denmark
10. Bazúa ER (2018) Personal communication
11. Mathias PM, Boston JF, Watanasiri S (1984) Effective utilization of equations of state for thermodynamic properties in process simulation. AIChE J 30(2): 182–186
12. Lucia A, Taylor R (1992) Complex iterative solutions to process model equations? Comput Chem Eng 16:S387–S394
13. Ma M, Chen S, Abedi J (2015) Binary interaction coefficients of asymmetric CH4, C2H6, and CO2 with High n-Alkanes for the simplified PC-SAFT correlation and prediction. Fluid Phase Equilib 405:114–123

14. Tsonopoulos C, Dymond JH, Szafranski AM (1989) Second virial coefficients of normal alkanes, linear 1-Alkanols and their binaries. Pure Appl Chem 61(8): 1387–1394
15. Carreón-Calderón B, Uribe-Vargas V, Ramírez-De-Santiago M, Ramírez-Jaramillo E (2014) Thermodynamic characterization of heavy petroleum fluids using group contribution methods. Ind Eng Chem Res 53(13)
16. Castellanos Díaz O, Modaresghazani J, Satyro MA, Yarranton HW (2011) Modeling the phase behavior of heavy oil and solvent mixtures. Fluid Phase Equilib 304(1–2):74–85
17. Agrawal P, Schoeggl FF, Satyro MA, Taylor SD, Yarranton HW (2012) Measurement and modeling of the phase behavior of solvent diluted bitumens. Fluid Phase Equilib 334:51–64
18. Gao G, Daridon JL, Saint-Guirons H, Xans P, Montel F (1992) A simple correlation to evaluate binary interaction parameters of the Peng-Robinson equation of state: binary light hydrocarbon systems. Fluid Phase Equilib 74:85–93
19. Wichterle I, Kobayashi R (1972) Vapor-Liquid equilibrium of methane—ethane system at low temperatures and high pressures. J Chem Eng Data 17(1):9–12
20. Simnick JJ, Sebastian HM, Lin H, Chao K (1979) Gas-Liquid equilibrium in binary mixtures of Methane + m-Xylene, and Methane +m-Cresol. Fluid Phase Equilib 3:146–151
21. Peters CJ, de Roo JL, de Swaan Arons J (1992) Measurements and calculations of phase equilibria in binary mixtures of Propane + Tetratriacontane. Fluid Phase Equilib 72(C): 251–266
22. Pedersen KS, Christensen PL, Shaikh JA (2015) Phase behavior of petroleum reservoir fluids, 2nd edn. CRC Press, Boca Raton, USA
23. Krejbjerg K, Pedersen KS (2006) Controlling VLLE equilibrium with a cubic EoS in heavy oil modeling, pp 1–15
24. Lopez-Echeverry JS, Reif-Acherman S, Araujo-Lopez E (2017) Peng-Robinson equation of state: 40 years through cubics. Fluid Phase Equilib 447:39–71
25. Smith JM, Van Ness HC, Abbott MM, Swihart MT (2018) Introduccion to chemical engineering thermodynamics, 8th edn. McGraw-Hill, New York, USA
26. Solórzano-Zavala M, Barragán-Aroche F, Bazúa ER (1996) Comparative study of mixing rules for cubic equations of state in the prediction of multicomponent vapor-liquid equilibria. Fluid Phase Equilib 122(1–2):99–116
27. Huron MJ, Vidal J (1979) New mixing rule in simple equations of state for representing vapour-liquid equilibria of strongly non-ideal mixtures. Fluid Phase Equilib 3:255–271
28. Renon H, Prausnitz JM (1968) Local composition in thermodynamic excess functions for liquid mixtures. AIChE J 14:135–144
29. Pedersen KS, Michelsen ML, Fredheim AO (1996) Phase equilibrium calculations for unprocessed well streams containing hydrate inhibitors. Fluid Phase Equilib 126(1):13–28
30. Michelsen ML (1990) A modified huron-vidal mixing rule for cubic equations of state. Fluid Phase Equilib 60(1–2):213–219
31. Dahl S, Michelsen ML (1990) High-Pressure vapor-liquid equilibrium with a UNIFAC-Based equation of state. AIChE J 36(12):1829–1836
32. Holderbaum T, Gmehling J (1991) PSRK: a group contribution equation of state based on UNIFAC. Fluid Phase Equilib 70(2–3):251–265
33. Fischer K, Gmehling J (1995) Further development, status and results of the PSRK method for the prediction of vapor-liquid equilibria and gas solubilities. Fluid Phase Equilib 112(1):1–22
34. Ahlers J, Gmehling J (2002) Development of a universal group contribution equation of state. 2. Prediction of vapor-liquid equilibria for asymmetric systems. Ind Eng Chem Res 41(14):3489–3498
35. Fredenslund A, Jones RL, Prausnitz JM (1975) Group-Contribution estimation of activity coefficients in nonideal liquid mixtures. AIChE J 21(61):1086–1099
36. Horstmann S, Jabłoniec A, Krafczyk J, Fischer K, Gmehling J (2005) PSRK group contribution equation of state: comprehensive revision and extension IV, including critical constants and α-Function parameters for 1000 components. Fluid Phase Equilib 227(2):157–164
37. Weidlich U, Gmehling J (1987) A modified UNIFAC model. 1. Prediction of VLE, HE, and ∞. Ind Eng Chem Res 26(7):1372–1381

38. Gmehling J, Li J, Schiller M (1993) A modified UNIFAC model. 2. Present parameter matrix and results for different thermodynamic properties. Ind Eng Chem Res 32(1):178–193
39. Chen J, Fischer K, Gmehling J (2002) Modification of PSRK mixing rules and results for vapor-liquid equilibria, enthalpy of mixing and activity coefficients at infinite dilution. Fluid Phase Equilib 200(2):411–429
40. Kontogeorgis GM, Vlamos PM (2000) An interpretation of the behavior of EoS/G(E) models for asymmetric systems. Chem Eng Sci 55(13):2351–2358
41. Schmid B, Gmehling J (2012) Revised parameters and typical results of the VTPR group contribution equation of state. Fluid Phase Equilib 317:110–126
42. Schmid B, Gmehling J (2016) Present status of the group contribution equation of state VTPR and typical applications for process development. Fluid Phase Equilib 425:443–450
43. Voutsas E, Magoulas K, Tassios D (2004) Universal mixing rule for cubic equations of state applicable to symmetric and asymmetric systems: results with the Peng-Robinson equation of state. Ind Eng Chem Res 43:6238–6246
44. Voutsas E, Louli V, Boukouvalas C, Magoulas K, Tassios D (2006) Thermodynamic property calculations with the universal mixing rule for EoS/GEmodels: results with the Peng-Robinson EoS and a UNIFAC model. Fluid Phase Equilib 241(1–2):216–228
45. Peneloux A, Abdoul W, Rauzy E (1989) Fluid phase equilibria. Fluid Phase Equilib 41:115–132
46. Abdoul W, Rauzy E, Péneloux A (1991) Group-Contribution equation of state for correlating and predicting thermodynamic properties of weakly polar and non-associating mixtures. Binary and multicomponent systems. Fluid Phase Equilib 68(C):47–102
47. Jaubert JN, Privat R (2010) Relationship between the binary interaction parameters (Kij) of the Peng-Robinson and those of the Soave-Redlich-Kwong equations of state: application to the definition of the PR2SRK model. Fluid Phase Equilib 295(1):26–37
48. Jaubert JN, Privat R, Mutelet F (2010) Predicting the phase equilibria of synthetic petroleum fluids with the PPR78 approach. AIChE J 56(12):3225–3235
49. Qian JW, Privat R, Jaubert JN (2013) Predicting the phase equilibria, critical phenomena, and mixing enthalpies of binary aqueous systems containing Alkanes, Cycloalkanes, Aromatics, Alkenes, and Gases (N2, CO2, H2S, H2) with the PPR78 equation of state. Ind Eng Chem Res 52(46):16457–16490
50. Nasrifar K, Rahmanian N (2018) Equations of state with group contribution binary interaction parameters for calculation of two-phase envelopes for synthetic and real natural gas mixtures with heavy fractions. Oil Gas Sci Technol. 73(7):1–9
51. Gregorowtez J, de Loos TW, do Swaan Arena J (1992) The system Propane + Eicosane: P, T, and x measurements in the temperature range 288–358 K. J Chem Eng Data 37(3):356–358
52. Liu Q (2017) Bubble pressure measurement and modeling for N-Alkane + Aromatic Hydrocarbon binary mixtures. University of Alberta, Canada
53. Péneloux A, Rauzy E, Fréze R (1982) A consistent correction for Redlich-Kwong-Soave volumes. Fluid Phase Equilib 8(1):7–23
54. Mathias PM, Naheiri T, Oh EM (1989) A density correction for the Peng-Robinson equation of state. Fluid Phase Equilib 47(1):77–87
55. Baled H, Enick RM, Wu Y, Mchugh MA, Burgess W, Tapriyal D, Morreale BD (2012) Prediction of hydrocarbon densities at extreme conditions using volume-translated SRK and PR equations of state fit to high temperature, high pressure PVT data. Fluid Phase Equilib 317:65–76
56. Carreón-Calderón B, Uribe-Vargas V, Ramírez-Jaramillo E, Ramírez-de-Santiago M (2012) Thermodynamic characterization of undefined petroleum fractions using group contribution methods. Ind Eng Chem Res 51:14188–14198
57. Carreón-Calderón B, Uribe-Vargas V, Ramírez-de-Santiago M, Ramírez-Jaramillo E (2014) Thermodynamic characterization of heavy petroleum fluids using group contribution methods. Ind Eng Chem Res 53:5598–5607
58. Ihmels EC, Gmehling J (2003) Extension and revision of the group contribution Method GCVOL for the prediction of pure compound liquid densities. Ind Eng Chem Res 42(2):408–412

59. Riazi MR, Bishara A (2004) The impact of characterization methods on properties of reservoir fluids and crude oils: options and restrictions. J Pet Sci Eng 42:195–207
60. Whitson CH, Brulé MR (2000) Phase behavior. Richarson, USA, SPE
61. Riazi MR (2005) Characterization and properties of petroleum fractions. Philadelphia, USA, ASTM
62. Kariznovi M, Nourozieh H, Abedi J (2010) Bitumen characterization and pseudocomponents determination for equation of state modeling. Energy Fuels 24(1):624–633
63. Lee BI, Kesler MG (1975) A generalized thermodynamic correlation based on three-parameters corresponding states. AIChE J 21:510–527
64. Khorasheh F, Khaledi R, Gray MR (1998) Computer generation of representative molecules for heavy hydrocarbon mixtures. Fuel 77(4):247–253
65. Hudebine D, Verstraete JJ (2004) Molecular reconstruction of LCO gasoils from overall petroleum analyses. Chem Eng Sci 59(22–23):4755–4763
66. Albahri TA (2005) Molecularly explicit characterization model (MECM) for light petroleum fractions. Ind Eng Chem Res 44(24):9286–9298
67. Verstraete JJ, Schnongs P, Dulot H, Hudebine D (2010) Molecular reconstruction of heavy petroleum residue fractions. Chem Eng Sci 65(1):304–312
68. Ahmad MI, Zhang N, Jobson M (2011) Molecular components-based representation of petroleum fractions. Chem Eng Res Des 89(4):410–420
69. De Oliveira LP, Vazquez AT, Verstraete JJ, Kolb M (2013) Molecular reconstruction of petroleum fractions: application to vacuum residues from different origins. Energy Fuels 27(7):3622–3641
70. Riizlcka V, Fredenslund A, Rasmussen P (1983) Representation of petroleum fractions by group contribution. Ind Eng Chem Process Des Dev 22(1):49–53
71. Xu X, Jaubert JN, Privat R, Duchet-Suchaux P, Braña-Mulero F (2015) Predicting Binary-interaction parameters of cubic equations of state for petroleum fluids containing pseudo-components. Ind Eng Chem Res 54(10):2816–2824
72. Twu CH (1984) An internally consistent correlations for predicting the critial properties and molecular weight of petroleum and coal-tar liquids. Fluid Phase Equilib 16:137–150
73. Uribe-Vargas V, Carreón-Calderón B, Ramírez-Jaramillo E, Ramírez-de-Santiago M (2016) Thermodynamic characterization of undefined petroleum fractions of gas condensate using group contribution. Oil Gas Sci Technol 71(1)
74. Haghtalab A, Moghaddam AK (2016) Prediction of minimum miscibility pressure using the UNIFAC group contribution activity coefficient model and the LCVM mixing rule
75. Aguilar-Cisneros H, Uribe-Vargas V, Carreón-Calderón B, Domínguez-Esquivel JM, Ramirez-De-Santiago M (2017) Hydrogen solubility in heavy undefined petroleum fractions using group contributions methods. Oil Gas Sci Technol 72(1)
76. Aguilar-Cisneros H, Carreón-Calderón B, Uribe-Vargas V, Domínguez-Esquivel JM, Ramirez-de-Santiago M (2018) Predictive method of hydrogen solubility in heavy petroleum fractions using EOS/GEand group contributions methods. Fuel 224(January):619–627
77. Joback KG, Reid RC (1987) Estimation of pure-component properties from group-contributions. Chem Eng Commun 57(233–243)
78. Marrero J, Gani R (2001) Group-Contribution based estimation of pure component properties. Fluid Phase Equilib 184:183–208

Chapter 4
PC-SAFT Equation of State

List of Abbreviations

HCR	Molar ratio of hydrogen to carbon
PC-SAFT	Perturbed-Chain SAFT
PNA	Paraffinic, Naphthenic, and Aromatic fractions
PR	Peng-Robinson equation of state
SAFT	Statistical Associating Fluid Theory
SAP	Saturate, Aromatic and Polyaromatic fractions
SARA	Saturates, Aromatics, Resins, and Asphaltenes fractions
SCN	Single carbon number
sPC-SAFT	Simplified PC-SAFT
SRK	Soave-Redlich-Kwong equation of state

Nomenclature

a	Slopes the linear functions
$\tilde{A}$	Molar Helmholtz free energy
$a_0i, a_{1i}, a_{2i}, b_{0i}, b_{1i},\ and\ b_{2i}$	Universal PC-SAFT constants
b	Intercepts of the linear functions
C	Total number of components
d	Average segment diameter
d_i	Effective hard-segment diameter of the chain segments
$f(\eta)$	Function of η
g^{hs}	Radial distribution function in the hard-sphere system for sPC-SAFT
g_{ii}^{hs}	Radial distribution function for segments of component i in the hard-sphere system

B. Carreón-Calderón et al., *Thermophysical Properties of Heavy Petroleum Fluids*, Petroleum Engineering, https://doi.org/10.1007/978-3-030-58831-1_4

k_{ij}	Binary interaction parameter between component i and j
M	Molecular Weight
m	Chain length
$\tilde{m}$	Average chain length
N_A	Avogadro number $\left(6.02214076 \times 10^{23}\,\text{mol}^{-1}\right)$
P	Pressure
R	The Universal Gas Constant (8.3144636 J/mol K)
S	Specific gravity
T	Temperature
T_b	Normal boiling temperature
$\tilde{V}$	Molar volume
x_i	Mole fraction of component i
Z	Compressibility factor
γ	Aromaticity factor
$\varepsilon^{A_iB_i}$	Energy of association
ε/k	Segment energy parameter
η	Reduced packing fraction
$\kappa^{A_iB_i}$	Volume of association
ρ	Total density of molecules
$\hat{\rho}_j$	Mass density
σ	Size parameter
υ_{ij}	Occurrence of a given functional group i in the petroleum fraction j
ϕ_i	Fugacity coefficient of component i
Ω	Generalized PC-SAFT parameter

Superscripts

0	Pure component or reference state
A	Aromatic
c	Chain contribution
dsp	Dispersion contribution
hs	Hard-sphere contribution
hsc	Hard-sphere chain contribution
ig	Ideal gas contribution
ig	Residual property
N	Naphthenic fraction
P	Paraffinic fraction
R	Residual property

Subscripts

A	Asphaltene or Alkane fractions
$A + R$	Aromatic + Resins fraction
$C+$	Plus petroleum fraction
G	Types of functional groups
i	Component i
j	Petroleum Fraction j
P	Polynuclear aromatic fractions

In the last two decades, Statistical Associating Fluid Theory (SAFT) equations have been proposed as alternative equation of state to the widely accepted cubic equations of state in the petroleum and chemical industries. The thermodynamic characterization of petroleum fractions using SAFT-type equations follows the same aim that in the case of the characterization using cubic equations of state, i.e., it seeks to provide the necessary information for performing equilibrium estimations starting from limited or incomplete experimental information as input data. Hence, the same general elements are present here: a distribution function between some petroleum fraction properties and composition, a splitting-lumping method of such distribution function, and schemes for estimating properties of the resulting pseudocomponents or cuts, where properties entering into equations of state are the most important ones.

There have been several modifications of the original SAFT equations that have been developed during the last three decades, which share common characteristics with each other, for example, the use of the same chain and association contributions. Most often, the dispersion contribution, which is a dispersive potential accounting for attraction between either spheres or chains, is the responsible for the presence of different SAFT variants. For more details, the reader is referred to the Kontogeorgis and Folas' manuscript [2], where they have provided a comprehensive review of SAFT-type equations. These equations are thermodynamic models capable of representing thermodynamic properties of systems where hydrogen bonding effects are not present (non-associating systems), and besides, they are capable of explicitly accounting for effects of hydrogen bonding in either pure components or mixtures (associating systems). In most versions of SAFT-type equations, a hard-sphere fluid is used as a reference fluid (repulsive contribution) to be perturbed by attractive forces (dispersion contribution); hence, it is said that these models are based on thermodynamic perturbation theories [3–6]. The model constants are usually determined by regression to molecular simulation data; however, for the Perturbed-Chain SAFT version (PC-SAFT) [7], the model constants are adjusted to pure component experimental data of n-alkanes. Because a hard-chain fluid is proposed to be the reference fluid rather than a hard-sphere fluid, the n-alkanes family is considered as the best option to serve as model molecules. The PC-SAFT equation is capable of modelling phase equilibrium problems in the presence of components with large molecular weights, for example, phase equilibrium in polymer systems [8–10]. This SAFT version has also shown to be an appropriate option for phase equilibrium calculations of highly asymmetric mixtures [11, 12] and for the liquid density and

Table 4.1 Deviation percent from cubic (SRK and PR) and PC-SAFT equations of state for hydrocarbon properties [1]

Properties	SRK	PR	PC-SAFT
Pure components			
Density[a]	0.93	1.00	0.65
Saturated liquid density[a]	6.03	3.93	2.03
Density[b]	8.95	6.30	1.67
Saturated liquid density[b]	11.90	5.73	1.31
Liquid density for heavy n-alkanes[b]	19.83	10.90	1.92
Binary mixtures			
Bubble point pressure	4.41	4.38	3.96
Vapor phase composition	1.01	1.00	0.87
Conventional reservoir fluids			
Saturation pressure	6.89	6.37	11.04
Density[b]	12.21	3.51	3.28

[a]Calculated with volume translation
[b]Calculated without volume translation
[c]Calculated with optimal interaction parameters

compressibility estimations in general. Although the PC-SAFT equation as well as all SAFT-type equations are theoretically more robust thermodynamic models than cubic equations of state, they are also unable of predicting critical points properly: they tend to overpredict the critical temperature and pressure; the larger the component molecular weight, the higher errors are obtained in predictions. Rescaling SAFT parameters is possible in order to match experimental critical points [13], but this usually leads to less accurate liquid density estimations, unless adjustable additional parameters are incorporated. In general, it has been shown that PC-SAFT parameters are a function of pressure [14]. Furthermore, for extreme temperatures (>373 K) and pressures (>7 MPa) a new set of pure-component parameters, different from those obtained at low pressures, may be necessary for estimating liquid density properly [15]. However, the PC-SAFT equation has been revealed as a possible alternative or replacement of cubic equations of state for systems containing hydrocarbons, particularly for those systems containing heavy hydrocarbons [16–19]. It has been shown that the PC-SAFT equation provides results comparable to those results obtained from cubic equations of state for pure components, binary mixtures and conventional (non-heavy) reservoir fluids [1, 20–24]. For instance, in Table 4.1 one of these comparisons is shown, where the performance of the Peng-Robinson (PR) and the Soave-Redlich-Kwong (SRK) cubic equations of state is compared with the performance of the PC-SAFT equation of state. According to this Table, the PR equation combined with volume translation has the best overall performance, whereas the PC-SAFT equation has the best performance in estimating the saturated liquid density and the liquid density of heavy n-alkanes. Like cubic equations of state, in these

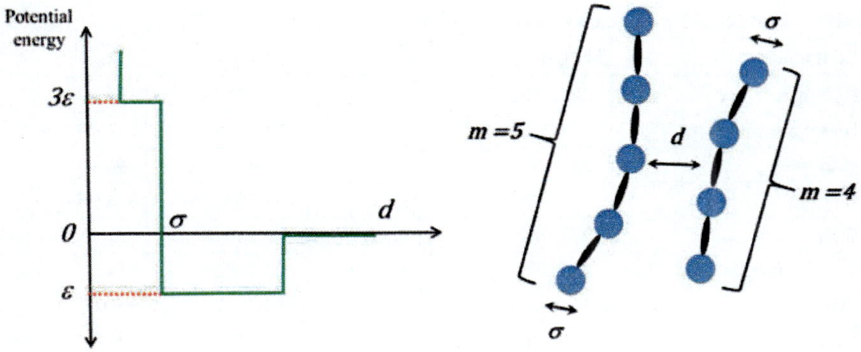

Fig. 4.1 Schematic meaning of non-associating PC-SAFT parameters: chain length m, size parameter σ and segment energy ε

assessments, it has been shown that PC-SAFT parameters must often be adjusted by regression to experimental pressure-volume-temperature data to achieve satisfactory results.

In SAFT-type equations, each pure non-associating component is characterized by a chain length m (the number of segments in a chain of component), a size parameter σ (the segment diameter in Å) and a segment energy parameter ε/k (the dispersion energy in K). In Fig. 4.1, the schematic meaning of these parameters is illustrated. These pure-component parameters are usually determined by regression to experimental vapor pressure and liquid density data. In Table 4.2, PC-SAFT parameters of non-associating components relevant to petroleum processes are shown. When a pure component is self-associating (molecules of pure components showing hydrogen bonding such as water and alcohols), there are two further parameters needed to characterize the component: the volume $\left(\kappa^{A_iB_i}\right)$ and energy $\left(\varepsilon^{A_iB_i}\right)$ of association. The associating contribution to the PC-SAFT and the corresponding parameters will be discussed in the following Chapter, where the phase equilibrium between heavy petroleum fractions and water will be illustrated. In this Chapter, hydrocarbon their mixtures, non-associating systems, will be analyzed.

One indication of the strong physical framework of SAFT-type equations is the linearity followed by their three non-associating parameters (m, σ and ε/k) with the molecular weight, especially when pure components are assembled according to their chemical family. For instance, Tihic et al. [12] determined a set of linear functions of molecular weight M for several families of components. The general forms of the proposed linear functions are [11]

$$m = a_m M + b_m \tag{4.1}$$

$$m\sigma^3 = a_{m\sigma} M + b_{m\sigma} \tag{4.2}$$

Table 4.2 PC-SAFT parameters of some non-associating components [7, 12, 25]

Component	M (g/mol)	m (−)	σ (Å)	ε/k (K)
Hydrogen	2.01588	1.0000	3.6752	56.27
Hydrogen sulfide	34.075882	1.6941	3.0214	226.79
Nitrogen	28.012	1.2053	3.3130	90.96
Oxygen	31.9988	1.1217	3.2098	114.96
Carbon monoxide	28.0094	1.3097	3.2507	92.15
Carbon dioxide	44.0088	2.0729	2.7852	169.21
Methane	16.04176	1.000	3.7039	150.03
Ethane	30.06764	1.6069	3.5206	191.42
Propane	44.09252	2.0020	3.6184	208.11
Butane	58.1194	2.3316	3.7086	222.88
Isobutane	58.1194	2.2616	3.7574	216.53
Pentane	72.14529	2.6896	3.7729	231.20
Isopentane	72.14529	2.5620	3.8296	230.75
Neopentane	72.14529	2.3543	3.9550	225.69
Cyclopentane	70.1294	2.3655	3.7114	265.83
n-Hexane	86.17116	3.0576	3.7983	236.77
2-methylpentane	86.17116	2.9317	3.8535	235.58
3-methylpentane	86.17116	2.8852	3.8605	240.48
2, 2-dimethylbutane	86.17116	2.6008	4.0042	243.51
2, 3-dimethylbutane	86.17116	2.6853	3.9545	246.07
Methylcyclopentane	84.15528	2.6130	3.8253	265.12
Cyclohexane	84.15528	2.5303	3.8499	278.11
Benzene	78.107635	2.4653	3.6478	287.35
Toluene	92.13352	2.8149	3.7169	285.69
n-Heptane	100.19704	3.4831	3.8049	238.40
n-Octane	114.22292	3.8176	3.8373	242.78
n-Nonane	128.24881	4.2079	3.8448	244.51
n-Decane	142.27469	4.6627	3.8384	243.87
n-Dodecane	170.32643	5.3060	3.8959	249.21
n-Tetradecane	198.3782	5.9002	3.9396	254.21
n-Hexadecane	226.42996	6.6485	3.9552	254.70
n-Octadecane	254.48174	7.3271	3.9668	256.20
n-Eicosane	282.5335	8.4092	3.9120	251.92
n-Octacosane	394.7405	11.083	3.9789	255.62
n-Hexatriacontane	506.94757	14.332	3.8836	252.89

Table 4.3 Slopes and Intercepts of the linear functions given by Eqs. (4.1), (4.2), and (4.3) for PC-SAFT parameters for different families of components 12

Family	a_m	b_m	$a_{m\sigma}$	$b_{m\sigma}$	$a_{m\varepsilon}$	$b_{m\varepsilon}$
Alkanes	0.0249	0.9711	1.6947	23.27	6.5446	177.92
Alkenes	0.0247	0.9173	1.7575	11.983	6.9348	119.25
Alkynes	0.0170	1.7585	1.7223	6.5264	6.1464	242.59
Cycloalkanes	0.0223	0.6646	1.7092	2.1664	6.4676	147.52
Ketones	0.0282	0.6646	1.6774	−0.8076	6.906	176.09
Esters	0.0309	0.6355	1.6799	−21.492	7.1101	167.66
Sulfides	0.0204	0.9247	1.4849	1.4107	6.5816	162.94
Polynuclear aromatic	0.0203	0.607	1.7024	−39.059	5.8932	325.34

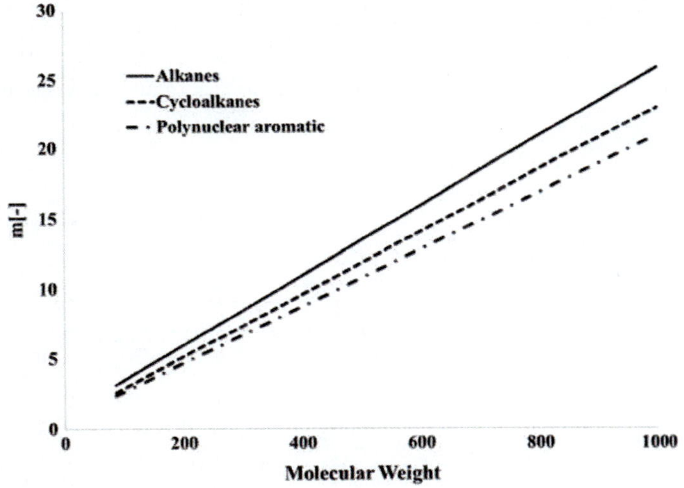

Fig. 4.2 The number of segments *m* from Eq. (4.1) for n-alkanes, cycloalanes and polynuclear aromatic families

$$m(\varepsilon/k) = a_{m\varepsilon}M + b_{m\varepsilon} \tag{4.3}$$

In Table 4.3, the corresponding slopes and intercepts of the linear functions are shown, while in Figs. 4.2, 4.3, and 4.4, the calculated parameters for n-alkanes, cycloalkane, and poly-nuclear aromatic families are illustrated. As observed from these figures, *m* is larger for the n-alkane family than for the aromatic one, i.e., there are more segments forming chains in the former family than in the latter, with cycloalkanes showing a middle behavior. It is worthy noticing that, in Fig. 4.3, σ shows an almost constant value regardless of the considered chemical family. This observation gives rise to the so-called simplified PC-SAFT version (sPC-SAFT) [11],

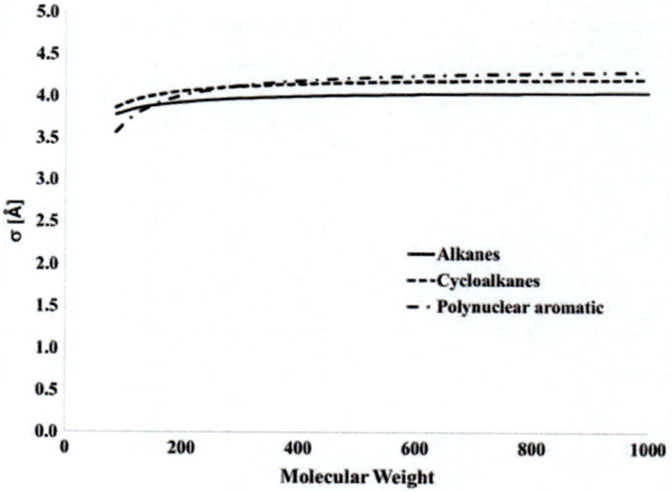

Fig. 4.3 The segment diameter σ in Å from Eq. (4.2) for n-alkanes, cycloalkanes and polynuclear aromatic families

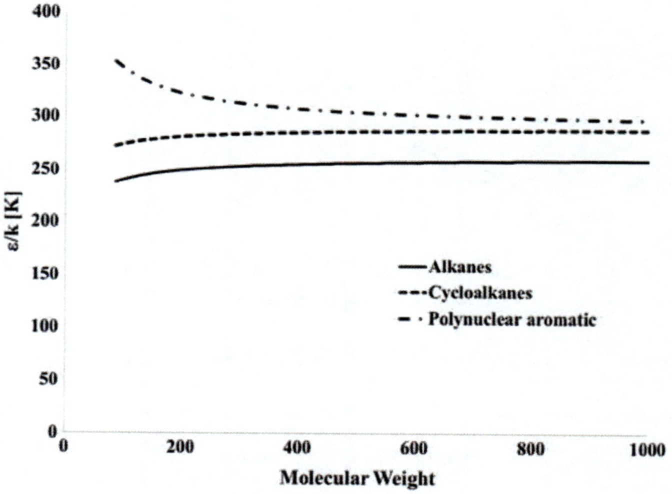

Fig. 4.4 The segment energy parameter ε/k in K from Eq. (4.3) for n-alkanes, cycloalkanes and polynuclear aromatic families

where an average segment diameter is proposed in order to simplify the original PC-SAFT version. Finally, in Fig. 4.4, we have that ε/k shows an opposite behavior in relation to m in Fig. 4.2, i.e., the aromatic family presents higher dispersion energies than the n-alkane family, with cycloalkanes showing a middle behavior again. The higher normal boiling temperatures in the aromatic family in relation to the n-alkane family is an example of these higher dispersion energies. The linear functions from Eqs. (4.1) to (4.3) allow PC-SAFT parameters of components with large molecular weight to be estimated when either the liquid density or the vapor pressure data are not available for regression purposes. This is a remarkable characteristic of the PC-SAFT equation concerning its application to heavy petroleum fractions, as will be illustrated later.

4.1 Compressibility Factor for Non-associating Systems

The PC-SAFT equation of state in terms of the compressibility factor Z is formed by three adding terms as follows [7]:

$$Z \equiv \frac{P\tilde{V}}{RT} = 1 + Z^{hsc} + Z^{dsp} \tag{4.4}$$

The symbols P, $\tilde{V}$, and T stand for pressure, molar volume, and temperature respectively, while R stands for the universal gas constant (8.3144725 J/mol K). The hard-sphere chain term Z^{hsc} accounts for the repulsion contribution, which arises from defining a reference fluid conceived as composed of hard spheres forming chains (see Fig. 4.1). On the other hand, the dispersion contribution Z^{disp} accounts for attractive interactions between chains. As observed in Eq. (4.4), Z^{disp} is a perturbation term to Z^{hsc}, i.e., a perturbation to the reference fluid. Similarly, Z^{hsc} and Z^{disp} may be seen as perturbation terms to the ideal-gas reference state $\left(Z^{ig} = 1\right)$. In terms of residual properties, which are defined as the difference between the actual property value and the ideal-gas property value, Z^{hsc} and Z^{disp} combined corresponds to the residual compressibility factor Z^R:

$$Z^R \equiv Z - Z^{ig} = Z^{hsc} + Z^{dsp} \tag{4.5}$$

In turn, Z^{hsc} is composed of the hard-sphere and chain contributions Z^{hs} and Z^c, respectively. These contributions are determined using an equation of state for hard-sphere chains from Chapman et al. [26, 27]:

$$Z^{hsc} = \tilde{m}Z^{hs} + Z^c \tag{4.6}$$

$$\tilde{m} = \sum_{i=1}^{C} x_i m_i \tag{4.7}$$

$$Z^{hs} = \frac{\xi_3}{1-\xi_3} + \frac{3\xi_1\xi_2}{\xi_0(1-\xi_3)^2} + \frac{3\xi_2^3 - \xi_3\xi_2^3}{\xi_0(1-\xi_3)^3} \tag{4.8}$$

$$\xi_k = \frac{\pi}{6}\rho \sum_{i=1}^{C} x_i m_i d_i^k \quad k \in \{0, 1, 2, 3\} \tag{4.9}$$

$$d_i = \sigma_i \left[1 - 0.12 e^{-3\frac{\varepsilon_i}{kT}}\right] \tag{4.10}$$

$$Z^c = \sum_{i=1}^{C} \frac{x_i(1 - m_i)}{g_{ii}^{hs}} \rho \frac{\partial g_{ii}^{hs}}{\partial \rho} \tag{4.11}$$

where x_i is the mole fraction of chains of component i, and d_i is an effective hard-segment diameter of the chain segments (Å), which depends on temperature and on the PC-SAFT parameters σ_i and (ε/k). ρ stands for the total density of molecules, i.e., the total number of molecules per unit volume (1/Å^3), whereas g_{ii}^{hs} is the radial distribution function for segments of component i in the hard-sphere system and is given by

$$g_{ij}^{hs} = \frac{1}{1 - \xi_3} + \frac{3\xi_2}{(1 - \xi_3)^2}\left(\frac{d_i d_j}{d_i + d_j}\right) + \frac{2\xi_2^2}{(1 - \xi_3)^3}\left(\frac{d_i d_j}{d_i + d_j}\right)^2 \tag{4.12}$$

$$\rho \frac{\partial g_{ij}^{hs}}{\partial \rho} = \frac{\xi_3}{(1 - \xi_3)^2} + \left(\frac{3\xi_2}{(1 - \xi_3)^2} + \frac{6\xi_2\xi_3}{(1 - \xi_3)^3}\right)\left(\frac{d_i d_j}{d_i + d_j}\right) + \left(\frac{4\xi_2^2}{(1 - \xi_3)^3} + \frac{6\xi_2^2\xi_3}{(1 - \xi_3)^4}\right)\left(\frac{d_i d_j}{d_i + d_j}\right)^2 \tag{4.13}$$

On the other hand, Z^{dsp} term is determined from

$$Z^{dsp} = -2\pi\rho \frac{\partial(\eta I_1)}{\partial \eta} \overline{m^2 \varepsilon \sigma^3} - \pi \rho \tilde{m} \left[C_1 \frac{\partial(\eta I_2)}{\partial \eta} + C_2 \eta I_2\right] \overline{m^2 \varepsilon^2 \sigma^3} \tag{4.14}$$

Together with

$$\frac{\partial(\eta I_1)}{\partial \eta} = \sum_{j=0}^{6} a_j(\tilde{m})(j + 1)\eta^j \tag{4.15}$$

$$\frac{\partial(\eta I_2)}{\partial \eta} = \sum_{j=0}^{6} b_j(\tilde{m})(j + 1)\eta^j \tag{4.16}$$

$$a_j(\tilde{m}) = a_{0j} + \frac{\tilde{m} - 1}{\tilde{m}} a_{1j} + \frac{(\tilde{m} - 1)(\tilde{m} - 2)}{\tilde{m}^2} a_{2j} \tag{4.17}$$

$$b_j(\tilde{m}) = b_{0j} + \frac{\tilde{m} - 1}{\tilde{m}} b_{1j} + \frac{(\tilde{m} - 1)(\tilde{m} - 2)}{\tilde{m}^2} b_{2j} \tag{4.18}$$

$$\overline{m^2\varepsilon^y\sigma^3} = \sum_{i=1}^{C}\sum_{j=1}^{C} x_i x_j m_i m_j \left(\frac{\varepsilon_{ij}}{kT}\right)^y \sigma_{ij}^3 \tag{4.19}$$

$$\sigma_{ij} = \frac{\sigma_i + \sigma_j}{2} \tag{4.20}$$

$$\varepsilon_{ij} = \left(1 - k_{ij}\right)\sqrt{\varepsilon_i \varepsilon_j} \tag{4.21}$$

$$C_1 = \left[1 + \tilde{m}\frac{8\eta - 2\eta^2}{(1-\eta)^4} + (1-\tilde{m})\frac{20\eta - 27\eta^2 + 12\eta^3 - 2\eta^4}{(1-\eta)^2(2-\eta)^2}\right]^{-1} \tag{4.22}$$

$$C_2 = -C_1^2\left[\tilde{m}\frac{-4\eta^2 + 20\eta + 8}{(1-\eta)^5} + (1-\tilde{m})\frac{2\eta^3 + 12\eta - 48\eta + 40}{(1-\eta)^3(2-\eta)^3}\right] \tag{4.23}$$

In Eqs. (4.17) and (4.18), the universal model constants $a_{0i}, a_{1i}, a_{2i}, b_{0i}, b_{1i}$, and b_{2i} were adjusted to vapor pressure, liquid, vapor and supercritical volumes of n-alkanes, because their molecules are elongated enough to be considered as representative of a hard-chain fluid. The obtained values are shown in Table 4.4. Like cubic equations of state, a binary interaction parameter k_{ij} is introduced in Eq. (4.21), which corrects the dispersion energies of dissimilar molecules. This binary parameter is used for describing phase behavior of mixtures and it must be estimated by regression to experimental data of phase equilibriums. Hence, Eqs. (4.20) and (4.21) are the conventional mixing rules for calculating parameters for couples of dissimilar molecules. Finally, the reduced packing fraction η is equal to ξ_3 in Eq. (4.9) and indicates the level of packing. Molecules in gases are lesser packed than molecules in liquids; hence, in the former case we have that $\eta \to 0$, whereas in the latter case we have that $\eta \gg 0$ with $\eta = \pi/\left(3\sqrt{2}\right)$ as the closest packing of molecules. On other words, $\eta \in \left[0, \pi/\left(3\sqrt{2}\right)\right]$ regardless of whether a liquid phase or vapor phase is present.

Taking as starting point that segments have similar diameters as illustrated in Fig. 4.3 regardless of the considered chemical family, von Solms et al. [11] proposed that the effective collision diameter d_i can be replaced by an average segment diameter d, that is,

$$d_i \approx d \quad \forall i \tag{4.24}$$

where the average diameter is defined through

$$\tilde{m}d^k = \sum_{i=1}^{C} x_i m_i d_i^k \quad k\varepsilon\{0, 1, 2, 3\} \tag{4.25}$$

Table 4.4 Universal model constants for Eqs. (4.17) and (4.18) in PC-SAFT equation [7]

i	a_{0i}	a_{1i}	a_{2i}	b_{0i}	b_{1i}	b_{2i}
0	0.9105631445	−0.3084016918	−0.0906148351	0.7240946941	−0.5755498075	0.0976883116
1	0.6361281449	0.1860531159	0.4527842806	2.2382791861	0.6995095521	−0.2557574982
2	2.6861347891	−2.5030047259	0.5962700728	−4.0025849485	3.8925673390	−9.155856153
3	−26.547362491	21.419793629	−1.7241829131	−21.003576815	−17.215471648	20.642075974
4	97.759208784	−65.25588533	−4.1302112531	26.855641363	192.67226447	−38.804430052
5	−159.59154087	83.318680481	13.77663187	206.55133841	−161.82646165	93.626774077
6	91.297774084	−33.74692293	−8.6728470368	−355.60235612	−165.20769346	−29.666905585

The only constraint imposed on this average is that the reduced packing fraction η of the actual mixture remains unchanged. Thus, bearing in mind that $\eta = \xi_3$, Eq. (4.25) transforms Eq. (4.9) into

$$\xi_k = \frac{\eta}{d^{3-k}} \quad k\varepsilon\{0, 1, 2, 3\} \tag{4.26}$$

From which, the hard-sphere and chain contributions to the compressibility factor given by Eq. (4.6) is expressed now in a simplified form as

$$Z^{hsc} = 2\tilde{m}\frac{2\eta - \eta^2}{(1-\eta)^3} + (1-\tilde{m})\frac{5\eta - 2\eta^2}{(1-\eta)(2-\eta)} \tag{4.27}$$

The hard-sphere chain contribution from Eq. (4.27), together with the dispersion contribution from Eq. (4.14) entering into Eq. (4.4), leads to the simplified PC-SAFT version (sPC-SAFT). For pure components, both PC-SAFT and sPC-SAFT version are identical; hence, PC-SAFT parameters remain unchanged for this simplified version. However, the interaction parameters k_{ij}, which are needed for describing phase equilibrium in multicomponent systems, differ from each other. When the performance of both the sPC-SAFT and original PC-SAFT versions has been compared for more complex systems than pure components, it has been shown that the simplifications introduced by means of Eq. (4.24) are not at expense of the accuracy of calculations, while the computational time and complexity are substantially reduced using the sPC-SAFT version [8, 9]. This makes the sPC-SAFT version a convenient choice over the PC-SAFT version for petroleum industry, where the computational time may become significant in calculations. It is important to underline that sPC-SAFT equation is not different from PC-SAFT version; rather, it is a simplified version using the mixing rule established by Eq. (4.25).

Irrespective of the PC-SAFT version, if pressure and temperature are known, then the estimation of the compressibility factor implies that the reduced packing fraction η must be determined iteratively from solving Eq. (4.4), which can be rewritten as

$$f(\eta) = 1 + Z^{hsc} + Z^{dsp} - \frac{P\tilde{V}}{RT} = 0 \tag{4.28}$$

where, Z^{disp} is given by Eq. (4.14) for both PC-SAFT versions, while Z^{hsc} is determined from Eq. (4.6) for the PC-SAFT version, and from Eq. (4.27) for the simplified one. As observed, the total density of molecules ρ appears in Eq. (4.14); hence, it is necessary to express ρ in terms of η so that Eq. (4.28) can be solved. The corresponding equation is Eq. (4.9) from the PC-SAFT version, which takes the form of

$$\eta = \xi_3 = \frac{\pi}{6}\rho\sum_{i=1}^{C} x_i m_i d_i^3 \tag{4.29}$$

In the case of the sPC-SAFT version, the needed relationship is derived from combing Eq. (4.25) with Eq. (4.29):

$$\eta = \frac{\pi}{6}\rho\tilde{m}d^3 \tag{4.30}$$

Finally, the molar volume $\tilde{V}$ in Eq. (4.28) must be expressed in terms of η as well. For the PC-SAFT version, we have that

$$\tilde{V} = 10^{-30}\frac{\pi N_A}{6\eta}\left(\sum_{i=1}^{C} x_i m_i d_i^3\right)[=]\frac{m^3}{mol} \tag{4.31}$$

whereas for the sPC-SAFT version, we have from Eqs. (4.25) to (4.31) that

$$\tilde{V} = 10^{-30}\frac{\pi N_A}{6\eta}\tilde{m}d^3[=]\frac{m^3}{mol} \tag{4.32}$$

where N_A stands for the Avogadro number. If one intends to calculate a vapor-phase molar volume, then the initial guess for η should be set equal to a small value like, for example, $\eta = 0.001$. On the other hand, if one intends to calculate a liquid-phase molar volume, then the initial guess for η should be set equal to a higher value like, for example, $\eta = 0.3$. Calculated values larger than the closest packing of molecules ($\eta > \pi/(3\sqrt{2})$) have no physical meaning and must be discarded. Using the sPC-SAFT version and parameters in Table 4.2, isotherms at the normal boiling temperature for three n-alkanes are illustrated in Fig. 4.5 as a function of

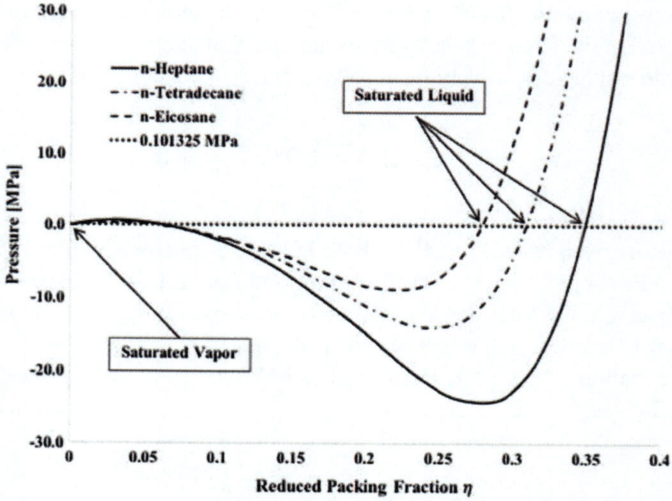

Fig. 4.5 PC-SAFT isotherms at the normal boiling temperature for three pure n-alkanes

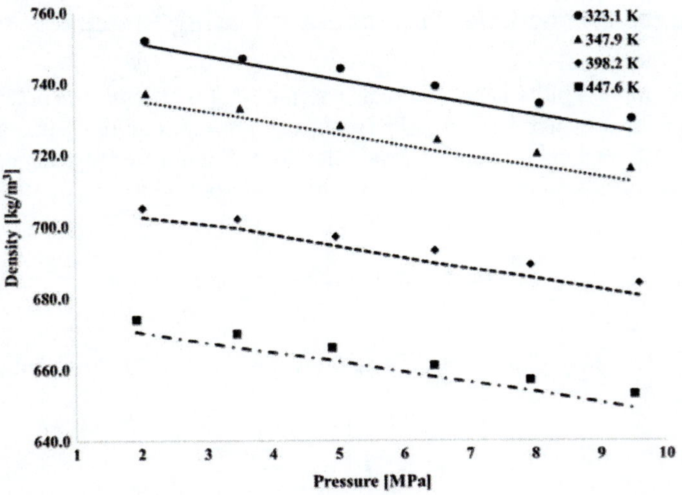

Fig. 4.6 Calculated (sPC-SAFT) and experimental liquid densities of Methane-n-Octadecane Mixtures $(k_{ij} = 0)$. Data from Kariznovi ei al. [28]

η. In this figure, the normal boiling pressure (0.101325 MPa) is also indicated to locate the corresponding η at saturated conditions. Saturated liquid-phase values are 0.3485, 0.3080 and 0.2782 for n-heptane, n-tetradecane and n-eicosane, respectively. These values indicate that the bigger the n-alkane, the smaller η, that is, the n-heptane molecules are more packed than the n-eicosane molecules in liquid phase. On the other hand, we have that saturated vapor-phase values are 0.0020, 0.0026, 0.0031 for n-heptane, n-tetradecane and n-eicosane, respectively. In contrast, these values indicate that the n-heptane molecules are lesser packed than the n-eicosane molecules in vapor phase. From these results, it is concluded that the bigger the component, the difference between the saturated liquid and vapor values is smaller. In the case of heavy petroleum fractions, this difference becomes as small as one order of magnitude, making phase equilibrium calculations difficult to converge. Finally, it is worthy noticing the similar shape shown by isotherms in Fig. 4.5 in comparison with those shown by cubic equations of state. At the normal boiling pressure, three values of η are found: the largest one corresponds to a saturated liquid-phase, i.e., the smallest molar volume (see Eq. 4.32), whereas the smallest value corresponds to a saturated vapor-phase, that is, the largest molar volume, with the middle value without physical meaning. Finally, for liquid saturated mixtures of methane and n-octadecane, Fig. 4.6 shows the comparison of experimental and calculated mass densities using $k_{ij} = 0$.

4.2 Fugacity Coefficient for Non-associating Systems

Once the compressibility factor has been determined, it is necessary to determine the fugacity coefficient in order to establish phase equilibrium conditions. Irrespective of the PC-SAFT version, to obtain the expression for calculating the fugacity coefficient ϕ_i of a component i in a C-component mixture, one start with

$$\frac{\tilde{A}^R}{RT} = 1 - Z + ln(Z) + \sum_{i=1}^{C} x_i ln(\phi_i) \tag{4.33}$$

which was derived in Chap. 2. By adding $ln(\phi_i)$ on both sides and remembering that

$$\left[\frac{\partial}{\partial x_i}\left(\frac{\tilde{A}^R}{RT}\right)\right]_{T,\rho,\not{x}} = ln(\phi_i) \tag{4.34}$$

we arrived at

$$ln(\phi_i) = \frac{\tilde{A}^R}{RT} + (Z - 1) - ln(Z) + \left[\frac{\partial}{\partial x_i}\left(\frac{\tilde{A}^R}{RT}\right)\right]_{T,\rho,\not{x}} - \sum_{k=1}^{C} x_k \left[\frac{\partial}{\partial x_k}\left(\frac{\tilde{A}^R}{RT}\right)\right]_{T,\rho,\not{x}} \tag{4.35}$$

The symbol $\not{x}$ points out that the differentiation of partial derivatives with respect to mole fraction x_i is performed keeping constant the mole fractions of all other components, without considering that the mole fraction of components must sum to unity ($\sum_{k=1}^{C} x_k = 1$). Like the compressibility factor, $\tilde{A}^R/RT$ is composed of two contributions: the hard-sphere-chain contribution $\tilde{A}^{hsc}/RT$ and the dispersion contribution $\tilde{A}^{dsp}/RT$, that is,

$$\frac{\tilde{A}^R}{RT} = \frac{\tilde{A}^{hsc}}{RT} + \frac{\tilde{A}^{dsp}}{RT} \tag{4.36}$$

If the original PC-SAFT version is intended to be used, then we have that the following expressions are required to determine the hard-sphere-chain contributions.

$$\frac{\tilde{A}^{hsc}}{RT} = \tilde{m}\frac{\tilde{A}^{hs}}{RT} + \sum_{k=1}^{C} x_k(1 - m_k) ln\left(g_{kk}^{hs}\right) \tag{4.37}$$

$$\frac{\tilde{A}^{hs}}{RT} = \frac{1}{\xi_0}\left[\frac{3\xi_1\xi_2}{1 - \xi_3} + \frac{\xi_2^3}{\xi_3(1 - \xi_3)^2} + \left(\frac{\xi_2^3}{\xi_3^2} - \xi_0\right) ln(1 - \xi_3)\right] \tag{4.38}$$

$$\left[\frac{\partial}{\partial x_i}\left(\frac{\tilde{A}^{hsc}}{RT}\right)\right]_{T,\rho,\Xi} = m_i\frac{\tilde{A}^{hs}}{RT} + \tilde{m}\left[\frac{\partial}{\partial x_i}\left(\frac{\tilde{A}^{hs}}{RT}\right)\right]_{T,\rho,\Xi} + \sum_{k=1}^{C}\frac{x_k(1-m_k)}{g_{kk}^{hs}}\left(\frac{\partial g_{kk}^{hs}}{\partial x_i}\right)_{T,\rho,\Xi} \quad (4.39)$$

$$\left[\frac{\partial}{\partial x_i}\left(\frac{\tilde{A}^{hs}}{RT}\right)\right]_{T,\rho,\Xi} = \frac{1}{\xi_0}\Bigg[-\xi'_{0i}\bar{a}^{hs} + \frac{3(\xi'_{1i}\xi_2 + \xi_1\xi'_{2i})}{1-\xi_3} + \frac{3\xi_1\xi_2\xi'_{3i}}{(1-\xi_3)^2} + \frac{3\xi_2^2\xi'_{2i}}{\xi_3(1-\xi_3)^2} + \frac{\xi_2^3\xi'_{3i}(3\xi_3-1)}{\xi_3^2(1-\xi_3)^3} + \left(\frac{3\xi_2^2\xi'_{2i}\xi_3 - 2\xi_2^3\xi'_{3i}}{\xi_3^3} - \xi'_{0i}\right)ln(1-\xi_3) + \left(\xi_0 - \frac{\xi_2^3}{\xi_3^2}\right)\frac{\xi'_{3i}}{1-\xi_3}\Bigg] \quad (4.40)$$

$$\left(\frac{\partial g_{kj}^{hs}}{\partial x_i}\right)_{T,\rho,\Xi} = \frac{\xi'_{3i}}{(1-\xi_3)^2} + \left[\frac{3\xi'_{2i}}{(1-\xi_3)^2} + \frac{6\xi_2\xi'_{3i}}{(1-\xi_3)^3}\right]\left(\frac{d_kd_j}{d_k+d_j}\right) + \left[\frac{4\xi_2\xi'_{2i}}{(1-\xi_3)^3} + \frac{6\xi_2^2\xi'_{3i}}{(1-\xi_3)^4}\right]\left(\frac{d_kd_j}{d_k+d_j}\right)^2 \quad (4.41)$$

For simplification purposes in the above expressions, a simplified notation for derivatives of Eq. (4.9) is introduced as follows:

$$\xi'_{ki} \equiv \left(\frac{\partial \xi_k}{\partial x_i}\right)_{T,\rho,\Xi} = \frac{\pi\rho m_i d_i^k}{6} \quad (4.42)$$

The dispersion contributions in Eq. (4.36) are in turn calculated from

$$\frac{\tilde{A}^{dsp}}{RT} = -\pi\rho\left(2I_1\overline{m^2\varepsilon\sigma^3} + \tilde{m}C_1I_2\overline{m^2\varepsilon^2\sigma^3}\right) \quad (4.43)$$

$$I_1 = \sum_{j=0}^{6} a_j(\tilde{m})\eta^j \quad (4.44)$$

$$I_2 = \sum_{j=0}^{6} b_j(\tilde{m})\eta^j \quad (4.45)$$

$$\left[\frac{\partial}{\partial x_i}\left(\frac{\tilde{A}^{dsp}}{RT}\right)\right]_{T,\rho,\Xi}$$
$$= -\pi\rho\left[2\overline{m^2\varepsilon\sigma^3}I_{1i}' + 2I_1\left(\overline{m^2\varepsilon\sigma^3}\right)_i' + \tilde{m}C_1I_2\left(\overline{m^2\varepsilon^2\sigma^3}\right)_i' + \left(m_iC_1I_2 + \tilde{m}I_2C_{1i}' + \tilde{m}C_1I_{2i}'\right)\overline{m^2\varepsilon^2\sigma^3}\right] \tag{4.46}$$

Here again, simplified notations for derivatives are introduced as follows:

$$I_{1i}' \equiv \left(\frac{\partial I_1}{\partial x_i}\right)_{T,\rho,\Xi} = \sum_{j=0}^{6}\left[ja_j(\tilde{m})\eta^{j-1}\xi_{3i}' + \eta^j a_{ji}'\right] \tag{4.47}$$

$$I_{2i}' \equiv \left(\frac{\partial I_2}{\partial x_i}\right)_{T,\rho,\Xi} = \sum_{j=0}^{6}\left[jb_j(\tilde{m})\eta^{j-1}\xi_{3i}' + \eta^j b_{ji}'\right] \tag{4.48}$$

$$a_{ji}' \equiv \left(\frac{\partial a_j}{\partial x_i}\right)_{T,\rho,\Xi} = \frac{m_i}{\tilde{m}^2}a_{1j} + \frac{m_i}{\tilde{m}^2}\left(3 - \frac{4}{\tilde{m}}\right)a_{2j} \tag{4.49}$$

$$b_{ji}' \equiv \left(\frac{\partial b_j}{\partial x_i}\right)_{T,\rho,\Xi} = \frac{m_i}{\tilde{m}^2}b_{1j} + \frac{m_i}{\tilde{m}^2}\left(3 - \frac{4}{\tilde{m}}\right)b_{2j} \tag{4.50}$$

$$\left(\overline{m^2\varepsilon^y\sigma^3}\right)_i' \equiv \left(\frac{\partial\overline{m^2\varepsilon^y\sigma^3}}{\partial x_i}\right)_{T,\rho,\Xi} = 2m_i\sum_{j=1}^{C}x_jm_j\left(\frac{\varepsilon_{ij}}{kT}\right)^y\sigma_{ij}^3 \tag{4.51}$$

$$C_{1i}' \equiv \left(\frac{\partial C_1}{\partial x_i}\right)_{T,\rho,\Xi} = C_2\xi_{3i}' - m_iC_1^2\left[\frac{8\eta - 2\eta^2}{(1-\eta)^4} - \frac{20\eta - 27\eta^2 + 12\eta^3 - 2\eta^4}{(1-\eta)^2(2-\eta)^2}\right] \tag{4.52}$$

When the sPC-SAFT version is intended to be used, then the dispersion contributions remain unchanged, but the hard-sphere-chain contributions change to simpler expressions by using again Eqs. (4.24) and (4.26), that is,

$$\frac{\tilde{A}^{hsc}}{RT} = \tilde{m}\frac{\tilde{A}^{hs}}{RT} + (1-\tilde{m})ln\left(g^{hs}\right) \tag{4.53}$$

$$\left[\frac{\partial}{\partial x_i}\left(\frac{\tilde{A}^{hsc}}{RT}\right)\right]_{T,\rho,\Xi} = m_i\left[\tilde{a}^{hs} - ln\left(g^{hs}\right)\right] + \tilde{m}\left[\frac{\partial}{\partial x_i}\left(\frac{\tilde{A}^{hs}}{RT}\right)\right]_{T,\rho,\Xi} + (1-\tilde{m})\left[\frac{\partial ln\left(g^{hs}\right)}{\partial x_i}\right]_{T,\rho,\Xi} \tag{4.54}$$

where the Helmholtz free energy of a hard-sphere fluid and the corresponding radial distribution function take the form

$$\frac{\tilde{A}^{hs}}{RT} = \frac{4\eta - 3\eta^2}{(1-\eta)^2} \tag{4.55}$$

$$g^{hs} = \frac{2-\eta}{2(1-\eta)^3} \tag{4.56}$$

$$\left[\frac{\partial}{\partial x_i}\left(\frac{\tilde{A}^{hs}}{RT}\right)\right]_{T,\rho,\Xi} = \frac{m_i d_i^3}{\tilde{m} d^3}\left[\frac{4\eta - 2\eta^2}{(1-\eta)^3}\right] \tag{4.57}$$

$$\left[\frac{\partial ln(g^{hs})}{\partial x_i}\right]_{T,\rho,\Xi} = \frac{m_i d_i^3}{\tilde{m} d^3}\left[\frac{5\eta - 2\eta^2}{(1-\eta)(2-\eta)}\right] \tag{4.58}$$

Substitution of Eqs. (4.55)–(4.58) into Eqs. (4.53) and (4.54) leads to the working expressions for the sPC-SAFT equation of state:

$$\frac{\tilde{A}^{hsc}}{RT} = \tilde{m}\frac{4\eta - 3\eta^2}{(1-\eta)^2} + (1-\tilde{m})ln\left[\frac{2-\eta}{2(1-\eta)^3}\right] \tag{4.59}$$

$$\begin{aligned}\left[\frac{\partial}{\partial x_i}\left(\frac{\tilde{A}^{hsc}}{RT}\right)\right]_{T,\rho,\Xi} &= m_i\left\{\frac{4\eta - 3\eta^2}{(1-\eta)^2} - ln\left[\frac{2-\eta}{2(1-\eta)^3}\right]\right. \\ &\left. + \left[\frac{4\eta - 2\eta^2}{(1-\eta)^3} + \left(\frac{1-\tilde{m}}{\tilde{m}}\right)\frac{5\eta - 2\eta^2}{(1-\eta)(2-\eta)}\right]\left(\frac{d_i}{d}\right)^3\right\}\end{aligned} \tag{4.60}$$

It is worth noticing that the effective hard-segment diameter of the chain segments d_i disappears explicitly from Eq. (4.59) because of the introduction of the average diameter d according to Eq. (4.25). It is import to remind that the partial derivative given by Eq. (4.60) is calculated regardless of the mole fraction constraint $(\sum_{k=1}^{C} x_k = 1)$. To take into account this constraint, we may choose the mole fraction of last component denoted by C as dependent variable on the remaining mole fractions. Thus, we have that [29]

$$\begin{aligned}\left[\frac{\partial}{\partial x_i}\left(\frac{\tilde{A}^{hsc}}{RT}\right)\right]_{T,\rho,\Sigma} &\equiv \left[\frac{\partial}{\partial x_i}\left(\frac{\tilde{A}^{hsc}}{RT}\right)\right]_{T,\rho,x_k,k\neq i=1,\ldots,C-1} \\ &= \left[\frac{\partial}{\partial x_i}\left(\frac{\tilde{A}^{hsc}}{RT}\right)\right]_{T,\rho,\Xi} - \left[\frac{\partial}{\partial x_C}\left(\frac{\tilde{A}^{hsc}}{RT}\right)\right]_{T,\rho,\Xi}\end{aligned} \tag{4.61}$$

As seen in this expression, the symbol Σ points out that the differentiation of partial derivatives with respect to mole fraction x_i is performed keeping constant the mole fractions of all other components, with exception of that corresponding to the last component. According to Eq. (4.61), we need to determine the partial derivative of component C; thus, from Eq. (4.60), the corresponding partial derivative is

$$\left[\frac{\partial}{\partial x_C}\left(\frac{\tilde{A}^{hsc}}{RT}\right)\right]_{T,\rho,\Sigma} = m_C\left\{\frac{4\eta - 3\eta^2}{(1-\eta)^2} - ln\left[\frac{2-\eta}{2(1-\eta)^3}\right] + \left[\frac{4\eta - 2\eta^2}{(1-\eta)^3} + \left(\frac{1-\tilde{m}}{\tilde{m}}\right)\frac{5\eta - 2\eta^2}{(1-\eta)(2-\eta)}\right]\left(\frac{d_C}{d}\right)^3\right\} \tag{4.62}$$

Introduction of Eqs. (4.60) and (4.62) into Eq. (4.61) leads to

$$\left[\frac{\partial}{\partial x_i}\left(\frac{\tilde{A}^{hsc}}{RT}\right)\right]_{T,\rho,\Sigma} = (m_i - m_C)\left\{\frac{4\eta - 3\eta^2}{(1-\eta)^2} - ln\left[\frac{2-\eta}{2(1-\eta)^3}\right] + \left[\frac{4\eta - 2\eta^2}{(1-\eta)^3} + \left(\frac{1-\tilde{m}}{\tilde{m}}\right)\frac{5\eta - 2\eta^2}{(1-\eta)(2-\eta)}\right]\frac{d_i^3 - d_C^3}{d^3}\right\} \tag{4.63}$$

One can observed that if $i = C$ (as in a pure-component system), then $m_i = m_C$ and the partial derivative given by Eq. (4.63) vanishes. On the other hand, writing Eq. (4.61) in terms of the dispersion contributions, similar results are obtained, because all the partial derivatives in Eq. (4.46) can be expressed as the product of m_i and a function of the other parameters. Therefore, we have that Eq. (4.35), considering the mole fraction constraint, is

$$ln(\phi_i) = \frac{\tilde{A}^R}{RT} + (Z-1) - ln(Z) + \left[\frac{\partial}{\partial x_i}\left(\frac{\tilde{A}^R}{RT}\right)\right]_{T,\rho,\Sigma} - \sum_{k=1}^{C-1} x_k\left[\frac{\partial}{\partial x_k}\left(\frac{\tilde{A}^R}{RT}\right)\right]_{T,\rho,\Sigma} \tag{4.64}$$

From which, the fugacity coefficient for a pure component ϕ_i^0 is given by

$$ln\left(\phi_i^0\right) = \frac{\tilde{A}_i^R}{RT} + (Z_i - 1) - ln(Z_i) \tag{4.65}$$

In terms of the properties of the mixture as one, this equation becomes

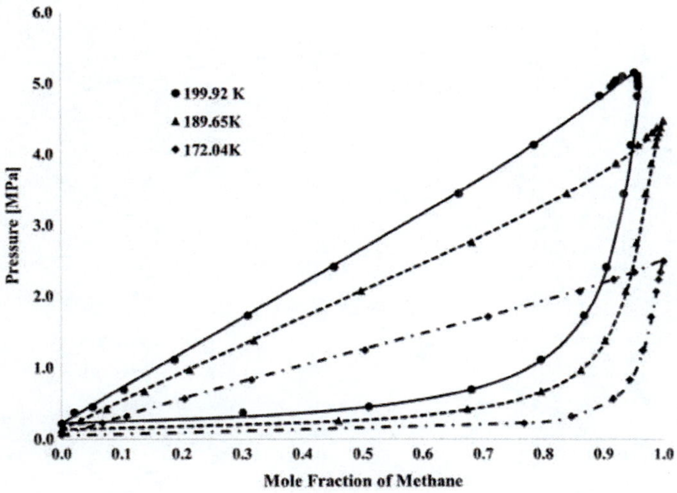

Fig. 4.7 Comparison between calculated and experimental vapor-liquid equilibriums of methane-ethane mixtures using sPC-SAFT $\left(k_{ij} = 0\right)$. Data from Wichterle and Kobayashi [30]

$$ln(\phi) = \frac{\tilde{A}^R}{RT} + (Z - 1) - ln(Z) \tag{4.66}$$

Although Eqs. (4.35) and (4.64) provide the same numerical results, Eq. (4.64) is preferred in mass transfer applications. From Figs. 4.7, 4.8, 4.9, and 4.10, vapor-liquid equilibriums are illustrated. As will be shown in the next Chapter, the calculated values are determined iteratively by solving the equilibrium condition for each component i in terms of ϕ_i, which are estimated using sPC-SAFT equation with $k_{ij} = 0$ and parameters from Table 4.2.

4.3 Estimation of PC-SAFT Parameters for Petroleum Fractions

In the application of SAFT-type equations to the thermodynamic characterization of petroleum fractions is assumed that such fractions are non-associating systems. This is the reason why association contributions are not considered in determining the compressibility factor and the fugacity coefficient from Eqs. (4.4) to (4.35) respectively. Hence, m, σ and ε/k are the only parameters estimated in the characterization procedures of petroleum fractions. This assumption is based on experimental analysis of asphaltenes precipitation: it has been found that polar interactions are not present for asphaltenes precipitation [34, 35]. Then, being asphaltenes the heaviest and most

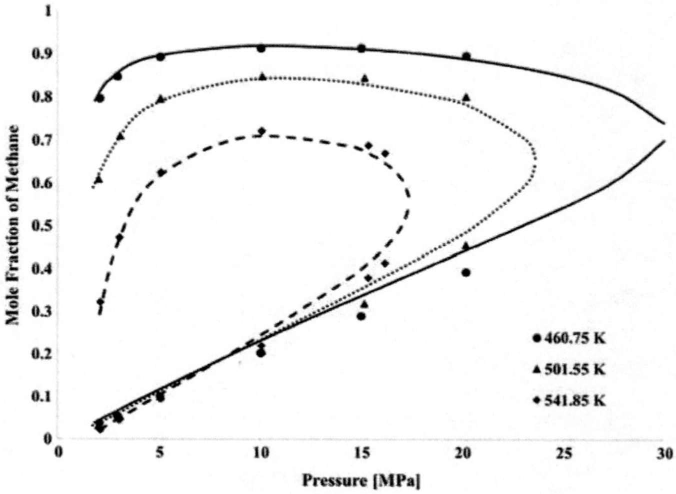

Fig. 4.8 Comparison between calculated and experimental vapor-liquid equilibriums of methane-m-Xylene mixtures using sPC-SAFT $(k_{ij} = 0)$. Data from Simnick et al. [31]

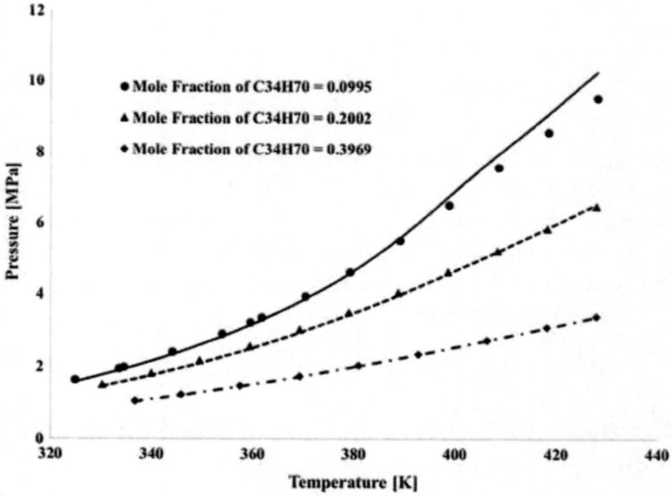

Fig. 4.9 Comparison between calculated and experimental vapor-liquid equilibriums of Propane-n-Tetratriacontane mixtures using sPC-SAFT $(k_{ij} = 0)$. Data from Peters et al. [32]

polarizable fraction, it is reasonable to assume that the remaining and lighter fractions can be properly treated as non-associating mixtures. Several approaches have been proposed to estimate the non-associating PC-SAFT parameters of petroleum fractions. In what follows, some of them are outlined according to their input data.

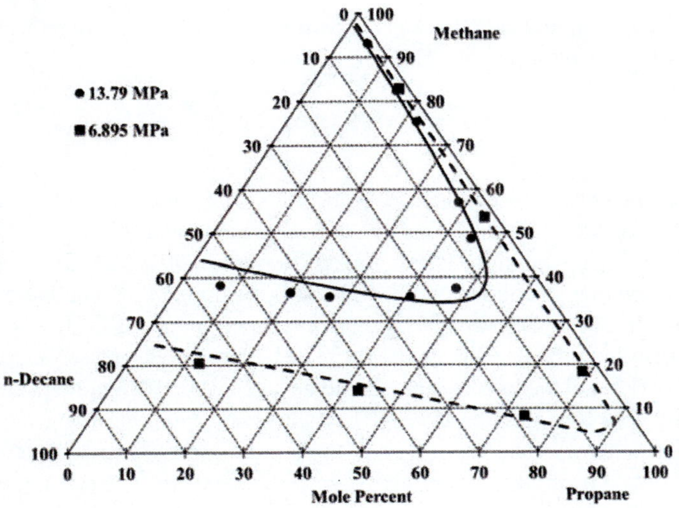

Fig. 4.10 Comparison between calculated and experimental vapor-liquid equilibriums of Methane–Propane–Decane mixtures using sPC-SAFT $(k_{ij} = 0)$. Data from Wiese et al. [33]

4.3.1 *Correlations in Terms of Molecular Weight and Compositional Analysis*

Taking into account the well-behaved tendencies with respect to molecular weight shown by component families, there have been estimating schemes of PC-SAFT parameters where a compositional analysis in terms of chemical families is required, for example, the Paraffinic, Naphthenic, and Aromatic (PNA) compositional analysis or the Saturates, Aromatics, Resins, and Asphaltenes (SARA) compositional analysis. The aim of these estimating methods is that suitable procedures can be developed for each chemical family so that the phase equilibrium problems of the petroleum fluid as a whole can be resolved properly. Thus, based on the expressions proposed previously [7, 12], Liang et al. [22] propose a new set of linear functions like those given by Eqs. (4.1), (4.2), and (4.3) for n-alkanes, cyclo-alkanes, and benzene derivatives. The respective slopes and intercepts are shown in Table 4.5.

Table 4.5 Slopes and intercepts of the linear functions given by Eqs. (4.1), (4.2), and (4.3) for PC-SAFT parameters for different families of components [22]

Family	a_m	b_m	$a_{m\sigma}$	$b_{m\sigma}$	$a_{m\varepsilon}$	$b_{m\varepsilon}$
Alkanes (paraffins)	0.02569	0.8709	1.7284	18.787	6.8248	141.14
Cycloalkanes (naphthenes)	0.02254	0.6827	1.7115	1.9393	6.4962	154.53
Benzenes derivatives (aromatics)	0.02576	0.2588	1.7539	−21.324	6.6756	172.4

The PC-SAFT parameters for a given petroleum fraction i are obtained by using the following mixing rule.

$$\Omega_i = \Omega_i^P x_i^P + \Omega_i^N x_i^N + \Omega_i^A x_i^A \tag{4.67}$$

where Ω represents the parameters m, $m\sigma^3$, and $m(\varepsilon/k)$ determined from the linear functions according to Table 4.5. x_i^P, x_i^N and x_i^A are the molar fractions of the PNA-compositional analysis, which is determined from some analytic estimation method. In the investigation performed by the authors, two estimation methods for the PNA content (the Riazi and Daubert [36] and the $n-d-M$ methods [37]) were assessed using 80 petroleum fluids over a wide range of temperatures, pressures and compositions. It was found that precautions should be taken when PNA content estimations are directly used for obtaining PC-SAFT parameters. Actually, among several characterization schemes analyzed by them, the highest errors were found from using the PNA content without an adjustment of model parameters to experimental normal boiling temperature or specific gravity of the specific petroleum fluid analyzed.

Pedersen et al. [38] suggested another set of correlations for estimating m and $(m\varepsilon/k)$ parameters of petroleum fractions (carbon number fractions in this case) in terms of the PNA content as well, while σ is estimated by regression to the liquid density at standard conditions. The proposed correlations are non-linear functions of the PNA content as well as of molecular weight and liquid density of the petroleum fraction analyzed. Similar to Liang et al. [22], the PNA content is estimated by using the $n-d-M$ method [37]. The PC-SAFT parameters of the PNA fractions are assumed equal to the corresponding parameters of the n-heptane (paraffins), cyclohexane (naphthenes) and benzene (aromatics), respectively.

As alternative to the PNA-compositional analysis, the SARA-compositional analysis may be preferred when asphaltenes precipitation is intended to be predicted [16]. From this analysis, three petroleum fractions are defined: saturates, aromatics + resins (A + R), and asphaltenes. The saturate fraction is composed of n-alkanes (n-paraffins), branched alkanes (iso-paraffins) and cyclo-alkanes (naphthenes). Because different families of components are combined into a single fraction, non-linear correlations in terms of molecular weight are proposed in order to account for the PC-SAFT parameters. Regarding the A + R fraction, its PC-SAFT parameters are determined from non-linear correlations in terms of molecular weight as well, but in this case the so-called aromaticity factor (γ_{A+R}) is used as an adjustable parameters in order to match the liquid density and bubble pressure simultaneously at saturation conditions of the petroleum fluid. Once γ_{A+R} has been determined, PC-SAFT parameters of the asphaltene fraction are adjusted to match the asphaltene precipitation pressure (the onset pressure) [17–19]. There are several available schemes: m, σ, and (ε/k) of the asphaltene fraction can be directly adjusted, or an aromaticity factor (γ_A), together with the corresponding molecular weight, can be used as adjustable parameters into the same set of correlations. One option of correlations for the saturate fraction is [19]

$$m = 0.0257M + 0.8444 \tag{4.68}$$

$$\sigma = 4.047 - 4.8013\frac{ln(M)}{M} \tag{4.69}$$

$$ln\left(\frac{\varepsilon}{k}\right) = 5.5769 - \frac{9.523}{M} \tag{4.70}$$

While the corresponding set of correlations for either the aromatic + resins ($\gamma = \gamma_{A+R}$) or asphaltene ($\gamma = \gamma_A$) fraction is

$$m = (1-\gamma)(0.0257M + 0.8444) + \gamma(0.0101M + 1.7296) \tag{4.71}$$

$$\sigma = (1-\gamma)\left(4.047 - 4.8013\frac{ln(M)}{M}\right) + \gamma\left(4.6169 - \frac{93.98}{M}\right) \tag{4.72}$$

$$\frac{\varepsilon}{k} = (1-\gamma)e^{5.5769-\frac{9.523}{M}} + \gamma\left(508 - \frac{234100}{M^{1.5}}\right) \tag{4.73}$$

As observed from these expressions, aromaticity provides an indication of the aromatic nature of the petroleum fraction. If the aromaticity is equal to zero, then the fraction is composed of saturate components, and if it is equal to one, then the fraction is composed of poly-nuclear aromatic components. Therefore, the adjusted aromaticity value must lie between zero and one.

4.3.2 *Correlations in Terms of Molecular Weight and Specific Gravity*

Encouraged by correlations used when cubic equations of state are employed in the characterization process, Assareh et al. [23] developed a set of similar correlations using the molecular weight and specific gravity as the only input data, which are typically available for reservoir petroleum fluids. One of the advantages of this set of correlations is that neither compositional analysis nor distillation curves are necessary for parameters estimation; the simple expressions suggested are

$$m = 33.58 + 0.08816M - 90.75S - 0.07727MS + 61.01S^2 \tag{4.74}$$

$$m\sigma^3 = -75.14 + 2.848M + 231.7S - 1.288MS - 186.9S^2 \tag{4.75}$$

Table 4.6 Regression constants of Eq. (4.77)

Ω	a	b	c	d	e	f	g
m	−0.122813	0.006268	0.563170	0.279508	−0.877908	0.702061	0.881099
$m\sigma^3$	−28.260228	−2.061987	−0.772729	2.701271	−0.642358	0.357911	2.156383
$m(\varepsilon/k)$	1.5669090	3.029941	0.226472	3.344329	−1.707719	0.479041	1.972062

$$m(\varepsilon/k) = 3372 + 11.24M - 8955S - 5.925MS + 6136S^2 \tag{4.76}$$

Another set of correlations was suggested by Hosseinifar et al. [24], which were derived from observing that the refractive index, combined with the molecular weight, modifies substantially the trends of the PC-SAFT parameters. Introducing an expression relating the refractive index to the liquid density, they arrived at a generalized expression in terms of the molecular weight and the specific gravity of the following form

$$\Omega = \left(a\left[\frac{3+2S}{3-S}\right]^{\frac{b}{2}} M^c + d\left[\frac{3+2S}{3-S}\right]^{\frac{e}{2}} M^f \right)^g \tag{4.77}$$

The values of the correlation constants are shown in Table 4.6. These constants were determined by minimizing the error between calculated and published parameters of pure hydrocarbons; the asterisk (∗) indicates a regression to parameters of pure hydrocarbons. Hence, to predict PC-SAFT parameters for petroleum fractions accurately, the calculated values using Eq. (4.77) must be modified by correction factors as follows:

$$m = m^* \theta_m \tag{4.78}$$

$$\sigma = \frac{(m\sigma)^*}{m^*} \theta_\sigma \tag{4.79}$$

$$\frac{\varepsilon}{k} = \frac{1}{m^*}\left(m\frac{\varepsilon}{k}\right)^* \theta_{(\varepsilon/k)} \tag{4.80}$$

$$\theta_{factor} = \left| a S_{C+}^{b} M_{C+}^{c} + d S_{C+}^{e} M_{C+}^{f} \right|^g \tag{4.81}$$

where, S_{C+} and M_{C+} are the specific gravity and the molecular weight of plus fraction. The corresponding values of the correlation constants of Eq. (4.81) are shown in Table 4.7.

Table 4.7 Regression constants of Eq. (4.81)

θ_{factor}	a	b	c	d	e	f	g
θ_m	−0.03741	−0.01070	0.718911	0.18353	0.201789	0.512177	1.237972
θ_σ	0.270725	−0.14515	0.316995	−0.0141	0.88177	0.662067	0.253528
$\theta_{(\varepsilon/k)}$	0.147172	−38.7519	−0.336230	−0.0835	1.915493	0.963658	0.028452

4.3.3 *Correlations in Terms of Normal Boiling Temperature and Specific Gravity*

Bearing in mind the two-step perturbation method used by Twu for critical properties [39], Yan et al. [1] developed correlations for PC-SAFT parameters in terms of specific gravity and normal boiling temperature. In this method, n-alkanes properties are used as reference properties in order to estimate actual properties. The PC-SAFT parameters of n-alkanes are estimated by using the following linear correlations:

$$m_0 = 0.02644M_0 + 0.835 \tag{4.82}$$

$$\left(m\sigma^3\right)_0 = 1.71638M_0 + 19.19189 \tag{4.83}$$

$$(m\varepsilon)_0/k = 6.90845M_0 + 139.3087 \tag{4.84}$$

The subscript 0 indicates properties of n-alkanes. As observed, these correlations are the same linear functions as those given by Eqs. (4.1), (4.2), and (4.3), but slopes and intercepts are different because of the used n-alkanes and temperature ranges. The corresponding molecular weight of n-alkanes (M_0) is determined at T_b using the Twu's correlation [39]

$$T_b = -13.7512\vartheta + 19.6197\vartheta^2 + e^{5.1264+2.71579\vartheta-0.28659\vartheta^2-\frac{39.8544}{\vartheta}-\frac{0.122488}{\vartheta^2}} \tag{4.85}$$

where $\vartheta = ln(M_0)$ and T_b is expressed in K. The previous equations contain M_0 as implicit variable; hence, in order to calculate M_0 from it knowing T_b, an iterative process is needed.

On the other hand, specific gravity (S_0) is estimated by the Soave's correlation [40]:

$$S_0 = (1.8T_b)^{\frac{1}{3}}\left(11.7372 + 3.336\times10^{-3}T_b - \frac{976.3}{T_b} + \frac{3.257\times10^5}{T_b^2}\right)^{-1} \tag{4.86}$$

Finally, PC-SAFT parameters for the actual petroleum fraction in terms of the specific gravity are calculated using

$$\sigma = \sigma_0 \tag{4.87}$$

$$\varepsilon = \varepsilon_0[1.1303391(S - S_0) + 1] \tag{4.88}$$

$$m = m_0\left[1.0460471(S - S_0)^2 - 1.6209973(S - S_0) + 1\right] \tag{4.89}$$

The purpose of any set of correlations is to provide generalized estimating methods in order to avoid a specific adjustment of model parameters, because the uncertainty on predictions beyond a restricted regression interval may be unpredictable. Even with the well-behaved linear functions from Eqs. (4.1) to (4.3) found for the PC-SAFT equation, as already been pointed out before, often, it is necessary to make use of some regression scheme for model parameters in order to achieve satisfactory results at least on a limited interval. In the case of T_b and S as independent variables, Liang et al. [22] found that good results are obtained by tuning m and σ to match T_b and S and by determining ε/k from the correlation

$$m(\varepsilon/k) = 6.8311M + 124.42 \tag{4.90}$$

4.3.4 Estimation of Parameters from Chemical Pseudostructures

As mentioned before, to identify all pure components forming petroleum fractions is either difficult or unfeasible. This is the reason why correlations in the methods described previously use bulk properties as input data (molecular weight, PNA content, etc.) in estimating the PC-SAFT parameters. When one intends to use rigorous procedures like free energy minimization, one deals with similar situations: components forming mixtures are unknown, and consequently their properties. However, as in the case of cubic equations of state (see Chap. 3), a free energy minimization can be used in the estimation of the PC-SAFT parameters, if the Gibbs free energy is written in terms of functional groups, which are structural units forming molecules in accordance with their chemical reactivity. The procedure starts by reminding that the residual Gibbs free energy of a pure-component system is related to the fugacity coefficient as indicated in Chap. 3. Assuming that petroleum fractions can be treated as pure components mathematically speaking, then the type and number of functional groups are the only responsible for minimizing the residual Gibbs free energy. Besides, as a system is at equilibrium state for pressure and temperature fixed when its Gibbs free energy is minimum, we search for a set of functional groups $(\upsilon_{1j}, \upsilon_{2j}, \ldots \upsilon_{Gj})$ that minimize

$$\min_{\upsilon_{1j},\upsilon_{2j},\ldots,\upsilon_{Gj}\geq 0} ln\left[\phi_j\left(\upsilon_{1j}, \upsilon_{2j}, \ldots, \upsilon_{Gj}\right)\right] \tag{4.91}$$

As been seen previously, some bulk properties are always needed in order to characterize a petroleum fraction. In the case of the minimization problem these bulk properties are established as constraints in the following form

$$\sum_{i=1}^{G} \upsilon_{ij} M_i - M_j = 0 \tag{4.92}$$

$$\sum_{i=1}^{G} \upsilon_{ij}\left[HCR_i - HCR_j\right] = 0 \tag{4.93}$$

$$\hat{\rho}_j - \frac{M_j}{\tilde{V}_j\left(\upsilon_{1j}, \upsilon_{2j}, \ldots, \upsilon_{Gj}\right)} = 0 \tag{4.94}$$

The subscript j indicates bulk properties of the petroleum fraction to be characterized, G is the types of functional groups considered to form such petroleum fraction, and υ_{ij} is the corresponding occurrence of a given functional group i in the petroleum fraction j; consequently, the subscript i indicates properties of such functional group. These constraints establish that the calculated set of functional groups must satisfy the experimental data M_j, HCR_j, and $\hat{\rho}_j$. In other words, Eq. (4.92) accounts for the molecular weight constraint, Eq. (4.93) for the molar ratio of hydrogen to carbon constraint, and Eq. (4.94) for the mass density constraint. Because $\hat{\rho}_j$ is usually available at standard conditions, ϕ_j and $\tilde{V}_j$ must be determined at the same temperature and pressure. Besides, a relationships between PC-SAFT parameters and functional groups are required in order to express the minimization problem as stated by Eq. (4.91) and the respective constraints in terms of $\upsilon_{1j}, \upsilon_{2j}, \ldots, \upsilon_{Gj}$. Tihic et al. [10] suggested relations between molecules and functional-groups parameters, which are used in estimating PC-SAFT parameters from chemical structure of molecules; the relationships are

$$m_j = \sum_{i=1}^{G} \upsilon_{ij} m_i \tag{4.95}$$

$$\left(m\sigma^3\right)_j = \sum_{i=1}^{G} \upsilon_{ij}\left(m\sigma^3\right)_i \tag{4.96}$$

$$\left(m\frac{\varepsilon}{k}\right)_j = \sum_{i=1}^{G} \upsilon_{ij}\left(m\frac{\varepsilon}{k}\right)_i \tag{4.97}$$

These relationships are also used by Burgess et al. [41] in order to suggest another set of values for the same and additional functional groups; the respective parameters are shown in Table 4.8.

One advantage of the current procedure is that the constraints given by Eqs. (4.92), (4.93), and (4.94) provided more flexibility to the characterization process, that is, the minimization problem may be solved by setting up two, three or even more constraints as active, depending on the experimental data available as input data and the prediction accuracy. The disadvantages of this method are that iterative calculations are needed, where it is not an easy task to find the global minimum of the Gibbs free energy. From Figs. 4.11, 4.12, and 4.13, the PC-SAFT parameters of pure components calculated from correlations and the minimization process are compared with data from Table 4.2.

4.4 PC-SAFT Parameters for Heavy Petroleum Fractions

In principle, all the previously described methods may be used for estimating the non-associating PC-SAFT parameters, depending on the experimental data available as input data and on whether the information is reliable or not for the heavy petroleum fraction under study. For instance, normal boiling temperatures are difficult to measure for heavy petroleum fractions because chemical reactions may occur, which may chemically decompose such fractions before the desired temperature is reached. Thus, when the normal boiling temperature is required by a procedure for estimating PC-SAFT parameters, extrapolation or prediction schemes are needed in order to estimate such input property. This is the reason why the methods previously discussed must be modified often when they are intended to be applied to heavy petroleum fractions. Accordingly, in the case of heavy petroleum fractions, specific methods have been suggested for estimating PC-SAFT parameters [17, 19, 42, 43].

In accordance with methods described previously, a generalized procedure for estimating the PC-SAFT parameters may consist of splitting out the heavy fraction into either two or three pseudocomponents. Then, the corresponding parameters are written in terms of correlations based on chemical families forming the chosen pseudocomponents. As options, correlations given by Eqs. (4.68)–(4.73) or those illustrated in Table 4.3 may be used. For instance, one may choose n-alkanes and polynuclear aromatics as chemical families forming the heavy fraction, with which we have that PC-SAFT parameters are determined from [42]

$$m = \gamma m_A + (1 - \gamma) m_P \tag{4.98}$$

$$\sigma = \gamma \sigma_A + (1 - \gamma) \sigma_P \tag{4.99}$$

$$\frac{\varepsilon}{k} = \gamma \left(\frac{\varepsilon}{k}\right)_A + (1 - \gamma) \left(\frac{\varepsilon}{k}\right)_P \tag{4.100}$$

Table 4.8 Functional-group parameters for minimization purposes of the free energy

Group	M_i	HCR_i	Tihic et al. [10]			Burgess et al. [41]		
			m_i	$(m\sigma^3)_i$ (Å^3)	$\left(m\frac{\varepsilon}{k}\right)_i$ (K)	m_i	$(m\sigma^3)_i$ (Å^3)	$\left(m\frac{\varepsilon}{k}\right)_i$ (K)
Paraffinic								
CH_3–	15.0348	3	0.644262	34.16955	129.3866	0.6960	35.425	144.63
CH_2–	14.0269	2	0.384329	24.33981	102.3238	0.4193	24.410	107.50
CH<	13.0189	1	0.043834	13.95391	68.2084	0.0665	10.784	47.48
>C<	12.0110	0	−0.492080	2.32542	−0.1098	−0.1123	1.473	−11.05
Aromatic								
AC	12.0110	0	0.010721	10.51468	87.9301	0.3789	20.109	113.39
ACH	13.0189	1	0.366330	19.81753	106.9481	0.9038	44.678	235.36
$ACCH_3$	27.0458	1.5	0.836861	43.57497	235.0636	0.3448	33.968	140.47
Naphthenic								
$cCCH_3$	27.0458	1.5	–	–	–	0.7045	50.859	176.07
$cCCH_2$–	26.0379	1	–	–	–	0.2322	41.733	117.31

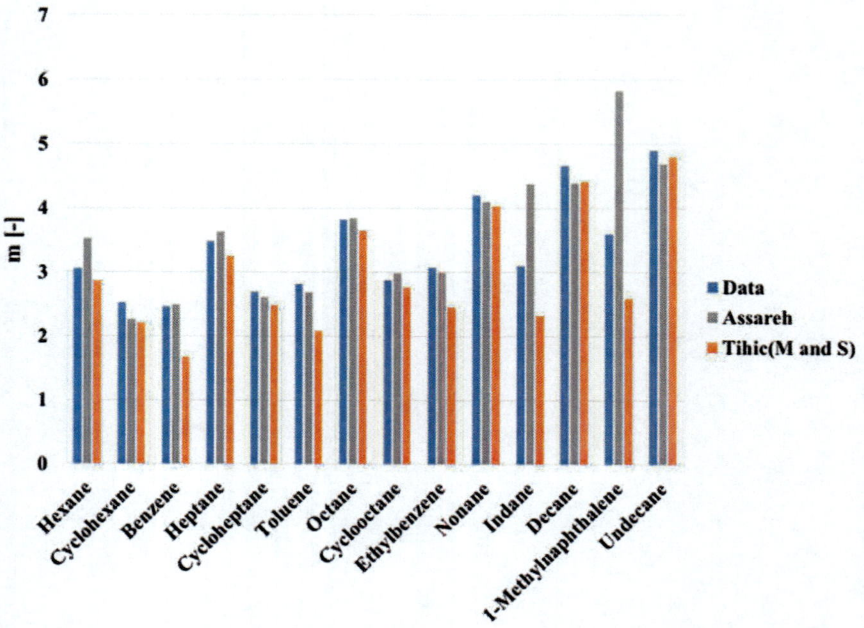

Fig. 4.11 Estimation of the number of segments of pure components using the Assareh correlation (Eq. 4.74) and the minimization energy setting M_j and $\hat{\rho}_j$ as constraints. Functional groups from Table 4.8 [10]

The subscripts A and P indicates n-alkanes and polynuclear aromatics parameters obtained from correlations in Table 4.3. In this case, the aromaticity factor (γ) represents the fraction of n-alkane family; if $\gamma = 0$ then the heavy fraction is composed of polynuclear aromatics alone, whereas if $\gamma = 1$ then such a fraction is formed by n-alkanes. Finally, the aromaticity factor is adjusted by regression to experimental data such as liquid density data.

4.4.1 Generalized Pseudocomponents

Bearing in mind the single carbon number (SCN) characterization, where a set of universal pseudocomponents are proposed in order to characterize any type of petroleum fluid using cubic equations of state, Saajanlehto et al. [43] proposed a set the universal pseudocomponents for heavy petroleum fractions. The main ideal behind this characterization procedure is also that such pseudocomponents can be used regardless of the type of the considered heavy petroleum fluid. However, the pseudocomponents are established here according to their normal boiling temperature rather than according to their carbon number as in the case of SCN characterization. An arbitrary range of temperatures is suggested for this purpose so that it

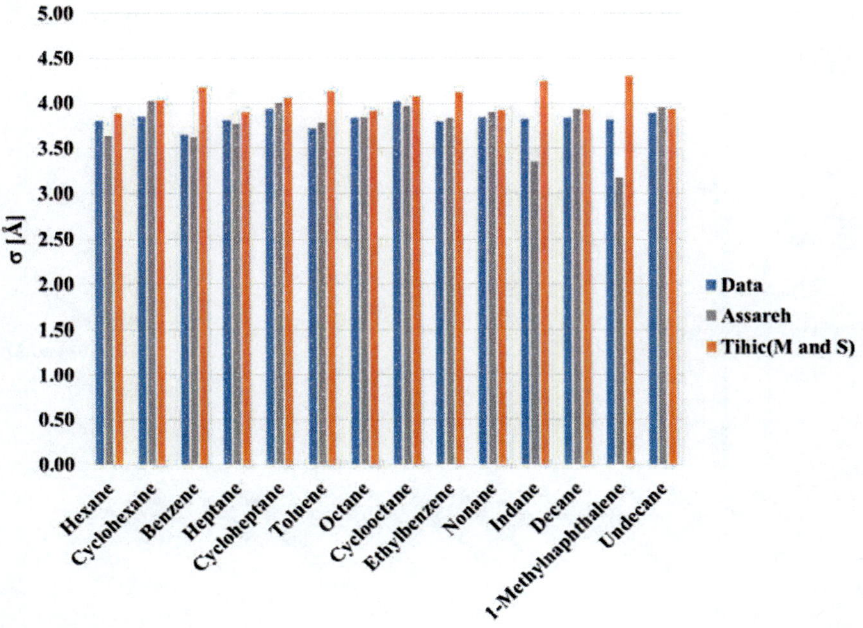

Fig. 4.12 Estimation of the segment diameter of pure components using the Assareh correlation (Eq. 4.75) and the minimization energy setting M_j and $\hat{\rho}_j$ as constraints. Functional groups from Table 4.8 [10]

contains most of measured distillation curves of heavy fractions. The range is set from 473 to 1023 K at 50 K intervals for each pseudocomponent, giving rise to twelve pseudocomponents plus an extra pseudocomponent at higher temperatures than 1023 K accounting for asphaltenes.

Once pseudocomponents have been set, each one of them is in turn split out into saturate, aromatic and polyaromatic fractions (SAP); splitting pseudocomponents into SAP fractions allows, in principle, PC-SAFT parameters to be estimated from correlations based on chemical families. However, the SAP fractions defined here are not chemical families strictly speaking, because the saturate fraction consists of all type of alkanes, normal and branched ones, while the aromatic fraction consists of monoaromatic and cycloalkanes components, and the polyaromatic fraction is assumed to be formed by polyaromatic components alone. An SAP-compositional analysis is chosen instead of other compositional analysis such as PNA or SARA because aromatic components are predominant in heavy fractions. Because of this splitting procedure into chemical families, the PC-SAFT parameters can be estimated by using methods as those described before that need molecular weight and compositional analysis as input data. Consequently, correlations from Eqs. (4.68) to (4.70) are used to the estimate PC-SAFT parameters for the saturate fraction, while the corresponding set of correlations for either the aromatic or polyaromatic fraction is:

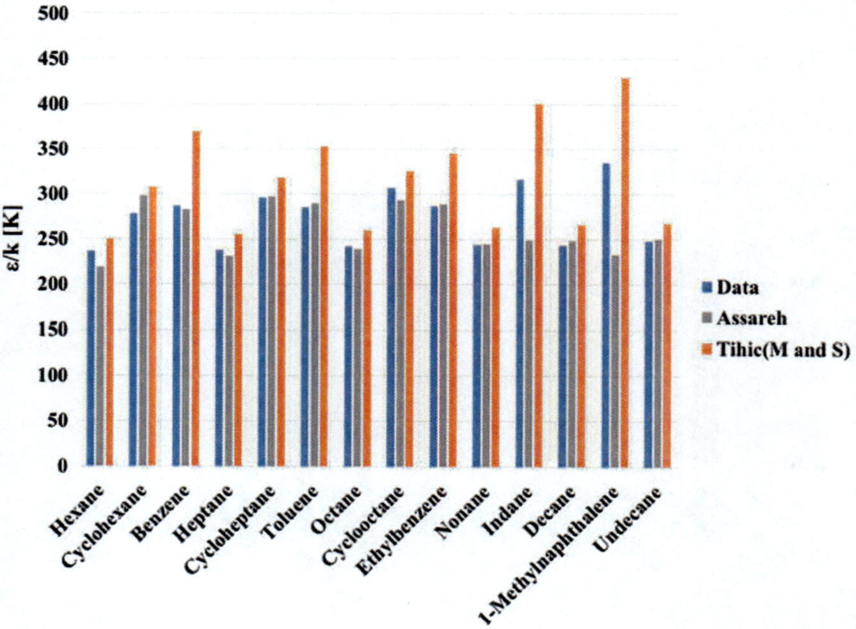

Fig. 4.13 Estimation of the dispersion energy of pure components using the Assareh correlation (Eq. 4.76) and the minimization energy setting M_j and $\hat{\rho}_j$ as constraints. Functional groups from Table 4.8 [10]

$$m = (1-\gamma)(0.0223M + 0.751) + \gamma(0.0101M + 1.7296) \tag{4.101}$$

$$\sigma = (1-\gamma)\left(4.1377 - \frac{38.1483}{M}\right) + \gamma\left(4.6169 - \frac{93.98}{M}\right) \tag{4.102}$$

$$\frac{\varepsilon}{k} = (1-\gamma)(0.00436M + 283.93) + \gamma\left(508 - \frac{234100}{M^{1.5}}\right) \tag{4.103}$$

Similarly, aromaticity γ gives an indication of the aromatic nature of the petroleum fraction between monoaromatic and polyaromatic in this case. If the aromaticity is equal to zero, then the fraction is composed of monoaromatic and cycloalkanes components, whereas if it is equal to one, then the fraction is composed of polyaromatic components. To obtain universal pseudocomponents, the corresponding PC-SAFT parameters must be universal as well, and consequently the aromaticity factor must acquire a universal value. Using experimental data from five heavy petroleum fluids, Saajanlehto et al. [43] found that $\gamma = 0.1$ and $\gamma = 0.6$ are suitable values for the aromatic and polyaromatic fractions, respectively.

Finally, the molecular weight and the corresponding PC-SAFT parameters are determined iteratively by solving the vapor-liquid equilibrium problem for a pure component at the normal boiling temperature and the atmospheric pressure. The

PC-SAFT parameters and bulk properties such as weight fractions w of the obtained universal-generalized pseudocomponents are shown in Tables 4.9, 4.10, and 4.11. As observed from these tables, we are taking about 37 pseudocomponents, which may represent a big number for some applications; however, these pseudocomponent can be lumped into a fewer number of pseudocomponents by using the lumping methods described later. One advantage of this generalized set of pseudocomponents is that good predictions are obtained as long as a suitable set of binary interaction parameters is used.

Table 4.9 PC-SAFT parameters of the generalized pseudocomponents (Saturate fractions)

$_b$ (K)	w	x	M (g/mol)	S (–)	m (–)	σ (Å)	ε/k (K)
473	0.3000	0.2691	160.3	0.7477	4.9646	3.8949	249.01
523	0.2725	0.2385	197.0	0.7698	5.9082	3.9183	251.78
573	0.2450	0.2089	239.7	0.7880	7.0057	3.9373	253.96
623	0.2175	0.1802	289.6	0.8030	8.2883	3.9530	255.70
673	0.1900	0.1526	348.3	0.8155	9.7969	3.9663	257.13
723	0.1625	0.1260	418.0	0.8261	11.5859	3.9777	258.30
773	0.1350	0.1005	501.4	0.8350	13.7293	3.9875	259.28
823	0.1075	0.0764	602.7	0.8427	16.3329	3.9960	260.11
873	0.0800	0.0538	728.1	0.8493	19.5569	4.0035	260.82
923	0.0525	0.0330	887.8	0.8550	23.6612	4.0103	261.43
973	0.0250	0.0144	1099.0	0.8600	29.0896	4.0164	261.97
1023	0.0000	0.0000	1390.0	0.8644	36.5681	4.0220	262.45

Table 4.10 PC-SAFT parameters of the generalized pseudocomponents (Aromatic fractions)

T_b (K)	w	x	M (g/mol)	S (–)	m (–)	σ (Å)	ε/k (K)
473	0.5700	0.5856	140.0	0.8897	3.7993	3.8732	292.75
523	0.5800	0.5887	169.9	0.9012	4.4300	3.9282	296.43
573	0.5900	0.5909	204.1	0.9123	5.1508	3.9713	299.11
623	0.6000	0.5924	243.1	0.9227	5.9734	4.0057	301.11
673	0.6100	0.5932	287.7	0.9323	6.9129	4.0336	302.67
723	0.6200	0.5932	338.7	0.9410	7.9878	4.0565	303.91
773	0.6300	0.5921	397.2	0.9489	9.2216	4.0755	304.94
823	0.6400	0.5899	464.6	0.9561	10.6434	4.0915	305.82
873	0.6500	0.5861	542.8	0.9627	12.2903	4.1050	306.62
923	0.6600	0.5804	633.8	0.9686	14.2103	4.1166	307.36
973	0.6700	0.5725	740.9	0.9741	16.4668	4.1266	308.08
1023	0.6800	0.5631	868.1	0.9791	19.1475	4.1352	308.83

Table 4.11 PC-SAFT parameters of the generalized pseudocomponents (Polyaromatic fractions)

T_b (K)	w	x	M (g/mol)	S (−)	m (−)	σ (Å)	ε/k (K)
473	0.1300	0.1453	128.7	0.9678	3.2659	3.8685	322.38
523	0.1475	0.1729	147.1	0.9866	3.5423	3.9383	339.93
573	0.1650	0.2003	168.4	1.0057	3.8606	3.9997	354.38
623	0.1825	0.2274	192.6	1.0249	4.2238	4.0533	366.17
673	0.2000	0.2542	220.1	1.0440	4.6348	4.0997	375.73
723	0.2175	0.2808	250.9	1.0628	5.0971	4.1397	383.47
773	0.2350	0.3073	285.5	1.0810	5.6144	4.1742	389.75
823	0.2525	0.3338	324.0	1.0985	6.1913	4.2041	394.85
873	0.2700	0.3602	366.8	1.1152	6.8335	4.2299	399.02
923	0.2875	0.3866	414.5	1.1310	7.5475	4.2524	402.45
973	0.3050	0.4131	467.5	1.1459	8.3414	4.2720	405.29
1023	0.3200	0.4369	526.5	1.1600	9.2247	4.2891	407.66
>1023	asphaltene		2250.0	–	40.0000	4.2000	370.00

4.4.2 Lumping Processes for PC-SAFT Parameters

As will be explained in the next chapter, once a distribution function between some properties and composition has been established for a petroleum fluid, a set of petroleum fractions (pseudocomponents, cuts, or single carbon numbers) is obtained. These fractions, together with pure components, may form a mixture of tens or even hundreds of components. Hence, it is advisable to reduce this number in order to decrease the computational time, which may become significant in some applications such as reservoir simulations. The lumping process is a method, which allows the number of petroleum fractions to be reduced into a smaller number so that bulk properties from previous characterization steps (the splitting method and the estimation of PC-SAFT parameters) remain unchanged as much as possible. Taking as starting point the lumping methods developed for critical properties entering into cubic equation of states, Pedersen et al. [38] suggested analogous mixing rules of the form of

$$m_j = \frac{\sum_{i=n_I}^{n_F} x_i m_i}{\sum_{i=n_I}^{n_F} x_i} \tag{4.104}$$

$$\sigma_j = \frac{\sum_{i=n_I}^{n_F} x_i \sigma_i}{\sum_{i=n_I}^{n_F} x_i} \tag{4.105}$$

$$\varepsilon_j = \frac{\sum_{i=n_I}^{n_F} \sum_{k=n_I}^{n_F} x_i x_k \sqrt{\varepsilon_i \varepsilon_k}}{\sum_{i=n_I}^{n_F} \sum_{k=n_I}^{n_F} x_i x_k} \tag{4.106}$$

In these expressions, n_I and n_F indicate the first and the last petroleum fraction in the corresponding group. It is worthy pointing out that the above mixing rules were suggested to be applied to petroleum fractions in terms of carbon numbers (the carbon number fractions). However, these mixing rules are not restricted to this type of fractions; in principle, any petroleum fraction such cuts or pseudocomponents in general can be incorporated into these mixing rules.

In comparison with cubic equation of state, SAFT-type equations show more complexity; hence, specific lumping approaches have been proposed. For instance, Assareh et al. [23, 44] suggested the following set of mixing rules for lumping purposes.

$$m_j = \frac{\sum_{i=n_I}^{n_F} x_i m_i}{\sum_{i=n_I}^{n_F} x_i} \tag{4.107}$$

$$T_1 = \frac{\frac{1}{8k}\sum_{i=n_I}^{n_F}\sum_{k=n_I}^{n_F} x_i x_k m_i m_k \sqrt{\varepsilon_i \varepsilon_k}(\sigma_i + \sigma_k)^3}{\sum_{i=n_I}^{n_F}\sum_{k=n_I}^{n_F} x_i x_k} \tag{4.108}$$

$$T_2 = \frac{\frac{1}{8k^2}\sum_{i=n_I}^{n_F}\sum_{k=n_I}^{n_F} x_i x_k m_i m_{jk} \varepsilon_i \varepsilon_k (\sigma_i + \sigma_k)^3}{\sum_{i=n_I}^{n_F}\sum_{k=n_I}^{n_F} x_i x_k} \tag{4.109}$$

When compared with Eq. (4.19), it is observed that Eqs. (4.108) and (4.109) are the normalized version of $\overline{m^2\varepsilon^y\sigma^3}$, where binary interaction parameters are set equal to zero. The segment diameter and a segment energy parameter are then obtained from

$$\left(\frac{\varepsilon}{k}\right)_j = \frac{T_2}{T_1} \tag{4.110}$$

$$\sigma_j^3 = \frac{T_1}{m_j^2(\varepsilon/k)_j} \tag{4.111}$$

In addition to mixing rules for lumping PC-SAFT parameters, mixing rules for lumping molecular weight and specific gravity may be necessary as well. Equations (4.112) and (4.113) provide such mixing rules; however, these equations are not empirical expressions; the former is a mass balance in terms of mole fractions, and the latter is the addition rule assuming ideal liquid mixtures, as explained in the Chap. 2. Hence, these two equations can be used together with both sets of expressions described above; the corresponding equations are

$$M_j = \frac{\sum_{i=n_I}^{n_F} x_i M_i}{\sum_{i=n_I}^{n_F} x_i} \tag{4.112}$$

$$S_j = \frac{\sum_{i=n_I}^{n_F} x_i M_i}{\sum_{i=n_I}^{n_F} \frac{x_i M_i}{S_i}} \tag{4.113}$$

Table 4.12 Properties and PC-SAFT parameters of lumped generalized pseudocomponents

Properties			Set 1			Set 2		
T_b (K)	M (g/mol)	S (−)	m (−)	σ (Å)	ε/k (K)	m (−)	σ (Å)	ε/k (K)
473	143.8	0.8502	4.0354	3.8783	284.76	4.0354	3.8747	282.21
523	172.4	0.8718	4.6290	3.9276	292.62	4.6290	3.9203	289.36
573	204.4	0.8915	5.2798	3.9699	299.89	5.2798	3.9586	295.76
623	240.0	0.9097	5.9928	4.0070	306.69	5.9928	3.9917	301.60
673	279.7	0.9269	6.7738	4.0401	313.10	6.7738	4.0208	307.00
723	324.0	0.9432	7.6292	4.0699	319.16	7.6292	4.0468	312.04
773	373.3	0.9588	8.5661	4.0970	324.95	8.5661	4.0704	316.81
823	428.2	0.9739	9.5921	4.1218	330.49	9.5921	4.0921	321.37
873	489.4	0.9886	10.7156	4.1446	335.81	10.7156	4.1123	325.78
923	557.4	1.0030	11.9458	4.1656	340.94	11.9458	4.1314	330.09
973	633.1	1.0172	13.2924	4.1851	345.89	13.2924	4.1495	334.35
1023	718.8	1.0305	14.8121	4.2025	350.32	14.8121	4.1662	338.21
>1023	2250.0	1.2305	40.0000	4.2000	370.00	40.0000	4.2000	370.00

In all the above lumping expressions, the properties of the lumped pseudocomponent are indicated by the subscript j.

As mentioned before, one drawback of the generalized pseudocomponents in the previous section is that they form a set of 37 pseudocomponents. One approach to reduce this number is to lump these generalized pseudocomponents into a smaller set by using the lumping expressions described above. In Table 4.12, the PC-SAFT parameters of lumped generalized pseudocomponents are shown. Set 1 refers to those calculated from Eqs. (4.104) to (4.106), while set 2 refers to those calculated from Eqs. (4.107) to (4.109); Eqs. (4.112) and (4.113) are used for lumping the respective molecular weight and specific gravity.

At this point, it is worth remembering that a lumping procedure searches for reducing the number of petroleum fractions so that bulk properties remain unaltered as much as possible. Hence, to verify whether lumped PC-SAFT parameters in Table 4.12 are suitable, normal boiling temperatures and specific gravity are calculated using such parameters and compared with values belonging to pseudocomponents from the same Table 4.12. The comparison is illustrated in Figs. 4.14 and 4.15. In these figures, calculations using the PC-SAFT parameters estimated from Eqs. (4.74) to (4.76) are also included. As observed in Fig. 4.14, the smallest relative deviations are provided by the set 2 for the entire temperature range, showing −0.4% as maximum deviation with respect to the pseudocomponent value. Therefore, these parameters satisfy what is expected from a lumping process. On the other hand, the relative deviations obtained from using the set 2 are larger, becoming even larger as the normal boiling point increases, that is, the heavier the pseudocomponent, the

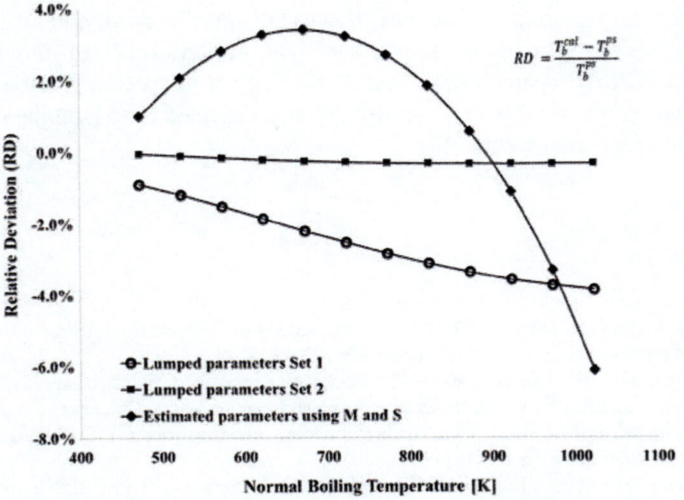

Fig. 4.14 Performance of the lumped parameters of the generalized pseudocomponents from Table 4.12 in estimating the normal boiling temperature of the same pseudocomponents

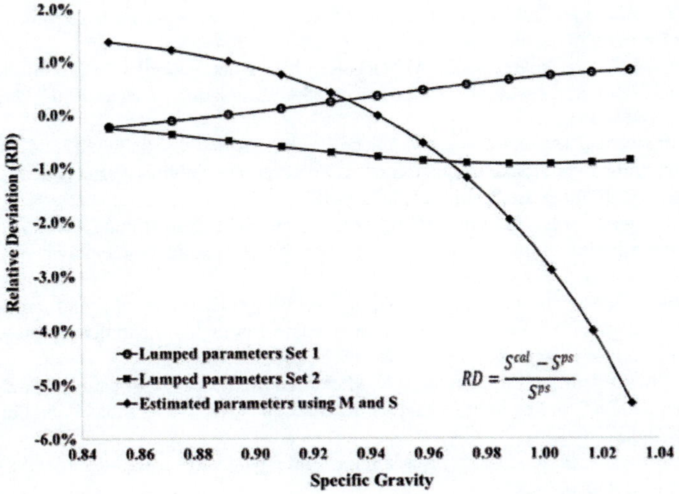

Fig. 4.15 Performance of the lumped parameters of the generalized pseudocomponents from Table 4.12 in estimating the specific gravity of the same pseudocomponents

lesser the efficiency of the lumping process. On the other hand, the relative deviations obtained by using parameters from correlations are larger with regard to the relative deviations from both sets of lumped parameters. In this case, it is worthy noticing that larger errors are obtained as the pseudocomponent becomes heavier. This is a usual result when procedures derived from pure component properties are applied to heavy petroleum fractions. Regarding specific gravity, similar conclusions can be drawn, except for similar and small deviations (<1%) from both sets of lumped PC-SAFT parameters in the whole interval of specific gravities, that is, lumped parameters from both sets satisfy what is expected from lumping processes in the case of specific gravity.

References

1. Yan W, Varzandeh F, Stenby EH (2015) PVT modeling of reservoir fluids using PC-SAFT EoS and Soave-BWR EoS. Fluid Phase Equilib 386:96–124
2. Kontogeorgis GM, Folas GK (2010) Thermodynamic models for industrial applications: from classical and advanced mixing rules to association theories. Wiley, Chichister, England
3. Wertheim MS (1984) Fluids with highly directional attractive forces. I. Statistical thermodynamics. J. Stat. Phys. 35(1):19–34
4. Wertheim, MS (1984) Fluids with highly directional attractive forces. II. Thermodynamic perturbation theory and integral equations. J Stat Phys 35:35–47
5. Wertheim MS (1986) Fluids with highly directional attractive forces. III. Multiple attraction sites. J Stat Phys 42:459–476
6. Wertheim MS (1986) Fluids with highly directional attractive forces. IV. Equilibrium polymerization. J Stat Phys 42:477–492
7. Gross J, Sadowski G (2001) Perturbed-Chain SAFT: an equation of state based on a perturbation theory for chain molecules. Ind Eng Chem Res 40(4):1244–1260
8. Von Solms N, Kouskoumvekaki IA, Michelsen ML, Kontogeorgis GM (2006) Capabilities, limitations and challenges of a simplified PC-SAFT equation of state. Fluid Phase Equilib 241(1–2):344–353
9. Kouskoumvekaki IA, Von Solms N, Michelsen ML, Kontogeorgis GM (2004) Application of the perturbed chain SAFT equation of state to complex polymer systems using simplified mixing rules. Fluid Phase Equilib 215(1):71–78
10. Tihic A, Kontogeorgis GM, von Solms N, Michelsen ML, Constantinou L (2008) A predictive group-contribution simplified PC-SAFT equation of state: application to polymer systems. Ind Eng Chem Res 47:5092–5101
11. Von Solms N, Michelsen ML, Kontogeorgis GM (2003) Computational and physical performance of a modified PC-SAFT equation of state for highly asymmetric and associating mixtures. Ind Eng Chem Res 42(5):1098–1105
12. Tihic A, Kontogeorgis GM, von Solms N, Michelsen ML (2006) Applications of the simplified perturbed-chain SAFT equation of state using an extended parameter table. Fluid Phase Equilib 248(1):29–43
13. Brignole EA, Mollerup J (2005) Rescaling of three-parameter equations of state: PC-SAFT and SPHCT, 234:108–121
14. Burgess WA, Tapriyal D, Morreale BD, Soong Y, Baled HO, Enick RM, Wu Y, Bamgbade BA, Mchugh MA (2013) Volume-translated cubic EoS and PC-SAFT density models and a free volume-based viscosity model for hydrocarbons at extreme temperature and pressure conditions. Fluid Phase Equilib 359:38–44
15. Burgess WA, Tapriyal D, Morreale BD, Wu Y, Mchugh MA, Baled H, Enick RM (2012) Prediction of fluid density at extreme conditions using the perturbed-chain SAFT equation correlated to high temperature, high pressure density data. Fluid Phase Equilib 319:55–66
16. Gonzalez DL, Hirasaki GJ, Creek J, Chapman WG (2007) Modeling of asphaltene precipitation due to changes in composition using the perturbed chain statistical associating fluid theory equation of state. Energy Fuels 21(3):1231–1242

17. Panuganti SR, Vargas FM, Gonzalez DL, Kurup AS, Chapman WG (2012) PC-SAFT characterization of crude oils and modeling of asphaltene phase behavior. Fuel 93:658–669
18. Panuganti SR, Tavakkoli M, Vargas FM, Gonzalez DL, Chapman WG (2013) SAFT model for upstream asphaltene applications. Fluid Phase Equilib 359:2–16
19. Punnapala S, Vargas FM (2013) Revisiting the PC-SAFT characterization procedure for an improved asphaltene precipitation prediction. Fuel 108:417–429
20. de Hemptine JC, Mougin P, Barreau A, Ruffine L, Tamouza S, Inchekel R (2006) Application to petroleum engineering of statistical thermodynamics–based equations of state. Oil Gas Sci Technol {l'IFP} 61(3):363–386
21. Leekumjorn S, Krejbjerg K (2013) Phase behavior of reservoir fluids: comparisons of PC-SAFT and cubic EOS simulations. Fluid Phase Equilib 359:17–23
22. Liang X, Yan W, Thomsen K, Kontogeorgis GM (2014) On petroleum fluid characterization with the PC-SAFT equation of state. Fluid Phase Equilib 375:254–268
23. Assareh M, Ghotbi C, Tavakkoli M, Bashiri G (2016) PC-SAFT modeling of petroleum reservoir fluid phase behavior using new correlations for petroleum cuts and plus fractions. Fluid Phase Equilib 408:273–283
24. Hosseinifar P, Assareh M, Ghotbi C (2016) Developing a new model for the determination of petroleum fraction PC-SAFT parameters to model reservoir fluids. Fluid Phase Equilib 412:145–157
25. Ma M, Chen S, Abedi J (2016) Modeling the solubility and volumetric properties of H_2 and heavy hydrocarbons using the simplified PC-SAFT. Fluid Phase Equilib 425:169–176
26. Chapman WG, Jackson G, Gubbins KE (1988) Phase equilibria of associating fluids chain molecules with multiple bonding sites. Mol Phys 65(5):1057–1079
27. Chapman WG, Gubbins KE, Jackson G, Radosz M, Chapman WG, Radosz M, Chapman WG, Jackson G (1990) New reference equation of state for associating liquids. Ind Eng Chem Res 29(8):1709–1721
28. Kariznovi M, Nourozieh H, Abedi J (2012) Measurement and equation of state prediction of vapor-liquid equilibrium and physical properties for the system Methane + n-Octadecane. Fluid Phase Equilib 314:102–106
29. Taylor R, Kooijman HA (1991) Composition derivatives of activity coefficient models (for the estimation of thermodynamic factors in diffusion). Chem Eng Commun 102:87–106
30. Wichterle I, Kobayashi R (1972) Vapor-liquid equilibrium of Methane–Ethane system at low temperatures and high pressures. J Chem Eng Data 17(1):9–12
31. Simnick JJ, Sebastian HM, Lin H, Chao K (1979) Gas-liquid equilibrium in binary mixtures of Methane + m-Xylene, and Methane +m-Cresol. Fluid Phase Equilib 3:146–151
32. Peters CJ, de Roo JL, de Swaan Arons J (1992) Measurements and calculations of phase equilibria in binary mixtures of Propane + Tetratriacontane. Fluid Phase Equilib 72(C):251–266
33. Wiese HC, Reamer HH, Sage BH (1970) Phase equilibriums in hydrocarbon systems. Phase behavior in the Methane–Propane-n-Decane system. J Chem Eng Data 15(1):75–82
34. Buckley JS, Hirasaki GJ, Liu Y, von Drasek S, Wang J-X, Gill BS (1998) Asphaltene precipitation and solvent properties of crude oils. Pet Sci Technol 16(3–4):251–285
35. Buckley JS (1999) Predicting the onset of asphaltene precipitation from refractive index measurements. Energy Fuels 13(2):328–332
36. Rlazl MR, Daubert TE (1986) Prediction of molecular-type analysis of petroleum fractions and coal liquids, 1009–1015
37. Van Nes K, Van Western HA (1951) Aspects of the constitution of mineral oils. Elsevier, New York, USA
38. Pedersen KS, Christensen PL, Shaikh JA (2015) Phase behavior of petroleum reservoir fluids, 2nd edn. CRC Press, Boca Raton, USA
39. Twu CH (1984) An internally consistent correlations for predicting the critial properties and molecular weight of petroleum and coal-tar liquids. Fluid Phase Equilib 16:137–150
40. Soave GS (1998) Estimation of the critical constants of heavy hydrocarbons for their treatment by the Soave–Redlich–Kwong equation of state. Fluid Phase Equilib 143:29–39

41. Burgess WA, Tapriyal D, Gamwo IK, Wu Y, Mchugh MA, Enick RM (2014) New group-contribution parameters for the calculation of PC-SAFT parameters for use at pressures to 276 MPa and temperatures to 533 K. Ind Eng Chem Res 53:2520–2528
42. Ma M, Chen S, Abedi J (2016) Modeling the density, solubility and viscosity of Bitumen/Solvent systems using PC-SAFT. J Pet Sci Eng 139:1–12
43. Saajanlehto M, Alopaeus V (2014) Heavy oil characterization method for PC-SAFT. Fuel 133:216–223
44. Assareh M, Ghotbi C, Pishvaie MR, Mittermeir GM (2013) An analytical delumping methodology for PC-SAFT with application to reservoir fluids. Fluid Phase Equilib 339:40–51

Chapter 5
Characterization of Petroleum Fractions

List of Abbreviations

CEoS	Cubic equations of state
EDF	Continuous exponential distribution function
GDF	Generalized distribution function
PC_SAFT	Perturbed-Chain Statistical Associating Fluid Theory

Nomenclature

A, B	Parameters from Eq. (5.18)
C_i	Carbon number
$E[g(Y)]$	Expected value
$F(M)$	Probability density function
$f(y)$	Distribution function
$F(y)$	Probability density function
$g(y)$	Single-valued function of the random variable Y
M	Molecular weight
m_i	Molecules from component i
n	Total number of moles
N_A	Avogadro number
p_i	Probabilities
PS	Pseudocomponents
r	Repetitions
RMS	Root Mean Square
S	Specific gravity
sd	Standard deviation
T_{bi}	Normal boiling temperature
u	Volume fraction

B. Carreón-Calderón et al., *Thermophysical Properties of Heavy Petroleum Fluids*, Petroleum Engineering, https://doi.org/10.1007/978-3-030-58831-1_5

w	Weight fraction
x	Mole fraction
β, τ	Parameters from Eq. (5.30)
Γ	Gamma function
ϵ_1	Tolerance
Ψ	Events
ψ	Sum of m_i $(m_i \leq r)$
Ω_a, Ω_b	CEoS parameters

Subscripts

+	Petroleum fraction
c	Cumulative
i	Each component
j	Subsystem

By definition, it is impossible to have complete experimental information of the unidentifiable fraction of petroleum fluids, because, otherwise, this would mean that we are able to identify every component forming such fraction. Therefore, the predictions of thermophysical properties of the unidentifiable petroleum fraction, and consequently the petroleum fluids as one, requires mathematical procedures based on physicochemical assumptions about its common behavior. One of these assumptions is to treat the petroleum fraction as a random system, where its bulk properties are average values of components that form the mixture. The application of this probabilistic treatment requires some concepts of the probability theory, which will be explained in what follows briefly.

5.1 Distribution Function and Its Properties

A distribution function indicates how the probability of a random variable is distributed; hence, a random variable Y can be defined by its distribution function $f(y)$ as follows:

$$f(y) \equiv p(Y \leq y) \tag{5.1}$$

Accordingly, $f(y)$ indicates the probability p that the random variable Y takes values between $-\infty$ and y, where y lies in the interval $(-\infty, \infty)$. $f(y)$ has the following properties: $f(-\infty) = 0$, $f(+\infty) = 1$, $f(y)$ is an increasing function of y, and $f(y)$ is continuous on the right.

When a random variable takes discrete values y_i, $(i = 1, 2, \ldots)$ with probabilities $p_i \equiv p(Y = y_i)$, then the distribution function is

$$f(y) = \sum_{y_i}^{y} p_i \tag{5.2}$$

On the other hand, a random variable is called continuous, if there is a function $F(y) > 0$ such that, the probability can be expressed as

$$f(y) = p(Y \leq y) = \int_{-\infty}^{y} F(y)dy \tag{5.3}$$

$F(y)$ is the so-called probability density function. As seen, $F(y)$ can alternatively be defined as $F(y) \equiv df(y)/dy$, where the continuous random variable Y has an everywhere continuous distribution function $f(y)$. As seen, $f(y)$ determines the cumulative probability of a random variable Y in the interval $(-\infty, y]$. This interval can be generalized to any finite interval $[a, b]$, for which the cumulative probability is

$$p(a \leq Y \leq b) = \int_{a}^{b} F(y)dy \tag{5.4}$$

If $g(y)$ is a single-valued function of the random variable Y, then $g(y)$ is also a random variable, whose expected value or expectation $E[g(Y)]$ is

$$E[g(Y)] = \sum_{k} p_k g(y_k) \tag{5.5}$$

when a discrete random variable is present. For the case of a continuous random variable,

$$E[g(Y)] = \int_{-\infty}^{\infty} F(y)g(y)dy \tag{5.6}$$

To express the previous probabilistic concepts in terms of properties of systems formed by molecules, the concept of an event is needed. In probability theory, all the possible outcomes of an experiment are called events, all of them forming a field of events Ψ with a total number of repetitions equal to r. Now, let ψ be a sum of m_i $(m_i \leq r)$ events with the same probability, and let each one of these events be the presence of one molecule from component i in a system formed by r molecules. Then, the probability for m_i molecules from component i to be present in the system is

$$p_i(\psi) = \frac{m_i}{r} \tag{5.7}$$

Multiplying and dividing by the Avogadro number N_A, we obtain that

$$p_i(\psi) = \frac{n_i}{n} = x_i \tag{5.8}$$

where n symbolizes the total number of moles. Therefore, $p_i(\psi)$ accounts for the mole fraction x_i of the component i in the system. Using this probability, for a C-component mixture, it follows from Eq. (5.2) that

$$f(C) = \sum_{i=1}^{C} x_i = 1 \tag{5.9}$$

which is known as the mole fraction constraint that must be satisfied by any system formed by C components.

Indeed, the number of molecules is always a discrete value, but, if the number of components is so big ($C \to \infty$) that $p_i(\psi)$ can be properly described by a continuous function $F(\psi)$, then

$$f(\infty) = \int_{-\infty}^{\infty} F(\psi)d\psi = 1 \tag{5.10}$$

where ψ is any property related to the mole fraction. This equation represents the mole fraction constraint for a continuous description of the components in the system.

According to Eq. (5.4), the cumulative probability can be determined once $F(y)$ is known; therefore, the cumulative mole fraction x_c in a finite interval $[a, b]$ is

$$x_c = \int_{a}^{b} F(\psi)d\psi \tag{5.11}$$

For a discrete number of components, x_c takes the form of

$$x_c = \sum_{i=a}^{b} x_i \tag{5.12}$$

Physically, x_c should correspond to the experimental mole fraction x_+ from the petroleum fraction; therefore, their difference may be used as an indication of the agreement between calculated and experimental values of x_i.

On the other hand, following the same reasoning, the expected values of bulk properties of the petroleum fraction can be determined from Eqs. (5.5) and (5.6) For instance, the molecular weight M is obtained as follows:

$$\langle M \rangle \equiv E[M] = \sum_{i=1}^{C} x_i M_i \tag{5.13}$$

$$\langle M \rangle \equiv E[M] = \int_{-\infty}^{\infty} F(\psi) M(\psi) d\psi \tag{5.14}$$

which should correspond to the experimental molecular weight M_+ of the petroleum fraction, and therefore, their difference may indicates the agreement between experimental and calculated values of both x_i and M_i. Finally, for a finite interval $[a, b]$, the expected values are

$$\langle M_c \rangle = \frac{1}{x_c} \sum_{i=a}^{b} x_i M_i \tag{5.15}$$

$$\langle M_c \rangle = \frac{1}{x_c} \int_{a}^{b} F(\psi) M(\psi) d\psi \tag{5.16}$$

$\langle M \rangle$ is also known as the number-average molar mass because the probability is defined in terms of the number of molecules present in the system. However, this is not the only way to define the probability of the presence of a given component in the system. For instance, let ψ be a sum of m_i events with the same probability again, but now let each event be the presence of one unit mass of component i in a system described in terms of mass as well. It can be easily shown that the probability $p_i(\psi)$ is equal to the mass fraction instead of the mole fraction. The resulting molecular weight is called the mass-average molar mass. The number-average and mass-average molar masses are different from each other generally. From an experimental point of view, the measurement of colligative properties is one of the classical methods for determining the molecular weight of heavy mixtures such as polymers and those present in petroleum fractions. It can be shown that the molecular weight measured by these techniques corresponds to the number-average molar mass [1], which is the molecular weight usually reported in petroleum applications.

5.2 Carbon Number Fractions

In the carbon number description, unidentifiable petroleum fractions are described by the random variable C_i, which takes discrete values resembling the carbon atoms in hydrocarbons. The simplest relationship between C_i and x_i is a linear expression of the form:

$$ln(x_i) = A + BC_i \tag{5.17}$$

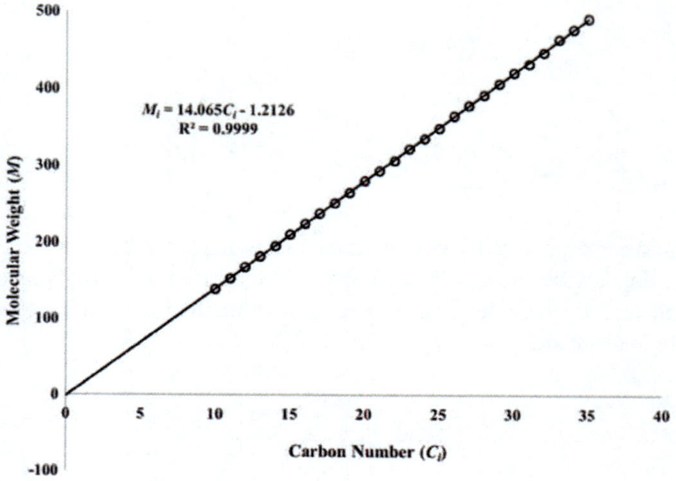

Fig. 5.1 Molecular weight distribution of Xinjian extra heavy oil with $S = 1.0352$ and $M = 574.5$. Data from Zhao et al. [2]

The intercept A and the slope B are specific for each petroleum mixture; therefore, they must be determined from experimental information from the same mixture, for example, the molecular weight.

The molecular weight of hydrocarbon families can be established as a linear function of the carbon atoms in the molecule, and a similar linear relationship is in general followed by unidentifiable petroleum fractions, as illustrated in Fig. 5.1. However, in this case, C_i represent hydrocarbon mixtures known as carbon number fractions, which are determined from the normal boiling temperatures of n-paraffins. For instance, C_{10} fraction consists of the hydrocarbons with normal boiling temperatures within 0.5 degrees below and above the normal boiling temperature of n-Decane (nC_{10}). Thus, assuming a linear relationship between C_i and M_i, Eq. (5.17) can be rewritten as:

$$\ln(x_i) = A + BM_i \tag{5.18}$$

Either of the two linear relationships allow the intercept A and the slope B to be calculated by regression to experimental compositional analysis, as illustrated in Fig. 5.2. A and B parameters so obtained allow the calculation of x_i from either C_i or M_i values not included in the experimental data. The cumulative mole fraction x_c and the expected molecular weight $\langle M_c \rangle$ can be then calculated from Eq. (5.12) and Eq. (5.15), respectively. In calculations, it has been suggested to set $C_{max} = 200$ as the heaviest carbon number fraction, when heavy petroleum fractions are present [3, 4]. When x_c and $\langle M_c \rangle$ differ from the experimental x_+ and M_+ as in the case of Fig. 5.2 ($x_c = 0.8656$ and $\langle M_c \rangle = 378.9$ against $x_+ = 1$ and $M_+ = 574.5$), then the parameters A and B must be changed so that calculated and experimental values

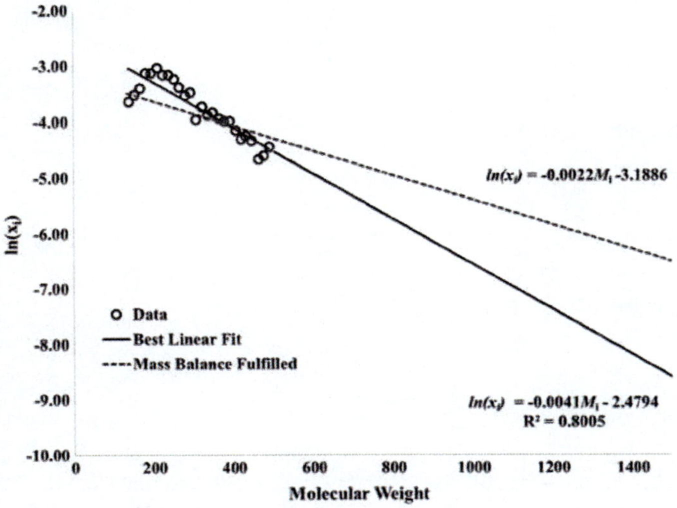

Fig. 5.2 Mol fraction distribution of Xinjian extra-heavy oil with $S = 1.0352$ and $M = 574.5$. Data from Zhao et al. [2]

coincide with each other. To accomplish such agreement, one option is to use x_+ and M_+ to calculate A and B directly. First, Eq. (5.12) is written in terms of x_+ as follows

$$x_+ = \sum_{i=+}^{C_{max}} x_i \tag{5.19}$$

where the subscript $"+"$ indicates the first carbon number fraction to be considered. In the case of Fig. 5.2, the subscript $"+"$ is set equal to 10 according to the provided experimental information [2]. By substituting x_i from Eq. (5.18) into the above equation, one obtains

$$x_+ = e^A \sum_{i=+}^{C_{max}} e^{BM_i} \tag{5.20}$$

from which

$$A = ln(x_+) - ln\left(\sum_{i=+}^{C_{max}} e^{BM_i}\right) \tag{5.21}$$

According to this equation, once B is known, A can be determined. To calculate B, Eq. (5.15) can be written in terms of properties of the petroleum fraction as

$$x_+M_+ = \sum_{i=+}^{C_{max}} x_i M_i \tag{5.22}$$

By substituting x_i from Eq. (5.18)

$$x_+M_+ = e^A \sum_{i=+}^{C_{max}} M_i e^{BM_i} \tag{5.23}$$

Combination of Eqs. (5.20) and (5.23) leads to

$$\sum_{i=+}^{C_{max}} (M_+ - M_i)e^{BM_i} = 0 \tag{5.24}$$

This equation must be solved iteratively in order to calculate B, which is in turn used to calculate A according to Eq. (5.21). If experimental values of M_i are not available to be used in Eqs. (5.21) and (5.24), then M_i can be estimated from [5]

$$M_i = 14C_i - 4 \tag{5.25}$$

In this equation, it is assumed that each C_i is linked to two hydrogen atoms, while the second term on the right-hand side accounts for the presence of aromatic structures. The characterization using Eqs. (5.21), (5.24) and (5.25) is illustrated in Fig. 5.2 as the mass balance fulfilled. As observed, the best linear fit differs from the mass balance fulfilled significantly. In general, such deviation is because the linear approximation given by Eq. (5.18) is not good enough for describing the experimental information properly $\left(R^2 = 0.8005\right)$.

As shown in previous chapters, the equation-of-state parameters of petroleum fractions can be estimated from correlations written in terms of two bulk properties among molecular weight, specific gravity and normal boiling temperature. Therefore, once the distribution function has been established through either Eq. (5.17) or Eq. (5.18), the characterization procedure requires to estimate either the specific gravity (S_i) or the normal boiling temperature (T_{bi}) for each carbon number fraction. In the former case, the calculated values of S_i should fulfill

$$\sum_{i=+}^{C_{max}} \frac{x_i M_i}{S_i} = \frac{x_+M_+}{S_+} \tag{5.26}$$

S_+ is the experimental specific gravity of the whole petroleum fraction. It can be shown that this expression arises from assuming an ideal-solution behavior for the total volume of the mixture as explained in Chap. 2. Therefore, Eq. (5.26) might be conceived as volume balance. To use Eq. (5.26) for estimating S_i, an expression relating S_i to M_i is necessary, which may be a distribution function of S_i similar to

Eq. (5.18). However, Søreide developed a correlation that has shown to be suitable for characterization purposes [6]

$$S_i = 0.2855 + \gamma(M_i - 66)^{0.13} \quad (5.27)$$

Typically, γ lies in the interval from 0.27 to 0.33; the heavier the petroleum fraction, the larger γ. Using this equation, Eq. (5.26) can be solved iteratively in order to find γ, and hence S_i for each carbon number fraction. Accordingly, x_i, M_i and S_i are now available in order to estimate the missing properties, becoming the unidentifiable petroleum fraction into a mixture of more or less two hundred identifiable carbon number fractions. To reduce such number, it is advisable to group the carbon number fractions into pseudocomponents so that each one contains approximately the same weight amount. The critical properties and acentric factor of such pseudocomponents are found as mass averages of the corresponding properties of the individual carbon number fractions [7]. In the case of the PC-SAFT parameters, the lumping procedure has been illustrated in the Chap. 4.

5.3 Continuous Description

To avoid the difficulty in establishing a heaviest carbon number fraction and the lumping procedure of carbon number fractions when a discrete characterization is considered, a continuous molar distribution can be used, which, in principle, should be a more accurate description, because petroleum fractions behave like continuous systems experimentally. Thus, according to Eq. (5.11), the cumulative mole fraction x_c in the finite interval of molecular weights $[M_{i-1}, M_i]$ is

$$x_c = \int_{M_{i-1}}^{M_i} F(M)dM \quad (5.28)$$

$F(M)$ accounts for the probability density function, which may be assumed to be an exponential function of the molecular weight as follows:

$$F(M) = \frac{1}{\beta} e^{\frac{\tau - M}{\beta}} \quad (5.29)$$

This expression comes from a more general three-parameter gamma distribution for describing molar distributions [8], where the parameter defining the form of the distribution is set equal to one. The other two parameters are related to each other by

$$\beta = M_+ - \tau \quad (5.30)$$

τ may be conceived as the minimum molecular weight M_{min} in the petroleum fraction; that is, $\tau = M_{min}$.

Alternatively, β and τ can be obtained by regression to experimental data as in the case of the discrete distribution, or its parameters, the intercept A and the slope B, can be used to calculate β and τ directly, according to

$$\beta = -\frac{1}{B} \tag{5.31}$$

$$\tau = \frac{1}{B} ln(B) - \frac{A}{B} \tag{5.32}$$

In any case, by using Eq. (5.29), the integration of Eq. (5.28) leads to

$$x_c = e^{\frac{\tau - M_{i-1}}{\beta}} - e^{\frac{\tau - M_i}{\beta}} \tag{5.33}$$

For given β and τ, this equation establishes the relationship between the mole fraction and the molecular weight limits for a given pseudocomponent of the petroleum fraction. For instance, one may choose to split out the petroleum fraction into pseudocomponents containing the same mole fraction x_c. Besides, one can determine or assume the M_0 value from experimental information. Thus, M_1 can be calculated from Eq. (5.34) with which the first pseudocomponent is established. M_1 can be now used to calculate M_2 from the same Eq. (5.34), and so on. Once the molecular weight limits has been calculated for the all pseudocomponents, the expected value of the molecular weight $\langle M_c \rangle$ is now required, which, according to Eq. (5.16), is derived from

$$\langle M_c \rangle = \frac{1}{x_c} \int_{M_{i-1}}^{M_i} MF(M)dM \tag{5.34}$$

By using the probability density function $F(M)$ given by Eq. (5.29), the above integration results in

$$\langle M_c \rangle = \frac{1}{x_c} \left[e^{\frac{\tau - M_{i-1}}{\beta}} (M_{i-1} + \beta) - e^{\frac{\tau - M_i}{\beta}} (M_i + \beta) \right] \tag{5.35}$$

Instead of pseudocomponents with the same mole fraction, one may prefer to split out the petroleum fraction into pseudocomponents containing the same weight fraction w_c, similar to the discrete distribution. That is,

$$w_c = \frac{1}{PS} \tag{5.36}$$

PS being the number of pseudocomponents. A useful relationship between w_c and x_c is

$$M_c x_c = M_+ w_c \tag{5.37}$$

which can be combined with Eq. (5.35) to obtain

$$g(M_i) = \tau - \beta ln\left[\frac{(M_{i-1} + \beta)e^{\frac{\tau - M_{i-1}}{\beta}} - M_+ w_c}{(M_i + \beta)}\right] - M_i = 0 \tag{5.38}$$

This equation is solved iteratively in order to calculate M_i, and hence x_c and $\langle M_c \rangle$ from Eq. (5.33) and Eq. (5.35), respectively. As before, calculations stars with M_0 in order to calculate M_1, and so on.

Often, the petroleum fluid is divided into two sets of components: the so-called identifiable or defined pure components such as methane and nitrogen, and the unidentifiable or undefined petroleum fraction. This means that x_+ can be less than unity ($x_+ < 1$), and therefore,x_c needs to be recalculated so that $x_+ = \sum_{c=+}^{PS} x_c < 1$. This is done by multiplying the calculated x_c by x_+; that is, $x_c = x_c x_+$. Finally, Eq. (5.26) is written in terms of the mole fraction and the molecular weight form the pseudocomponents

$$g(\gamma) = \sum_{c=1}^{PS} \frac{x_c M_c}{S_c} - \frac{x_+ M_+}{S_+} = 0 \tag{5.39}$$

with which the γ parameter is calculated iteratively, and hence $\langle S_c \rangle$ from Eq. (5.27). Figure 5.3 shows schematically the procedure of the characterization method by

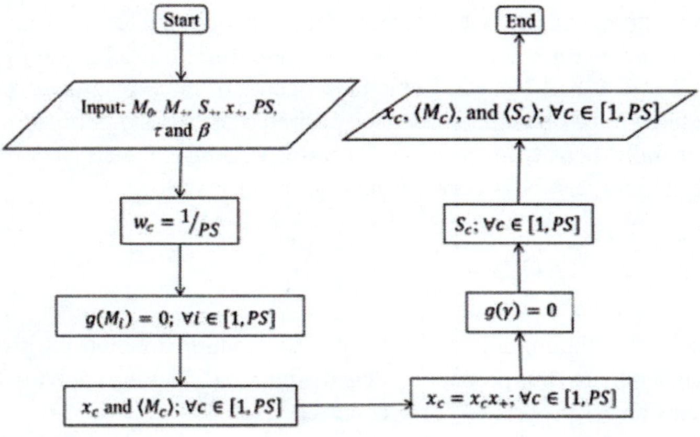

Fig. 5.3 Determination of properties of pseudocomponents containing the same weight fraction from the continuous exponential distribution function (EDF)

using the continuous exponential distribution function. Notice that if a molecular weight distribution is not available in order to calculate the *A* and *B* parameters from a regression procedure, then the only required data of the unidentifiable petroleum fraction are x_+, M_+, and S_+ as long as β is determined from Eq. (5.30). This means that M_{min} has to be known. In the case of compositional analysis for petroleum reservoir applications, M_{min} is set equal to the molecular weight of the n-heptane, that is $M_{min} = 100.204$.

The continuous exponential distribution function as explained before is limited to the molecular weight distribution of the petroleum fraction. A generalized distribution function (GDF), which can be applied to the distribution of the normal boiling temperature, molecular weight, and specific gravity, takes the form of [9, 10]

$$P^* = \left[\frac{A}{B} ln\left(\frac{1}{\chi^*}\right)\right]^{\frac{1}{B}} \tag{5.40}$$

where

$$P^* \equiv \frac{P - P_0}{P_0} \tag{5.41}$$

and

$$\chi^* = 1 - \chi_c \tag{5.42}$$

χ_c is the accumulative mole, weight or volume fraction of the property distribution. *P* can be any property among T_b, *M*, and *S*; consequently, P_0 is specific for each property and may be conceived as the initial value of the property distribution (T_{b0}, M_0 *or* S_0). Similarly, *A* and *B* are specific parameters for each property; that is, we have A_T, A_M or A_S, and B_T, B_M or B_S. *A* is specific for each unidentifiable petroleum fraction, but *B* may be the same for all unidentifiable petroleum fractions regardless of the petroleum fluid where they come from. It has been suggested that $B_T = 1.5$, $B_M = 1$, and $B_S = 3$. Using these values of the *B* parameter, only *A* and P^0 are required to be determined for each desired property and petroleum fraction.

The probability density function $F(P^*)$ corresponding to the distribution function given by Eq. (5.40) can be obtained from Eq. (5.11) as follows

$$F(P^*) \equiv \frac{d\chi_c}{dP*} = \frac{B^2}{A}(P^*)^{B-1} e^{-\frac{B}{A}(P^*)^B} \tag{5.43}$$

As in the case of the continuous exponential distribution function, the probabilistic concepts are applied to this probability density function. Thus, by applying Eq. (5.14) to the dimensionless property P^*, the expected value is

$$\langle P^* \rangle = \int_0^\infty P^* F(P^*) dP^* = \left(\frac{A}{B}\right)^{\frac{1}{B}} \Gamma\left(1 + \frac{1}{B}\right) \tag{5.44}$$

The corresponding expected value of the P property arises from the combination of Eq. (5.41) and Eq. (5.44)

$$\langle P \rangle = P_0\left(1 + \langle P^* \rangle\right) \tag{5.45}$$

In Eq. (5.44), Γ is the gamma function, whose values can be found in mathematics textbooks [11], or it can be evaluated from mathematics software. Microsoft Excel has available the $Gamma(x)$ function, with which Eq. (5.44) can be evaluated directly.

One useful characteristic of Eq. (5.24) is that it can be written in a linear form

$$Y = C_1 + C_2 X \tag{5.46}$$

where $Y = ln(P^*)$, $X = ln[ln(1/\chi)]$, $B = 1/C_2$, and $A = Be^{BC_1}$. Therefore, the parameters A and B are available by carrying out a linear regression procedure as long as P_0 is a known value. It can be assumed that P_0 represents the initial value of the distribution function. For instance, in the case of compositional analysis in petroleum reservoirs applications, P_0 can be set equal to $M_0 = 100.204$, $T_{b0} = 371.58\ K$ or $S_0 = 0.7557$ depending on whether a molecular-weight, normal boiling temperature or specific gravity distribution is available as experimental information.

A more convenient way in order to establish P_0 is by means of the Root Mean Square (RMS)

$$RMS(P_0) = \left[\frac{1}{N}\sum_{i=1}^{N}\left(P_i^{cal} - P_i^{exp}\right)^2\right]^{\frac{1}{2}} \tag{5.47}$$

N is the number of experimental points, while the superscripts *cal* and *exp* indicate calculated and experimental values, respectively, Here, P_0 is determined by a trial-and-error procedure so that $RMS(P_0)$ is minimized. During this procedure, the parameters A and B are calculated from a linear regression according to Eq. (5.46). From Figs. 5.4 to 5.7, experimental and calculated distribution functions are illustrated for three different heavy petroleum fluids. Figure 5.4 shows a molecular-weight distribution, while Figs. 5.5 and 5.6 show normal-boiling-temperature distributions. In these figures, P_0 is calculated by minimizing Eq. (5.47).

Similar to the continuous exponential distribution function, the generalized distribution function can be written in terms of pseudocomponents. Thus, remembering that χ_c can be the cumulative mole, weight or volume fraction, we have that

$$\chi_c = \int_{P_{i-1}^*}^{P_i^*} F(P^*) dP^* \tag{5.48}$$

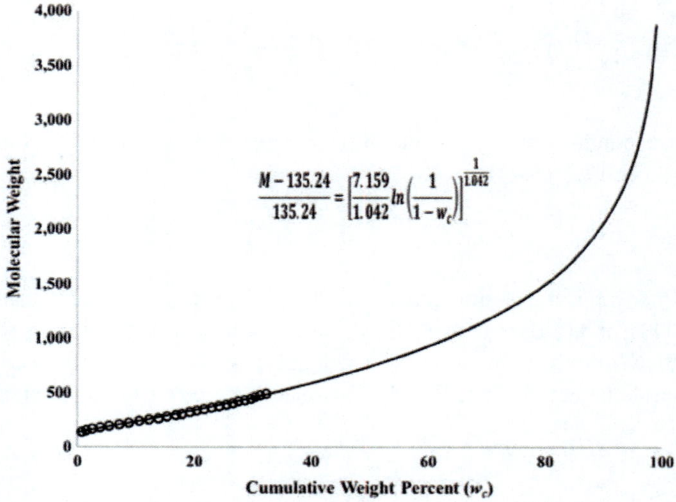

Fig. 5.4 Molecular weight distribution of Xinjian extra heavy oil with $S = 1.0352$ and $M = 574.5$. Data from Zhao et al. [2]

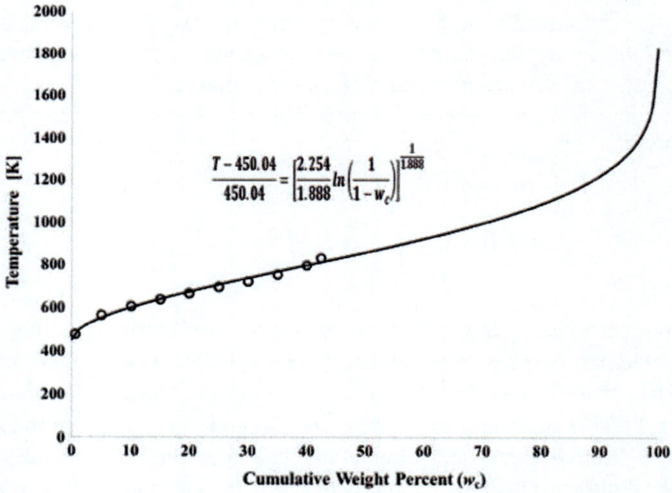

Fig. 5.5 Experimental and calculated simulated distillation of Athabasca Bitumen with $S = 1.009$ and $M = 522$. Data from Lal et al. [12]

in the finite interval $\left[P^*_{i-1}, P^*_i\right]$ of the dimensionless property P^*. Introducing the probability density function $F(P^*)$ given by Eq. (5.43), we arrive at

$$\chi_c = e^{-\frac{B}{A}\left(P^*_{i-1}\right)^B} - e^{-\frac{B}{A}\left(P^*_i\right)^B} \tag{5.49}$$

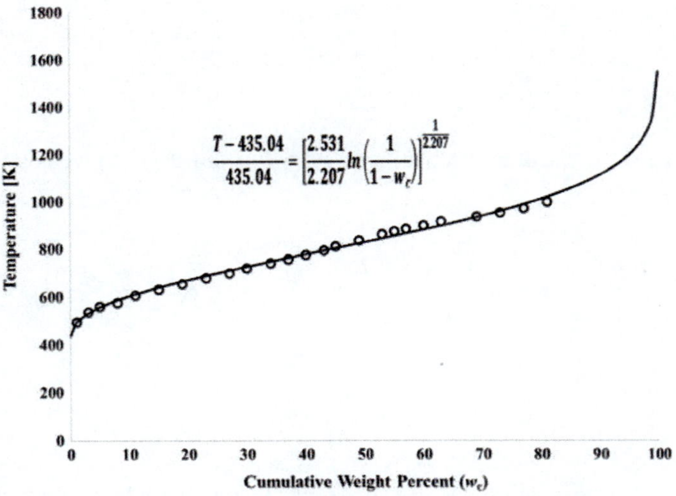

Fig. 5.6 Experimental and calculated simulated distillation of MacKay River Bitumen with $S = 1.0004$ and $M = 513$. Data from Zirrahi et al. [13]

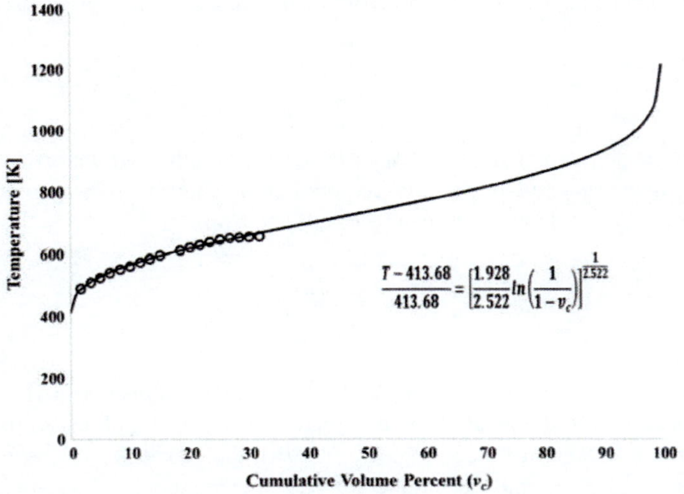

Fig. 5.7 Experimental and calculated simulated distillation of WC-B-B2 Bitumen with $S = 1.015$ and $T_b = 743.72$ K. Data from Johnston et al. [14]

By assuming pseudocomponents containing the same composition (the same molar, weight or volume fraction), this expression can be used to calculated P_i^* as long as A, B, and P_{i-1}^* are available. With P_i^*, the expected value $\langle P_c^* \rangle$ is obtained from

$$\langle P_c^* \rangle = \frac{1}{\chi_c} \int_{P_{i-1}^*}^{P_i^*} P^* F(P^*) dP^* \tag{5.50}$$

Here again, the use of the probability density function $F(P^*)$ given by Eq. (5.43) leads to

$$\langle P_c^* \rangle = \frac{1}{\chi_c} \left(\frac{A}{B} \right)^{\frac{1}{B}} \left[\Gamma\left(1 + \frac{1}{B}, q_{i-1}\right) - \Gamma\left(1 + \frac{1}{B}, q_i\right) \right] \tag{5.51}$$

with

$$q_i = \frac{B}{A} \left(P_i^*\right)^B \tag{5.52}$$

$\Gamma\left(1 + \frac{1}{B}, q_i\right)$ corresponds to the so-called upper incomplete gamma function, which can be evaluated as explained in mathematics textbooks. In Microsoft Excel, the complement of the $Distr.Gamma.N\left(q_i, 1 + \frac{1}{B}, 1, true\right)$ function can be used to evaluate $\Gamma\left(1 + \frac{1}{B}, q_i\right)$.

Finally, the expected value of the generalized property P of a given pseudocomponent c is

$$\langle P_c \rangle = P_0\left(1 + \langle P_c^* \rangle\right) \tag{5.53}$$

Usually, experimental distributions are in terms of weight fractions instead of mole fractions; however, the latter are the required compositions for phase equilibrium calculations. The mole fractions can be calculated from

$$x_c = \frac{\frac{w_c}{\langle M_c \rangle}}{\sum_{j=1}^{PS} \frac{w_j}{\langle M_j \rangle}} \tag{5.54}$$

Figure 5.8 shows a schematic diagram for the determination of properties of pseudocomponents from the generalized distribution function. As observed the parameters of the distribution function (A, B, and P_0), the properties of the petroleum fraction(x_+ and P_+), and the number of pseudocomponents (PS) are the input data. Assuming pseudocomponents containing the same weight fraction, the upper limit P_i^* is calculated for each pseudocomponent by means of Eq. (5.49). Using these values, the upper incomplete gamma function is evaluated. Then, the expected values are determined for each pseudocomponent from Eqs. (5.51) and (5.53). If the generalized property P corresponds to the molecular weight, the mole fraction is calculated according to Eq. (5.54). Otherwise, the expected value of the whole petroleum fraction is evaluated as a function of the calculated properties from the pseudocomponents. In the case of the molecular weight, this means

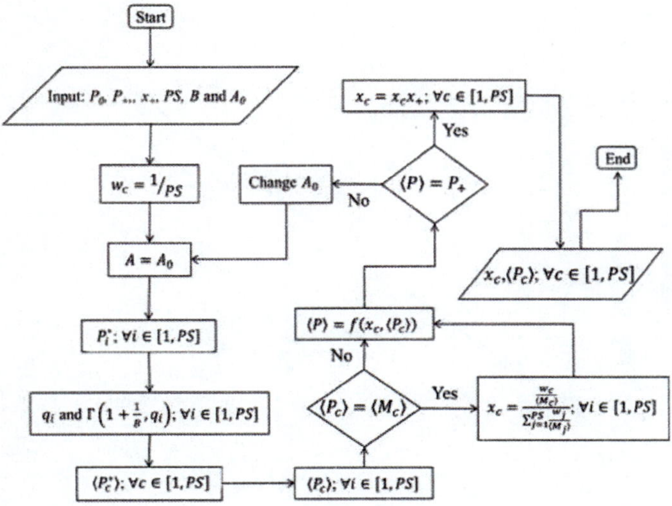

Fig. 5.8 Determination of properties of pseudocomponents from the generalized distribution function (GDF)

$$M = \sum_{c=1}^{PS} x_c \langle M_c \rangle \tag{5.55}$$

while in the case of the specific gravity

$$\langle S \rangle = \frac{\sum_{c=1}^{PS} x_c \langle M_c \rangle}{\sum_{c=1}^{PS} \frac{x_c \langle M_c \rangle}{\langle S_c \rangle}} = \frac{1}{\sum_{c=1}^{PS} \frac{w_c}{\langle S_c \rangle}} \tag{5.56}$$

In principle, $\langle P \rangle$ should be equal to P_+, the experimental value of the Petroleum fraction. Otherwise, the parameter A from the GDF must be modified until such agreement is reached. From Figs. 5.9 to 5.11, x_c, $\langle M_c \rangle$, and $\langle S_c \rangle$ of 13 pseudocomponents for an Extra Heavy Oil [2] are shown. In these figures, the x-coordinate corresponds to the properties calculated by using the regression parameters from Fig. 5.4, and by setting $B_S = 3$ for calculating the distribution function of specific gravity because there are not available experimental information on which a regression procedure can be performed. The y-coordinate corresponds to the properties calculated by setting $B_M = 1$ and $B_S = 3$. Besides, the values calculated from EDF according to Fig. 5.3 are also shown on the y-coordinate. It is worth pointing out that all distribution functions illustrated in these figures reproduce the experimental properties, x_+, M_+, and S_+, of the whole petroleum fraction. It has been

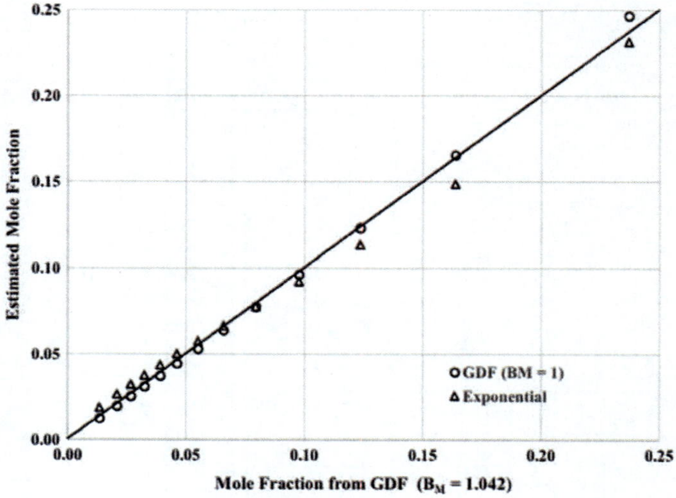

Fig. 5.9 Composition of 13 pseudocomponents for Xinjian extra-heavy oil from Fig. 5.4 as function of the continuous distribution function. The smallest fractions correspond to the heaviest pseudocomponent in Figs. 5.10 and 5.11

illustrated that phase equilibrium calculations are a weak function of the used distribution function when heavy petroleum fluids are involved [15, 16]. Therefore, the generalized distribution function is recommended because of its relative simplicity and generality.

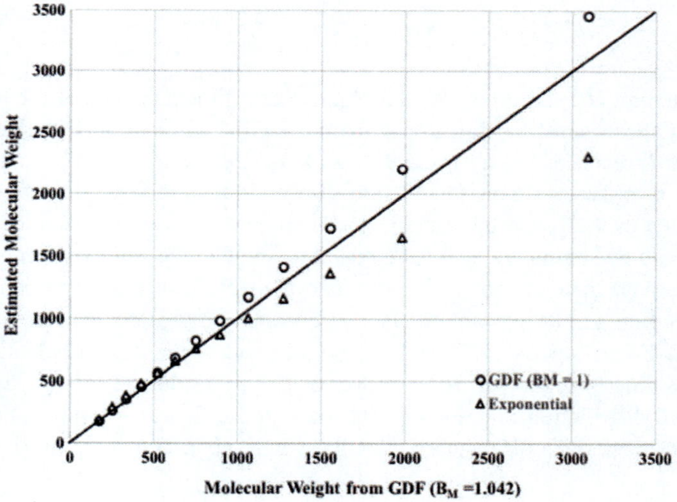

Fig. 5.10 Molecular weight of 13 pseudocomponents for Xinjian extra-heavy oil from Fig. 5.4 as function of the continuous distribution function

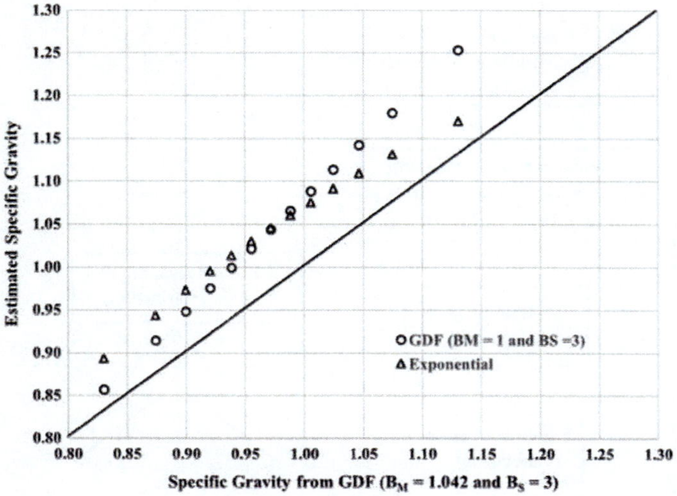

Fig. 5.11 Specific gravity of 13 pseudocomponents for Xinjian extra-heavy oil from Fig. 5.4 as function of the continuous distribution function

5.4 Strategies of Characterization

Bearing in mind that parameters entering equations of state, critical properties for cubic equations of state (CEoS) or non-associating parameters for Perturbed-Chain Statistical Associating Fluid Theory (PC-SAFT), can be estimated using the correlations, Kariznovi et al. [17] suggested a thermodynamic characterization procedure of heavy petroleum fractions, which is schematically illustrated in Fig. 5.12. Originally, it was developed for CEoS, but can be extended to PC-SAFT as well. First, experimental data are required, which may consist of Compositional Analysis, Distillation Curve, and Density and Molecular Weight distributions. It is worthy pointing out that not all the above experimental information is required at input data. The total composition of the petroleum fraction in the mixture, and the corresponding molecular weight and density at standard conditions are usually enough. By using the procedures illustrated before, the unidentifiable petroleum fraction is split out in pseudocomponents. The properties estimated from such procedures are individual composition and at least two of properties from $\langle M_c \rangle$, $\langle S_c \rangle$ and $\langle T_{bc} \rangle$ for each pseudocomponent c, because the third one can be estimated by using one of the correlations in terms of M, S, and T_b from the previous Chapters. Using these bulk properties, the critical properties and the acentric factor can be calculated from the Lee-Kesler correlations when a CEoS is considered. In the case of the PC_SAFT equation of state, it has not been defined yet which set of correlations gives suitable outcomes for regression purposes. Once the equation-of state parameters have been estimated roughly, we can use the selected equation of state to carry out phase equilibrium calculations, where either the saturation temperature or pressure are estimated. If

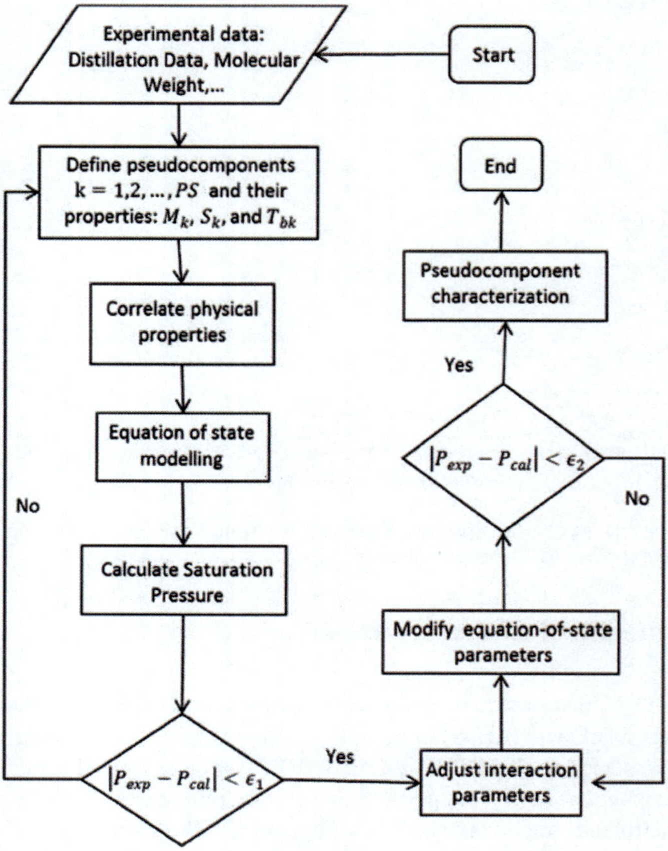

Fig. 5.12 Schematic diagram for characterization procedure using correlations for equation-of-state parameters [17]

these saturation values are different from the experimental ones, properties of pseudocomponents may be modified by means of an adjustment procedure so that the deviation of the calculated and experimental saturation values becomes smaller than a tolerance ϵ_1. Once this convergence criterion has been fulfilled, the calculated values are improved by fitting acentric factors, binary interaction parameters of the classical van der Waals mixing rule, and if necessary CEoS parameters Ω_a and Ω_b, when a CEoS is considered. The fitting procedure is limited to the binary interaction parameters when the PC-SAFT equation of state is considered. Thus, the characterization procedure of a petroleum fraction is complete.

We see that a thermodynamic characterization procedure provides the necessary information for performing equilibrium estimations using equations of state beginning with limited or incomplete experimental information as input data. According to Fig. 5.12 a representative characterization procedure for heavy petroleum fractions, where correlations are involved, implies adjustment procedures to phase equilibrium data by fitting pseudocomponent properties, binary interaction parameters and even equation-of-state parameters. The considerable extent of adjustment is a consequence of the chemical complexity of heavy petroleum fractions, which is beyond of properties and conditions from which the correlations are derived, as discussed before. An alternative option in order to reduce the extent of adjustment necessities is by using the concept of chemical pseudostructures, which allows that the non-classical mixing rules described previously can be used for the thermodynamic characterization of heavy petroleum fractions. Figure 5.13 illustrates the characterization procedure of petroleum fractions using chemical pseudostructures. In comparison with the use of correlations as depicted in Fig. 5.12 adjustment to phase equilibrium data is carried out by searching for the best combination of CEoS and mixing rule so that the available experimental data are predicted. If such a combination is not found, the match of equilibrium data is carried out by adjusting the molecular weight of the

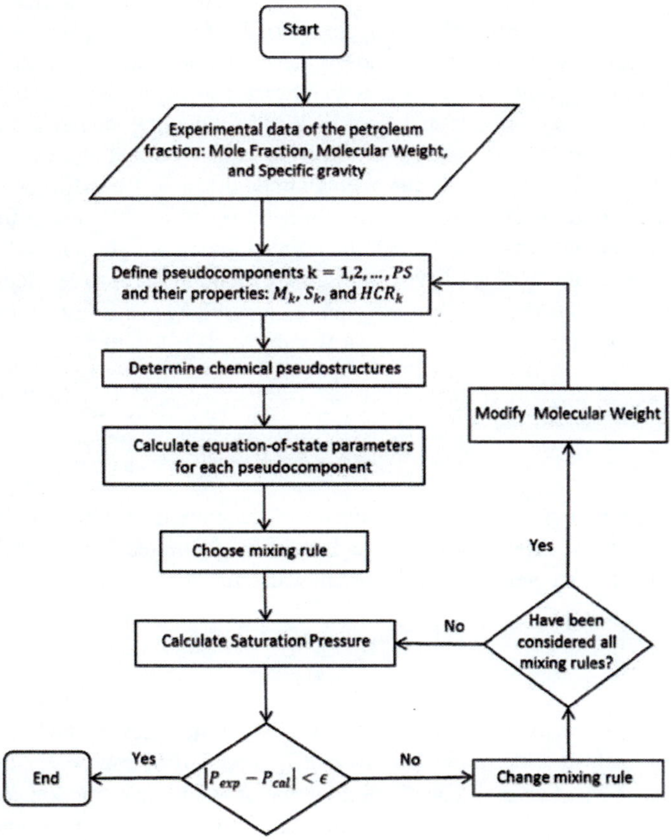

Fig. 5.13 Schematic diagram for characterization procedure using chemical pseudostructures for estimating equation of state parameters

whole petroleum fraction. CEoS parameters like Ω_a and Ω_b or the binary interaction parameters are not adjusted in this case; instead, the predictive mixing rules explained previously are used, keeping their corresponding binary interaction parameters fixed. The aim of such a procedure is to hold the predictive capabilities of such mixing rules for the predictions of phase equilibrium problems involving heavy petroleum fractions. In principle, extrapolations beyond of the range of available experimental data should be better in this case. Unlike CEoS, there is only one mixing rule for the PC-SAFT equation of state, whose binary interaction parameters are not well established for predictive purposes. Therefore, the binary interaction parameters are in general adjusted, especially for asymmetric mixtures.

As seen from Figs. 5.12 to 5.13, measured properties of the unidentifiable petroleum fraction are susceptible to be changed in order to calculations coincide with experimental equilibrium data. Among such properties, the molecular weight is the suggested one to be modified, because the experimental techniques are not particularly accurate [18, 19]. In general, the experimental uncertainty for a given property is about twice the standard deviation (sd). This means that there is a 95% probability that the measured M_+ lies in the interval $\langle M\rangle - 2sd < \langle M\rangle < \langle M\rangle + 2sd$ [20]. This experimental uncertainty leads to inaccurate experimental mole fractions, because the composition is measured as weight fractions and not as mole fractions. Hence, it has been suggested to make regressions by adjusting the molecular weight up to $\pm$ 10% for either the heavy fraction from compositional analysis or for the stable whole sample [21]. It has even been claimed that a 20% experimental error is most likely inherent to the molecular weight determination [22]. When the molecular weight is intended to be adjusted, the corresponding weight fractions are held constant. Aguilar and McCain have proposed a method in order to recalculate mole fractions keeping weight fractions constant as a result of adjusting the molecular weight [22]. First, the apparent molecular weight M_a is calculated

$$M_a = \sum_{i=1}^{+} x_i M_i \tag{5.57}$$

where the summation includes the identifiable components and the unidentifiable fraction. x_i and M_i represent the mole fractions and molecular weight reported experimentally. Then weight fractions are now calculated

$$w_i = \frac{x_i M_i}{M_a} \tag{5.58}$$

w_i are the weight fractions to be held constant, while the mole fraction x_i has to be recalculated each time the molecular weight is modified. It can be easily verified that changes in M_a of either the whole fluid or the unidentifiable fraction provide the same outcomes. Therefore, it is more convenient to modify M_a of the whole fluid according to

$$M_a^* = M_a(1 \mp \epsilon) \tag{5.59}$$

As mentioned before, $\epsilon \leq 0.1$. Thus, the new mole fractions of each identifiable component in the mixture are given by

$$x_i^* = \frac{w_i M_a^*}{M_i} \tag{5.60}$$

with which the new mole fraction of the unidentifiable fraction is

$$x_+^* = 1 - \sum_{i=1}^{+-1} x_i^* \tag{5.61}$$

Finally, the corresponding molecular weight (M_+^*) is calculated with

$$x_+^* M_+^* = M_a^* - \sum_{i=1}^{+-1} x_i^* M_i \tag{5.62}$$

As explained before, regression to phase equilibrium data should be performed by adjusting the molecular weight as the first option, because of its high experimental uncertainty. If the agreement is not satisfactory, then the equation of state or its parameters have to be adjusted, as indicated in Figs. 5.12 and 5.13.

A regression procedure of equation of state parameters to experimental data is not a simple task. It requires specialized strategies, which has been widely discussed and illustrated with the use of CEoS [21, 23]. However, for both CEoS and PC-SAFT equations of state, the strategies may consist of adjusting the pure-component parameters of the pseudocomponents or the binary interaction parameters. For CEoS, the pure-component parameters are the critical properties and the acentric factor, whereas, for the PC-SAFT equation of state, they are the non-associating parameters. In the latter case, some strategies are briefly described in the corresponding Chapter. In principle, these properties can be modified because they do not account for actual properties of the pseudocomponents. They are mathematical approximations, with which pseudocomponents can be treated as pure components from the mathematical point of view, in spite of the fact that they are mixtures. On the other hand, the adjustment of the binary interaction parameters is the most usual option, because they are in principle the most evident deficiency of equations of state. Thus, they are in practice empirical parameters for regression purposes. In the case of CEoS, the adjustment procedures are in general restricted to the binary parameters of the classical mixing rule of Van der Waals. The other mixing rules are not commonly used in phase equilibrium calculations involving petroleum fluids because, they, as explained in Chap. 3, need information that is not usually available for such mixtures.

Once the adjustment procedure is concluded, calculated and experimental values of phase equilibriums should match each other within acceptable defined deviations. In principle, this means the equations of state are ready to be used to predict phase equilibriums at conditions different from those used in the regression process. If conditions lie in the composition, pressure and pressure intervals of the regression process, then an interpolation is performed. Otherwise, one is taking about an extrapolation. From a mathematical point of view, the extrapolation error increases away from the points of interest; therefore, predictions should be used with caution outside the regression intervals. Besides, there is no guarantee that the adjusted parameters are valid for other thermophysical properties such as thermodynamic potentials or transport properties. This is the reason why efforts have been carried out to develop functions for estimated binary interaction parameters of the classical mixing rule with some theoretical basis [24–27].

References

1. Castellan WG (1983) Physical chemistry, 3rd edn. Addison-Wesley, New York, USA
2. Zhao Q, Li Z, Wang S, Lai F, Li H (2019) Phase behavior measurements and modeling for N 2 /CO 2 /extra heavy oil mixtures at elevated temperatures. Ind Eng Chem Res 58(1):428–429
3. Pedersen KS, Milter J, Sørensen H (2004) Cubic equations of state applied to HT/HP and highly aromatic fluids. SPE J 9(02):186–192
4. Krejbjerg K, Pedersen KS (2006) Controlling VLLE equilibrium with a Cubic EoS in heavy oil modeling 1–15
5. Pedersen KS, Thomassen P, FredenslundA (1984) Thermodynamics of petroleum mixtures containing heavy hydrocarbons. 1. Phase envelope calculations by use of the Soave-Redlich-Kwong Equation of state. Ind Eng Chem Process Des Dev 23(1):163–170
6. Soreide I (1989) Improved phase behavior predictions of petroluem reservoir fluids from a cubic equation of state. Norwegian Inst. of Technology
7. Pedersen KS, Thomassen P, Fredenslund A (1985) Thermodynamics of petroleum mixtures containing heavy hydrocarbons. 3. efficient flash calculation procedures using the SRK equation of state. Ind Eng Chem Process Des Dev 24(4):948–954
8. Whitson CH, Brulé MR (2000) Phase behavior. Richarson, USA, SPE
9. Riazi MR (1989) Distribution model for hydrocarbon plus fractions. Ind Eng Chem Res 28:1731–1735
10. Riazi MR (1997) A Continuous model for C 7 + Fraction characterization of petroleum fluids. Ind Eng Chem Res 36(10):4299–4307
11. Bronshtein IN, Semendyayev KA, Musiol G, Muehlig H (2004) Handbook of mathematics. Springer, Berlin
12. Lal D, Otto FD, Mather AE (1999) Solubility of hydrogen in Athabasca Bitumen. Fuel 78(12):1437–1441
13. Zirrahi M, Hassanzadeh H, Abedi J (2016) Experimental and modeling studies of MacKay River bitumen and water. J Pet Sci Eng 2017(151):305–310
14. Johnston KA, Schoeggl FF, Satyro MA, Taylor SD, Yarranton HW (2017) Phase behavior of bitumen and n-pentane. Fluid Phase Equilib 442:1–19
15. Carreón-Calderón B, Uribe-Vargas V, Ramírez-de-Santiago M, Ramírez-Jaramillo E (2014) Thermodynamic characterization of heavy petroleum fluids using group contribution methods. Ind Eng Chem Res 53:5598–5607
16. Ma M, Chen S, Abedi J (2016) Modeling the density, solubility and viscosity of bitumen/solvent systems using PC-SAFT. J Pet Sci Eng 139:1–12
17. Kariznovi M, Nourozieh H, Abedi J (2010) Bitumen characterization and pseudocomponents determination for equation of state modeling. Energy Fuels 24(1):624–633

18. Peramanu S, Pruden BB, Rahimi P (1999) Molecular weight and specific gravity distributions for Athabasca and Cold Lake bitumens and their saturate, aromatic, resin, and asphaltene fractions. Ind Eng Chem Res 38(8):3121–3130
19. Acevedo S, Gutierrez LB, Negrin G, Pereira JC, Mendez B, Delolme F, Dessalces G, Broseta D (2005) Molecular weight of petroleum asphaltenes: a comparison between mass spectrometry and vapor pressure osmometry. Energy Fuels 19(4):1548–1560
20. Satya S, Roehner RM, Deo MD, Hanson FV (2007) Estimation of properties of crude oil residual fractions using chemometrics. Energy Fuels 21(2):998–1005
21. Pedersen KS, Christensen PL, Shaikh JA (2015) Phase behavior of petroleum reservoir fluids, Second. CRC Press, Boca Raton, USA
22. Rafael AA, William DM (2002) An efficient tuning strategy to calibrate cubic EOS for compositional simulation. SPE Annu Tech Conf Exhib 77382:1–14
23. Danesh A (2003) PVT and phase behaviour of petroleum reservor fluids. Elsevier, The Netherlands
24. Jaubert JN, Privat R, Mutelet F (2010) Predicting the phase equilibria of synthetic petroleum fluids with the PPR78 approach. AIChE J 56(12):3225–3235
25. Qian JW, Privat R, Jaubert JN (2013) Predicting the phase equilibria, critical phenomena, and mixing enthalpies of binary aqueous systems containing Alkanes, Cycloalkanes, Aromatics, Alkenes, and Gases (N_2, CO_2, H_2S, H_2) with the PPR78 equation of state. Ind Eng Chem Res 52(46):16457–16490
26. Xu X, Jaubert JN, Privat R, Duchet-Suchaux P, Braña-Mulero F (2015) Predicting binary-interaction parameters of cubic equations of state for petroleum fluids containing pseudo-components. Ind Eng Chem Res 54(10):2816–2824
27. Nasrifar K, Rahmanian N (2018) Equations of state with group contribution binary interaction parameters for calculation of two-phase envelopes for synthetic and real natural gas mixtures with heavy fractions. Oil Gas Sci Technol 73(7):1–9

Chapter 6
Phase Equilibrium

List of Abbreviations

AAD%	Absolute average deviation percent
AS	Number of association sites on each molecule
CEoS	Cubic equation of state
CPA	Cubic-plus-association equation of state
EDF	Exponential distribution function
GCVOL	Volume group contribution method
GDF	Generalized distribution function
GFEM	Gibbs free energy minimization
GP	Generalized pseudocomponents
HCR	Molar ratio of hydrogen to carbon
MG	Marreno-Gani group contribution Method
MHV1	Modified-Huron-Vidal-of-order-one mixing rule
ND	Number of data
PCES	Predicative cubic equation of state
PC-SAFT	Perturbed-chain statistical associating fluid theory equation of state
PPR78	Predicative Peng-Robinson equation of state
PR	Peng-Robinson equation of state
PS	Pseudocomponents
RR	Rachford-Rice equation
sPC-SAFT	Simplified version of the PC-SAFT equation of state
SRK	Soave-Redlich-Kwong equation of state
UMR	Universal mixing rule
UNIFAC	Universal quasi-chemical functional group solution model
vdW	Van der Waals
VPO	Vapor phase osmometry

B. Carreón-Calderón et al., *Thermophysical Properties of Heavy Petroleum Fluids*, Petroleum Engineering, https://doi.org/10.1007/978-3-030-58831-1_6

Nomenclature

ΔG	Total Gibbs free energy of mixing
$\Delta \tilde{G}$	Molar Gibbs free energy of mixing
°API	API gravity
a	Attractive forces of CEoS
$\tilde{A}^R$	Molar residual Helmholtz free energy
A and B	Parameters of generalized distribution function
a_i	Activity of component i
b	Co-volume of CEoS
C	Number of components
C6	Hexanes petroleum fraction
C7+	Heptane-plus petroleum fraction
$D_\perp$	Tangent plane distance
$d_\perp$	Dimensionless chemical potential difference
E_i	Inverse value of ϕ_{it}
F	Objective function to be minimized
f_{ij}	Fugacity of the component i in the phase j
G	Total Gibbs free energy
g	Hard sphere radial distribution function
$\tilde{G}$	Total molar Gibbs free energy
$\tilde{G}_t^R$	Molar residual Gibbs free energy of the whole system
$G_\perp$	Tangent plane to G at n_{it}
H	Total enthalpy
$\tilde{H}$	Total molar enthalpy
k	Boltzmann Constant $\left(1.380549 \times 10^{-23}\ \mathrm{J/K}\right)$
k_0 and k_1	Parameters of the Tsonopoulos mixing rule
K_{bi}	Boiling-point elevation constant of solvent i
K_{fi}	Freezing-point depression constant
K_i	Two-phase equilibrium ratio for a component i
K_{ij}	Two-phase equilibrium ratio for a component i in the phase j
k_{ij}	Binary interaction parameter between components i and j
$L(x, \lambda)$	Lagrangian function
l_j	Molality of solute j
l_t	Total molality
M	Molecular weight
m	Number of segments in a chain of component
N_A	Avogadro number $\left(6.02214076 \times 10^{23}\ \mathrm{mol}^{-1}\right)$
n_{ij}	Number of moles of the component i in the phase j
n_{it}	Total number of moles the component i in the system
n_t	Total number of moles
P	Pressure
P^0	Reference pressure (0.101325 MPa)
P_{ci}	Critical pressure of component i

Q	Expression to be minimized
q_1	Parameter of the MHV1 mixing rule
R	The universal gas constant (8.3144636 J/molK)
S	Specific gravity or total entropy
$\tilde{S}$	Total molar entropy
T	Temperature
T_b	Normal boiling temperature
T_{ci}	Critical temperature of component i
T_{fi}	Freezing temperature of pure solvent i
$\tilde{V}$	Molar volume
w_c	Cumulative weight fraction
w_i	Amount of mass of solvent i
W_s	Weight ratio of solute to solvent
X_{A_i}	Mole fraction of molecules from a component i not bonded at site A
X_1	Monomer fraction of associating components
x_i	Mole fraction of the component i
X_{ij}	Modified mole fraction of the component i in the phase j
Z^{ass}	Association compressibility factor
z_i	Total mole fraction of the system
α_j	Molar phase fraction
α_{nm}, β_{nm}, and χ_{nm}	Parameters of the UNIFAC solution model
$\Delta^{A_iB_j}$	Association strength
δ_{jr}	Kronecker delta
$\Delta\tilde{G}_i^{VL}$	Molar Gibbs energy of vaporization of solvent i
$\Delta\tilde{H}_i^{LS}$	Molar enthalpy of fusion of pure solvent i
$\Delta\tilde{H}_i^{VL}$	Molar enthalpy of vaporization of solvent i
$\Delta\tilde{S}_i^{LS}$	Molar entropy of fusion of pure solvent i
$\Delta\tilde{S}_i^{VL}$	Molar entropy of vaporization of solvent i
$\varepsilon^{A_iB_j}$	Cross-association energy
ε/k	The dispersion energy
η	Reduced packing fraction
θ_{bi}	Elevation of the boiling temperature
θ_{fi}	Elevation of the freezing temperature
$\kappa^{A_iB_j}$	Cross-association volume
λ	Lagrange multiplier
μ_i^0	Chemical potential of component i at pure state
μ_{ij}	Chemical potential of the component i in the phase j
μ_{it}	Chemical potential of component i evaluated at the total composition of the system n_{it}
$\hat{\rho}_i$	Mass density of component i
σ	Segment diameter or number of phases
σ_{ij}	Cross-association segment diameter

ϕ_{ij}	Fugacity coefficient of the component i in the phase j
ϕ_{it}	Fugacity coefficient of component i evaluated at the total composition of the system n_{it}
ϕ_t	Fugacity coefficient evaluated at the total composition of the system n_t
ω_i	Acentric factor of component i

Superscripts

0	Pure component or reference state
ass	Association contribution
Cal	Calculated value
dsp	Dispersion contribution
Exp	Experimental value
hsc	Hard-sphere-chain contribution
ig	Ideal gas
phy	Physical contribution
R	Residual property

Subscripts

0	Initial point
i	Component i
j	Phase j
L	Liquid phase
M	Molecular weight
r	Reference phase
s	Solute
S	Specific gravity
T	Temperature
V	Vapor phase

As explained in Chap. 2, the total Gibbs free energy *G* is the thermodynamic potential from which, it can be established when a system formed by σ phases and *C* components is in equilibrium at constant temperature *T* and pressure *P*. Although there are other thermodynamic potentials that can be used depending on the available experimental variables, *G* is chosen because *T* and *P* are the available ones in experiments usually.

For such a system, *G* depends on the number of moles n_{ij} and the chemical potential μ_{ij} of the component *i* in the phase *j* according to the additivity rule

$$G = \sum_{j=1}^{\sigma} \sum_{i=1}^{C} n_{ij} \mu_{ij} \tag{6.1}$$

From a practical point of view, we are interested in knowing how many phases are present and how the total mass in the composed system is distributed among such phases at equilibrium conditions. In other words, by minimizing G according to Eq. (6.1), we search for σ and n_{ij} from the knowledge of a given temperature and pressure.

Considerable effort has been expended in developing mathematical methods for the minimization problem of G. A detailed discussion of such methods is beyond of the scope of the current manuscript. The reader is referred to the references of this chapter, and specialized manuscripts [1], where such mathematical methods are treated in detailed. Here, the main concepts and working expressions are derived and analyzed. Only those methods that have been implement for phase equilibrium calculations in the presence of heavy petroleum fluids of this manuscript are explained in depth.

6.1 Fundamental Relationships

In addition to the minimum value of G, the thermodynamic equilibrium of a system must satisfy some physical constraints. Specifically, a mass balance must be satisfied, because the whole system is assumed as a closed system. Therefore, the total mass of the system, which is usually expressed as the total number of moles n_t, remains constant. The mass balance can be expressed in several ways, as will be shown next. For example, the total number of moles the component i in the system n_{it}, which also remains constant, is

$$n_{it} = \sum_{j=1}^{\sigma} n_{ij}; \forall i = 1, \ldots, C \tag{6.2}$$

From the mathematical point of view, this mass balance becomes the optimization problem of finding the minimum value of G from an unconstrained problem into a constrained one. Besides, the minimization problem is subjected to the constraints

$$n_{ij} \geq 0 \tag{6.3}$$

which states that it does not make any physical sense negative masses. Thus, the minimization problem of G consists of the objective function given by Eq. (6.1) subjected to the constraints expressed by Eqs. (6.2) and (6.3).

It is well known that the necessary condition for any function to be on a stationary point (minimum, maximum, or saddle point), is that the corresponding partial derivatives vanish. Therefore, it may be advisable to determine such derivatives so that they can be used in searching for the minimum value of G instead of starting from Eq. (6.1) directly. The partial derivatives of Eq. (6.1) can be obtained from the corresponding total differential

$$dG = \sum_{j=1}^{\sigma}\sum_{i=1}^{C}\left(n_{ij}d\mu_{ij} + \mu_{ij}dn_{ij}\right) \tag{6.4}$$

To simplify this expression, we make use of the Gibbs-Duhen equation derived in Chap. 2, but written for σ phases, that is,

$$\sum_{j=1}^{\sigma}\sum_{i=1}^{C} n_{ij}d\mu_{ij} = 0 \tag{6.5}$$

which reduces Eq. (6.4) to

$$dG = \sum_{j=1}^{\sigma}\sum_{i=1}^{C} \mu_{ij}dn_{ij} \tag{6.6}$$

To include the mass balance constraint in this total differential, it is necessary to choose a reference phase, which is indicated by the subscript r, as follows

$$dG = \sum_{i=1}^{C}\left[\sum_{j=1}^{\sigma}(1-\delta_{jr})\mu_{ij}dn_{ij} + \mu_{ir}dn_{ir}\right] \tag{6.7}$$

The symbol δ_{jr} is the Kronecker delta introduced in Chap. 2, and likewise, it allows any phase to be chosen as reference, because if $j = r$, then $\delta_{jr} = 1$ discarding the phase r from the summation. On the contrary, if $j \neq r$, then $\delta_{jr} = 0$ including the phase j in the summation. On the other hand, the total mass balance that has to be satisfied takes the form of

$$n_t = \sum_{j=1}^{\sigma}\sum_{i=1}^{C} n_{ij} \tag{6.8}$$

Because n_t is constant, its differential form is

$$0 = \sum_{j=1}^{\sigma}\sum_{i=1}^{C} dn_{ij} \tag{6.9}$$

This equation can be written in term of the chosen phase r as well

$$\sum_{i=1}^{C}\left[\sum_{j=1}^{\sigma}(1-\delta_{jr})dn_{ij}+dn_{ir}\right]=0 \tag{6.10}$$

For a given component *i*, the following mass balance must be satisfied:

$$dn_{ir}=-\sum_{j=1}^{\sigma}(1-\delta_{jr})dn_{ij} \tag{6.11}$$

The elimination of dn_{ir} in Eq. (6.7) by using this equation leads to

$$dG=\sum_{i=1}^{C}\sum_{j=1}^{\sigma}(\mu_{ij}-\mu_{ir})dn_{ij} \tag{6.12}$$

Finally, from the definition of a total differential from any function, the partial derivatives of *G* with respecto to n_{ij} must vanish on a stationary point as mentioned before, that is,

$$\left(\frac{\partial G}{\partial n_{ij}}\right)_{T,P,n_t}=\mu_{ij}-\mu_{ir}=0 \tag{6.13}$$

Because the phase *r* is chosen arbitrarily, this equation is valid regardless of the considered component *i* or phase *j*. Therefore, the equality of chemical potentials between phases is the necessary condition for a system to be in equilibrium $(\mu_{ij}=\mu_{ir})$. This conclusion is reached in Chap. 2 as well, but by performing a hypothetical mass-transfer experiment among the phases forming the whole system. There, it was shown that mass transfer vanishes when the chemical potentials of each component in all phases are equal to each other.

To express thermodynamic potentials and chemical potentials in terms of measured variables, the fugacity concept was introduced in the same Chap. 2. There, it is shown that μ_{ij} is related to the fugacity f_{ij} through

$$\mu_{ij}=\mu_i^0+RTln\left(\frac{f_{ij}}{P^0}\right) \tag{6.14}$$

P^0 is the reference pressure whose value is setting equal to 0.101325 MPa. This pressure represents the point from which the chemical potential is calculated. On the other hand, μ_i^0 is the chemical potential of component *i* at pure state, whose value depends on P^0 and *T* only. As observed, if a component behaves as a pure component at P^0 and *T*, then the fugacity is equal to the pressure $(f_{ij}=P^0)$. Equation (6.14) together the mass balance given by Eq. (6.2) allows Eq. (6.1) to be written as

$$\frac{G}{RT} = \sum_{i=1}^{C} n_{it}\left[\frac{\mu_i^0}{RT} - ln(P^0)\right] + \sum_{j=1}^{\sigma}\sum_{i=1}^{C} n_{ij} ln(f_{ij}) \tag{6.15}$$

Because n_{it}, P^0, and T are held constant during the minimization process, the first term is also constant; therefore, the objective function to be minimized reduces to

$$F = \sum_{j=1}^{\sigma}\sum_{i=1}^{C} n_{ij} ln(f_{ij}) \tag{6.16}$$

Similar to G, it can be verified that the partial derivatives corresponding to F are

$$\left(\frac{\partial F}{\partial n_{ij}}\right)_{n_{it}, n_{mn;n\neq p}} = ln(f_{ij}) - ln(f_{ir}) \tag{6.17}$$

which must vanish on stationary points as well. Therefore, the necessary equilibrium conditions analogous to those given by Eq. (6.13), but in terms of the fugacity take the form of

$$f_{i1} = f_{i2} = \ldots = f_{i\sigma}; \forall i = 1, \ldots, C \tag{6.18}$$

As illustrated in Chaps. 3 and 4, the fugacity coefficient ϕ_{ij} rather than the fugacity are determined from equations of state. Hence, it is advisable to write the above equilibrium conditions in terms of ϕ_{ij} instead of f_{ij}. This can be done by means of the relationship

$$\phi_{ij} \equiv \frac{f_{ij}}{x_{ij} P_j} \tag{6.19}$$

Thus, the substitution of f_{ij} from this equation into the necessary equilibrium conditions as expressed by Eq. (6.18) results in

$$x_{i1}\phi_{i1} = x_{i2}\phi_{i2} = \ldots = x_{i\sigma}\phi_{i\sigma}; \forall i = 1, \ldots, C \tag{6.20}$$

where the equilibrium mechanic condition from the Chap. 2 ($P_1 = P_2 = \ldots = P_\sigma = P$) has been used in order to eliminate P_j. As mentioned before, Eqs. (6.13), (6.18), and (6.20) represent the necessary equilibrium conditions, but they are not sufficient for a system formed by σ phases and C components to be in a real equilibrium, because all stationary points (minimum, maximum or saddle points) fulfilled the equality between either chemical potentials or fugacities. Besides, the equilibrium state requires that the mathematical solution given by the necessary conditions corresponds to the global minimum for G. Considerable effort has been devoted to ensure that such solution corresponds to the actual physical equilibrium, that is, to find the global minimum of G [2–7]. One of the most common methods consists of combining the solution of the necessary equilibrium conditions with the so-called phase stability analysis [8–12]. This analysis determines whether a system may undergo the formation of a new phase so that the system reaches a lesser free energy.

6.2 Stability Analysis

Let the system consist of a single phase ($\sigma = 1$), where its temperature, and pressure are held constant during all possible state changes. We are interested in changes, where an infinitesimal amount of any phase may emerge from the system. The small changes imply that the free energy of the system will not be far from the original one, and therefore it can be estimated through a Taylor series expansion about its original state, that is,

$$G_{\perp} = G(\boldsymbol{n_t}) + \sum_{i=1}^{C} \left(\frac{\partial G}{\partial n_i}\right)_{T,P,n_t}\Bigg|_{n_i=n_{it}} (n_i - n_{it}) \tag{6.21}$$

In this expansion, the first-order terms are the only ones retained. In other words, $G_{\perp}$ accounts for the Gibbs free energy of the system as a result of state changes infinitesimally small. Geometrically, $G_{\perp}$ accounts for a tangent plane to G at $\boldsymbol{n_{it}}$. From Eq. (6.13), $G_{\perp}$ can be written as

$$G_{\perp} = G(\boldsymbol{n_t}) + \sum_{i=1}^{C} (\mu_{it} - \mu_{rt})(n_i - n_{it}) \tag{6.22}$$

where μ_{it} stands for the chemical potential of component i evaluated at the total composition of the system $\boldsymbol{n_{it}}$. On the other hand, from the additivity rule, $G(\boldsymbol{n_t})$ is given by

$$G(\boldsymbol{n_t}) = \sum_{i=1}^{C} \mu_{it} n_{it} \tag{6.23}$$

which simplifies Eq. (6.22) to

$$G_{\perp} = \sum_{i=1}^{C} \mu_{it} n_i \tag{6.24}$$

or

$$\tilde{G}_{\perp} = \sum_{i=1}^{C} \mu_{it} x_i \tag{6.25}$$

The difference between Gibbs free energy of any arbitrary phase and $\tilde{G}_{\perp}$ is the so-called Tangent Plane Distance ($D_{\perp}$)

$$D_{\perp}(\boldsymbol{x}) \equiv \frac{\tilde{G} - \tilde{G}_{\perp}}{RT} = \sum_{i=1}^{C} \frac{(\mu_i - \mu_{it})}{RT} x_i \tag{6.26}$$

As observed, if $D_{\perp}(\boldsymbol{x}) < 0$ for at least one composition, there is at least one phase with a smaller free energy; consequently, the original phase may split up into other phases, that is, it is not stable. In contrast, if $D_{\perp}(\boldsymbol{x}) \geq 0$ for the whole range of composition, there is not any other phase with a smaller free energy; consequently, the original phase will not split up into other phases, that is, it is stable. Therefore, if the minimum value of $D_{\perp}(\boldsymbol{x})$ is found and this minimum value is nonnegative, it can be guaranteed that the original phase is stable. To find such a minimum value, it is advisable first to obtain the total differential of $D_{\perp}$

$$dD_{\perp} = \sum_{i=1}^{C} \frac{(1 - \delta_{ir})(\mu_i - \mu_{it})}{RT} dx_i + \frac{(\mu_r - \mu_{rt})}{RT} dx_r + \sum_{i=1}^{C} \frac{d(\mu_i - \mu_{it})}{RT} x_i \tag{6.27}$$

Here again, the properties of a reference component have been indicated by the symbol r. In addition, the mass balance constraint in terms of the mole fraction must be fulfilled

$$\sum_{i=1}^{C} x_i = 1 \tag{6.28}$$

which allows the mole fraction of the reference component to be written in terms of the remaining ones

$$dx_r = -\sum_{i=1}^{C} (1 - \delta_{ir}) dx_i \tag{6.29}$$

The substitution of this expression into Eq. (6.27) results in

$$dD_{\perp} = \sum_{i=1}^{C} \left[\frac{(\mu_i - \mu_{it})}{RT} - \frac{(\mu_r - \mu_{rt})}{RT} \right] dx_i + \sum_{i=1}^{C} \frac{d(\mu_i - \mu_{it})}{RT} x_i \tag{6.30}$$

This total differential can be simplified, on the one hand, by using the Gibbs-Duhem equation

$$\frac{1}{RT}\sum_{i=1}^{C} x_i d\mu_i = 0 \quad (6.31)$$

and on the other hand, by using the fact that $d\mu_{it} = 0$, because μ_{it} is the chemical potential evaluated at n_t, which is held constant. Thus, the total differential of $D_\perp$ is

$$dD_\perp = \sum_{i=1}^{C}\left[\frac{(\mu_i - \mu_{it})}{RT} - \frac{(\mu_r - \mu_{rt})}{RT}\right]dx_i \quad (6.32)$$

From this expression, the partial derivatives with respect to the composition x_i are obtained directly. Besides, remembering that the minimum value of $D_\perp$ has to be mathematically a stationary point, we finally obtain that

$$\left(\frac{\partial D_\perp}{\partial x_i}\right)_{T,P,x_{j;i\neq j,i\neq r}} = \frac{(\mu_i - \mu_{it})}{RT} - \frac{(\mu_r - \mu_{rt})}{RT} = 0 \quad (6.33)$$

Therefore, the difference between chemical potentials is equal to a constant value on any stationary point, regardless of the considered component, that is,

$$d_\perp \equiv \frac{(\mu_i - \mu_{it})}{RT} = \frac{(\mu_r - \mu_{rt})}{RT} \quad (6.34)$$

According to Eq. (6.26), this difference, which is constant regardless of the component, transforms $D_\perp$ into

$$D_\perp(\boldsymbol{x}) = \sum_{i=1}^{C} d_\perp x_i = d_\perp(\boldsymbol{x}) \quad (6.35)$$

which states that the difference between the thermodynamic state of any arbitrary phase and the original one is equal to the constant $d_\perp(\boldsymbol{x})$. In other words, the stability criteria for a given phase can be establishes in terms of $d_\perp(\boldsymbol{x})$. Hence, if $d_\perp(\boldsymbol{x}) < 0$ for the entire range of composition, the original phase is not stable. On the contrary, if $d_\perp(\boldsymbol{x}) \geq 0$ the original phase it is stable. It is worth pointing out that $d_\perp(\boldsymbol{x}) < 0$ is a sufficient condition for a given phase to be thermodynamically unstable, but it does mean that a new phase emerges spontaneously from such a phase stabilizing the whole system. An extra energy is necessary in order to create the surface between the new phase and the original one, the so-called interface. In this context, it can be demonstrated that $d_\perp(\boldsymbol{x})$ is the volume contribution to the so-called nucleation work of a new phase emerging from a unstable one [13].

To find the minimum value $D_\perp$, it is advisable to write Eq. (6.26) in a more practical expression. This can be done by using Eqs. (6.14) and (6.19). Thus, we can express $D_\perp$ is terms of fugacity coefficients and composition

$$D_\perp(\boldsymbol{x}) = \sum_{i=1}^{C}[ln(\phi_i) + ln(x_i) - ln(\phi_{it}) - ln(z_i)]x_i \tag{6.36}$$

in which, the total mole fraction of the system z_i is determined from

$$z_i = \frac{n_{it}}{n_t} \tag{6.37}$$

The minimization problem of $D_\perp$ is subjected to the mass balance constraint given by Eq. (6.28), and the non-negative mole fraction constraint

$$x_i \geq 0 \tag{6.38}$$

The objective function given by Eq. (6.36) with its corresponding constraints is the minimization problem usually found in practice, even though activity coefficients can be found rather than fugacity coefficients, depending on the used thermodynamic model for the system. Typically, a nonlinear constrained optimization problem is solved by setting up the Lagrangian function $L(\boldsymbol{x}, \lambda)$ of the objective function [14]. In this case, $L(\boldsymbol{x}, \lambda)$ corresponding to $D_\perp(\boldsymbol{x})$ is

$$L(\boldsymbol{x}, \lambda) = \sum_{i=1}^{C}[ln(\phi_i) + ln(x_i) - ln(\phi_{it}) - ln(z_i)]x_i - \lambda\left(\sum_{i=1}^{C} x_i - 1\right) \tag{6.39}$$

where λ is the Lagrange multiplier corresponding to constraint given Eq. (6.28). The logarithm dependence on mole fractions ensures that all mole fractions are positive. Again, on the stationary points, the corresponding partial derivatives must vanish

$$\left(\frac{\partial L}{\partial x_k}\right)_{\lambda, x_j; k\neq j} = ln(\phi_k) + ln(x_k) - ln(\phi_{kt}) - ln(z_k) + 1 - \lambda = 0 \tag{6.40}$$

where the Gibbs-Duhem equation has been used. Besides, we have that

$$\left(\frac{\partial L}{\partial \lambda}\right)_{x_i} = 1 - \sum_{i=1}^{C} x_i = 0 \tag{6.41}$$

The evaluation of $L(\boldsymbol{x}, \lambda)$ on the stationary points given by these partial derivatives leads to

$$L(\boldsymbol{x}, \lambda) = \lambda - 1 \tag{6.42}$$

On the other hand, it was demonstrated before that $D_{\perp}(\boldsymbol{x}) = d_{\perp}(\boldsymbol{x})$ on the stationary points of $D_{\perp}(\boldsymbol{x})$. Therefore,

$$d_{\perp}(\boldsymbol{x}) = \lambda - 1 \tag{6.43}$$

In other words, the stability criteria explained before can be established in terms of the Lagrange multiplier usually present in the mathematical methods for solving optimization problems. Accordingly, the stability criteria can now be restated as follows: if $\lambda < 1$ for the whole range of composition, then the original phase is unstable. On the contrary, if $\lambda \geq 1$, the original phase is stable.

The stability criteria can be applied to a system where more than one phase is present ($\sigma > 1$). In this case, it is useful to define the Gibbs free energy of mixing ΔG as

$$\Delta G = \sum_{j=1}^{\sigma} \sum_{i=1}^{C} n_{ij}\mu_{ij} - \sum_{i=1}^{C} n_{it}\mu_{it} \tag{6.44}$$

Here, the first term accounts for the free energy because of the presence of σ phases within the system, while the second term accounts for the total free energy of the system as one, irrespective of its internal nature. Besides, it is advisable to write ΔG as a function of mole fractions x_{ij} and molar phase fractions α_j. This can be done by using the following relationships

$$x_{ij} = \frac{n_{ij}}{n_j} \tag{6.45}$$

and

$$\alpha_j \equiv \frac{n_j}{n_t} \tag{6.46}$$

Thus, ΔG turns into

$$\Delta \tilde{G} = \sum_{j=1}^{\sigma} \sum_{i=1}^{C} \alpha_j x_{ij}\mu_{ij} - \sum_{i=1}^{C} z_i \mu_{it} \tag{6.47}$$

Again, this expression can be written in terms of fugacity coefficients by means of Eqs. (6.14) and (6.19)

$$\frac{\Delta \tilde{G}}{RT} = \sum_{j=1}^{\sigma} \alpha_j \sum_{i=1}^{C} x_{ij}\left[ln\left(\phi_{ij}\right) + ln\left(x_{ij}\right)\right] - \sum_{i=1}^{C} z_i[ln(\phi_{it}) + ln(z_i)] \tag{6.48}$$

As before, the mass balance must be fulfilled. Equation (6.2) is used here, but expressed in terms of α_j and x_{ij}, that is,

$$z_i = \sum_{j=1}^{\sigma} \alpha_j x_{ij} \tag{6.49}$$

which transforms Eq. (6.48) into

$$\frac{\Delta \tilde{G}}{RT} = \sum_{j=1}^{\sigma} \alpha_j \sum_{i=1}^{C} x_{ij}\left[ln(\phi_{ij}) + ln(x_{ij}) - ln(\phi_{it}) - ln(z_i)\right] \tag{6.50}$$

According to Eq. (6.40), the expression inside brackets is equal to $\lambda - 1$; consequently,

$$\frac{\Delta \tilde{G}}{RT} = \sum_{j=1}^{\sigma} \alpha_j (\lambda_j - 1) = \sum_{j=1}^{\sigma} \alpha_j d_{\perp j} \tag{6.51}$$

Considering that $\alpha_j \geq 0$, if $d_{\perp j} > 0$ for all possible phases, then $\Delta \tilde{G} > 0$ and therefore the system is stable with respect to the formation of any other phase, that is, the system consists of just one single phase. On the other hand, $d_{\perp j} = 0$ for all possible phases, then $\Delta \tilde{G} = 0$ and therefore the system is formed by σ phases in equilibrium with each other. Finally, if $d_{\perp j} < 0$ for at least phase j, then the system is unstable with respect to the formation of such a phase, and therefore, the system is not in equilibrium, even though $\Delta \tilde{G} \geq 0$. Mathematically, this means that a local minimum of $\tilde{G}$ has been found; therefore, the global minimum of $\tilde{G}$ requires that $d_{\perp j} \geq 0$ for all possible phases. This is the reason why mathematical methods that search for all solutions of a system of nonlinear equations have been suggested in order to find the global minimum of $\tilde{G}$ [2, 6, 8–10, 15, 16]. From Figs. 6.1, 6.2 and 6.3, an unstable phase, a stable phase and two phases in equilibrium for a binary mixture are schematically illustrated in terms of $d_{\perp}(\boldsymbol{x})$.

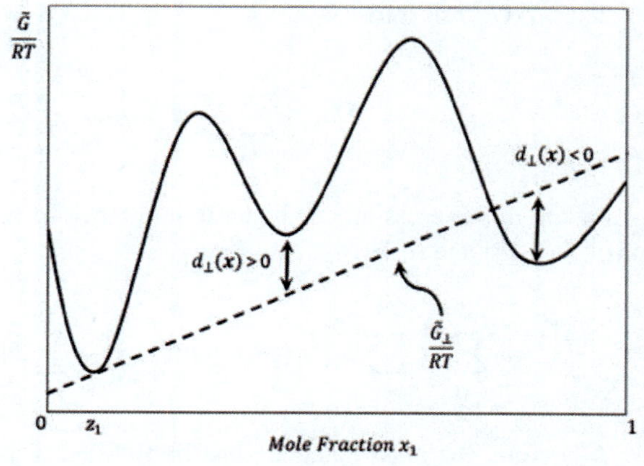

Fig. 6.1 Schematic diagram of an unstable binary mixture with composition z_1, ($d_{\perp}(\boldsymbol{x}) < 0$)

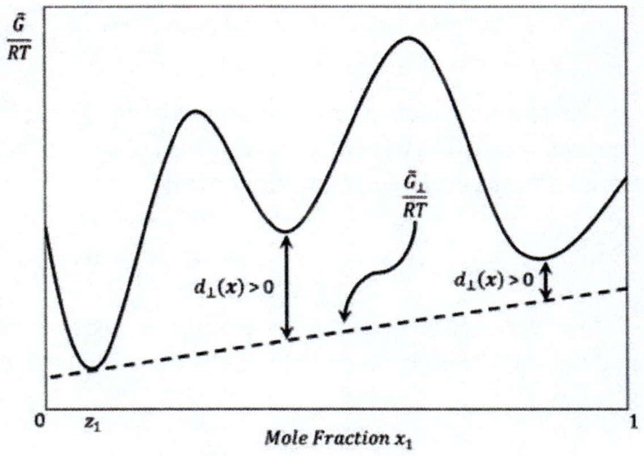

Fig. 6.2 Schematic diagram of a stable binary mixture with composition z_1, $(d_\perp(x) > 0)$

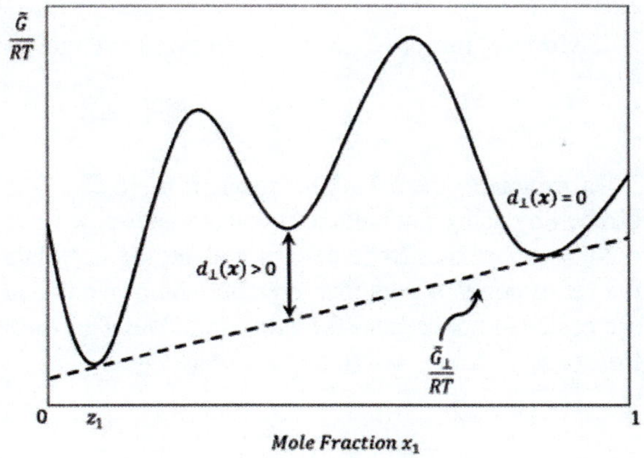

Fig. 6.3 Schematic diagram of a stable binary mixture with composition z_1 in equilibrium with a second phase $(d_\perp(x) = 0)$

6.3 Two-Phase Equilibrium

For the two-phase equilibrium problem, the necessary equilibrium conditions, as expressed by Eq. (6.20), turn out to be

$$\begin{aligned} x_{11}\phi_{11} &= x_{12}\phi_{i2} \\ \vdots \quad & \vdots \quad \vdots \\ x_{C1}\phi_{C1} &= x_{C2}\phi_{C2} \end{aligned} \tag{6.52}$$

Instead of using fugacity coefficients as such in the two-phase calculations, it is a common practice to use the equilibrium ratio K_i for a component i. K_i is defined as the ratio of the mole fraction of component i in the phase 1 to the corresponding mole fraction in the phase 2, that is,

$$K_i \equiv \frac{x_{i1}}{x_{i2}} = \frac{\phi_{i2}}{\phi_{i1}} \tag{6.53}$$

The second equality comes from using Eq. (6.52) In addition to K_i, each component must satisfied the component mass balance as expressed by Eq. (6.49), which, for a two-phase system, takes the form of

$$z_i = \alpha_1 x_{i1} + \alpha_2 x_{i2} \tag{6.54}$$

In general, for a σ-phase system, it can easily verified from the same Eq. (6.49) that the total mass balance in terms of the molar phase fractions α_j is given by

$$\sum_{j=1}^{\sigma} \alpha_j = 1 \tag{6.55}$$

Therefore, for a two-phase system, we have that

$$\alpha_1 + \alpha_2 = 1 \tag{6.56}$$

In principle, the set of equations Eqs. (6.53), (6.54), and (6.56) can be solved directly by using a suitable numerical method, where x_{i1}, x_{i2}, and α_1 are the $2C + 1$ unknown variables to be determined. However, it is a common practice to split up the set of coupled nonlinear equations into two decoupled sets of equations, which are easier to handle and therefore to solve. The former set arises from combining Eqs. (6.53), (6.54), and (6.56) in order to give

$$x_{i1} = \frac{z_i K_i}{1 + \alpha_1 (K_i - 1)} \tag{6.57}$$

$$x_{i2} = \frac{z_i}{1 + \alpha_1 (K_i - 1)} \tag{6.58}$$

Once α_1 and K_i have been either estimated or calculated, these equations allow x_{i1} and x_{i2} to be calculated explicitly. This is a necessary characteristic for solving nonlinear equations by successive substitution methods. To avoid unrealistic mole fractions during calculations, the mole fractions calculated from Eqs. (6.57) and (6.58) should be always positive $\left(x_{ij} \geq 0\right)$, even though the mass balance constraint might be not fulfilled. The purpose of this constraint is to avoid negative mole fractions $\left(x_{ij} < 0\right)$, for which fugacity and activity coefficients do not exist and, at the same time, to have a wider search interval where iterative calculations can proceed, even if $x_{ij} > 1$. To ensure that $x_{ij} > 0$, we have that the following conditions has to be satisfied

$$1 + \alpha_1 (K_i - 1) > 0 \tag{6.59}$$

For the largest value of K_i, the minimum value of α_1 is

$$\alpha_1^{min} = \frac{1}{1 - K_i^{max}} \tag{6.60}$$

Similarly, for the smallest value of K_i, the maximum value of α_1 is

$$\alpha_1^{max} = \frac{1}{1 - K_i^{min}} \tag{6.61}$$

On the other hand, the latter decoupled equation arises from combining Eqs. (6.57) and (6.58) in order to give rise to the well-known Rachford-Rice equation [17–19]

$$RR(\alpha_1) = \sum_{i=1}^{C} (x_{i1} - x_{i2}) = \sum_{i=1}^{C} \frac{z_i(K_i - 1)}{1 + \alpha_1(K_i - 1)} = 0 \tag{6.62}$$

where, α_1 is the only unknown variable to be numerically found by keeping the set of K_i values constant. For solving the RR equation by a numerical method such as the Newton-Raphson method, it is necessary to determined its derivative

$$\frac{d(RR)}{d\alpha_1} = -\sum_{i=1}^{C} \frac{z_i(K_i - 1)^2}{[1 + \alpha_1(K_i - 1)]^2} < 0 \tag{6.63}$$

which is always negative. This means that the RR equation is decreasing on the interval $\left(\alpha_1^{min}, \alpha_1^{max}\right)$, as illustrated in Fig. 6.4.

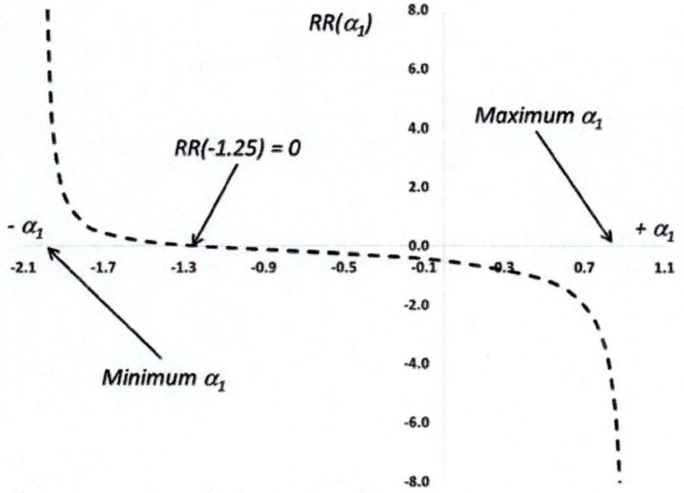

Fig. 6.4 Schematic representation of the Rachford-Rice function

As observed from Fig. 6.4, ensuring realistic mole fractions $\left(x_{ij} \geq 0\right)$ can result in unrealistic negative values of α_1 $(\alpha_1 < 0)$. However, these negative values can be used as stability criteria, as will be demonstrated next.

From Eq. (6.40), we have that

$$\sum_{i=1}^{C} z_i \frac{\phi_{it}}{\phi_i} = \frac{1}{e^{\lambda - 1}} \tag{6.64}$$

By defining $K_i = \phi_{it}/\phi_i$, this equation turns out to be

$$\sum_{i=1}^{C} z_i K_i = \frac{1}{e^{\lambda - 1}} \tag{6.65}$$

As discussed before, If $\lambda < 0$, then the original homogenous phase is not stable, with which the above expression reduces to

$$RR(0) = \sum_{i=1}^{C} K_i z_i - 1 < 0 \tag{6.66}$$

On the contrary, If $\lambda \geq 0$, then the original homogenous phase is stable, and one obtains that

$$RR(0) = \sum_{i=1}^{C} K_i z_i - 1 \geq 0 \tag{6.67}$$

Similarly, by defining $K_i = \phi_i/\phi_{it}$, Eq. (6.64) becomes

$$\sum_{i=1}^{C} \frac{z_i}{K_i} = \frac{1}{e^{\lambda - 1}} \tag{6.68}$$

Following the same arguments, we arrive at

$$RR(1) = \sum_{i=1}^{C} \frac{z_i}{K_i} - 1 < 0 \tag{6.69}$$

and

$$RR(1) = \sum_{i=1}^{C} \frac{z_i}{K_i} - 1 \geq 0 \tag{6.70}$$

Therefore, the phases 1 and 2 are both stable as long as Eqs. (6.67) and (6.70) are fulfilled simultaneously. This means that a solution exists in the interval [0, 1]. On the other hand, the phase 1 is unstable when Eq. (6.66) is satisfied, which means that $\alpha_1 < 0$ because of the RR equation is a monotonic function. If the phase 1 is unstable, then phase 2 is stable, that is, Eq. (6.70) is fulfilled and $\alpha_2 > 1$. Finally, the phase 1 is stable and the phase 2 is unstable as long as $\alpha_1 > 1$ and $\alpha_2 < 0$, Eqs. (6.67) and (6.69). In Table 6.1 are summarized these stability criteria for the vapor-liquid and liquid-liquid calculations. In the case of vapor-liquid equilibrium, the vapor phase is assigned as the phase 1 and the liquid phase as the phase 2, generally. For the liquid-liquid equilibrium, both liquid phases can be assigned as either phase 1 or phase 2. A method, where negative molar fractions are used as stability criteria is called Negative Flash [17, 18, 20].

In Fig. 6.5, an algorithm for determining the vapor-liquid equilibrium of multicomponent systems is shown. In principle, it may be used for any petroleum fluid, but its use is specially suggested for heavy petroleum fluids using cubic equations of state. The initial values of K_i are estimated from the Wilson's approximation.

$$ln(K_i) = ln\left(\frac{P_{ci}}{P}\right) + 5.373(1 + \omega_i)\left(1 + \frac{T_{ci}}{T}\right) \quad (6.71)$$

which is a function of the critical properties, pressure P_{ci} and temperature T_{ci}, and the acentric factor ω_i. The phase extrapolation may be omitted; its use is recommended as an initialization strategy, that is, after a few iterations, it is no longer needed [21]. Finally, to accelerate the successive substitution of the mole fractions, the dominant eigenvalue method is suggested [1, 22].

Instead of decoupling the set of nonlinear equations, one may prefer to solve the phase equilibrium problem directly by using a numerical solver package. In this case, Eq. (6.53) are written in a more suitable way

$$g_i(\alpha_1, \boldsymbol{K}) = ln(K_i) + ln(\phi_{i1}) - ln(\phi_{i2}) = 0; \forall i = 1, \ldots, C \quad (6.72)$$

Table 6.1 Stability criteria for two-phase calculations

$RR(0) = \sum_{i=1}^{C} K_i z_i - 1$		$RR(1) = \sum_{i=1}^{C} \frac{z_i}{K_i} - 1$	
< 0	≥ 0	< 0	≥ 0
$\alpha_1 < 0$	$\alpha_1 \geq 0$	$\alpha_1 > 1$	$\alpha_1 \leq 1$
Phase 2	Phase 1 and possibly Phase 2	Phase 1	Phase 2 and possibly Phase 1
Vapor(1)–Liquid(2)			
Liquid	Vapor and possibly Liquid	Vapor	Liquid and possibly Vapor
Liquid(1)–Liquid(2)			
Liquid(2)	Liquid(1) and possibly Liquid(2)	Liquid(1)	Liquid(2) and possibly Liquid(1)

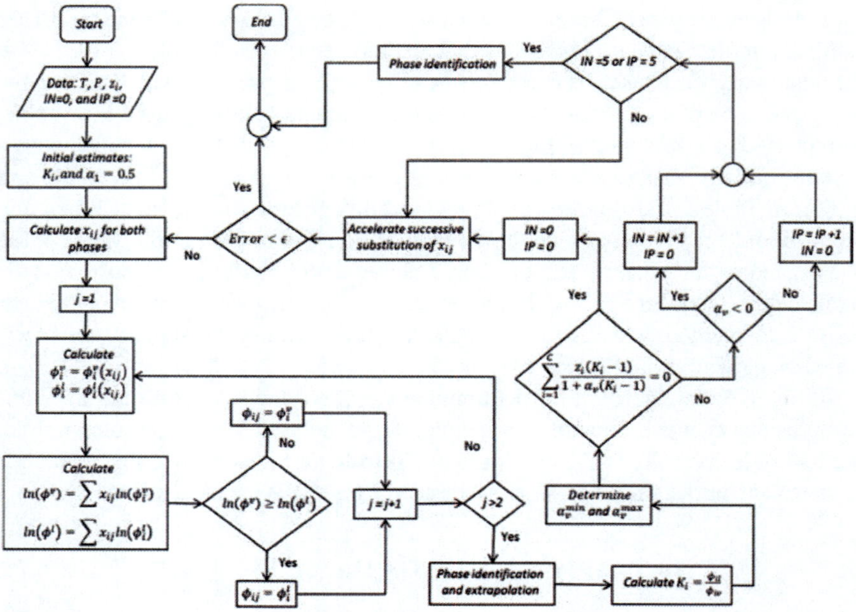

Fig. 6.5 Flowchart of the vapor-liquid equilibrium algorithm using cubic equations of state

which are C nonlinear equations with $C+1$ unknown variables $(\alpha_1, \boldsymbol{K})$. The missing one is the RR equation

$$g_{C+1}(\alpha_1, \boldsymbol{K}) = \sum_{i=1}^{C} \frac{z_i(K_i - 1)}{1 + \alpha_1(K_i - 1)} = 0 \tag{6.73}$$

From the knowledge of α_1 and K_i, the mole fractions can be calculated by using Eqs. (6.57) and (6.58), and consequently, the fugacity coefficients with an equation of state.

6.4 Multiphase Equilibrium

When more than two phases are present inside the system and the system is stable, we have from Eq. (6.51) that $d_{\perp j} = 0$, that is,

$$x_{ij}\phi_{ij} = z_i\phi_{it} \tag{6.74}$$

From the combination of this equation with Eq. (6.49), we obtained that

$$z_i = \sum_{j=1}^{\sigma} \alpha_j x_{ij} = \sum_{j=1}^{\sigma} \alpha_j \frac{\phi_{it}}{\phi_{ij}} z_i \tag{6.75}$$

The parameter E_i is defined as the inverse value of ϕ_{it}

$$E_i \equiv \frac{1}{\phi_{it}} = \sum_{j=1}^{\sigma} \frac{\alpha_j}{\phi_{ij}} \tag{6.76}$$

Thus, according to Eq. (6.74), the necessary equilibrium conditions for a system formed by σ phases can be rewritten as

$$x_{i1}\phi_{i1} = x_{i2}\phi_{i2} =, \ldots, = x_{i\sigma}\phi_{i\sigma} = \frac{z_i}{E_i} \tag{6.77}$$

with the mole fractions given by

$$x_{ij} = \frac{1}{\phi_{ij}} \frac{z_i}{E_i} \tag{6.78}$$

As observed, the condition $E_i > 0$ must be fulfilled in order to $x_{ij} > 0$. Therefore, E_i is the multiphase parameter analogous to the denominator given by Eq. (6.59) from the RR equation. On the other hand, the molar residual Gibbs free energy $\tilde{G}_t^R$ of the whole system is

$$\frac{\tilde{G}_t^R}{RT} = ln(\phi_t) = \sum_{i=1}^{C} z_i ln(\phi_{it}) \tag{6.79}$$

which can be expressed in terms of the parameters E_i as follows

$$\frac{\tilde{G}_t^R}{RT} = -\sum_{i=1}^{C} z_i ln\left(\frac{1}{\phi_{it}}\right) = -\sum_{i=1}^{C} z_i ln(E_i) \tag{6.80}$$

The total molar Gibbs free energy $\tilde{G}_t$ of the whole system can be determined as the summation of the ideal and residual contributions, that is,

$$\frac{\tilde{G}_t}{RT} = \frac{\tilde{G}_t^{ig}}{RT} + \frac{\tilde{G}_t^R}{RT} = ln\left(\frac{P}{P^0}\right) + \sum_{i=1}^{C} z_i \left[\frac{\mu_i^0}{RT} + ln(z_i)\right] - \sum_{i=1}^{C} z_i ln(E_i) \tag{6.81}$$

Because temperature, pressure and the total composition are held constant, the residual contribution is the only one which is a function of the mole fractions and the molar phase fractions; therefore,

$$\min \frac{\tilde{G}_t}{RT} = \min \frac{\tilde{G}_t^R}{RT} \tag{6.82}$$

As shown before for the two-phase case, the mole fractions are numerically calculated by using the successive substitution method. In the multiphase case, it is performed by using Eq. (6.78). Thus, we may assume $\tilde{G}_t^R$ to be a function of the molar phase fractions $\boldsymbol{\alpha}$ only. As will be illustrated next, the partial derivatives of $\tilde{G}_t^R$ with respect to $\boldsymbol{\alpha}$ are equal to the summation of the mole fractions, which are never equal to zero as required on a stationary point. Hence, a more suitable expression to be minimized is that suggested by Michelsen [23], which is

$$Q(\boldsymbol{\alpha}) = \sum_{j=1}^{\sigma} \alpha_j - \sum_{i=1}^{C} z_i ln(E_i) \tag{6.83}$$

where the only constraint is

$$\alpha_j \geq 0; \forall j \tag{6.84}$$

This minimization problem has been taken as starting point for multiphase equilibrium algorithms [18, 24, 25]. To ensure that the function $Q(\boldsymbol{\alpha})$ always exits during the minimization process, the following form of $\tilde{G}_t^R$ is preferred instead

$$\frac{\tilde{G}_t^R}{RT} = \frac{1}{2} \sum_{i=1}^{C} z_i ln(\phi_{it}^2) = -\frac{1}{2} \sum_{i=1}^{C} z_i ln(E_i^2) \tag{6.85}$$

As observed, E_i can take negative values and $\tilde{G}_t^R$ still exits during iterative calculations. The partial derivatives of $Q(\boldsymbol{\alpha})$ are

$$\left(\frac{\partial Q}{\partial \alpha_k}\right)_{\alpha_j; k \neq j} = 1 - \sum_{i=1}^{C} \frac{z_i}{E_i} \frac{1}{\phi_{ik}} = 1 - \sum_{i=1}^{C} x_{ik} \tag{6.86}$$

which must vanish on stationary points. As in the case of two-phase systems, we can perform a stability analysis in the multiphase case as well. Instead of Eq. (6.74), we start with

$$x_{ij} = z_i \frac{\phi_{it}}{\phi_{ij}} e^{\lambda_j - 1} \tag{6.87}$$

From which, we have that

$$\sum_{i=1}^{C} z_i \frac{\phi_{it}}{\phi_{ik}} = \frac{1}{e^{\lambda_j - 1}} \sum_{i=1}^{C} x_{ij} \tag{6.88}$$

Often, it is advisable to define a modified summation of the mole fractions as follows:

$$\sum_{i=1}^{C} X_{ij} \equiv \frac{1}{e^{\lambda_j - 1}} \sum_{i=1}^{C} x_{ij} \tag{6.89}$$

This modified summation of the mole fractions can be also used as a phase stability measurement. Therefore, Eq. (6.86) may be written as

$$\left(\frac{\partial Q}{\partial \alpha_k}\right)_{\alpha_j;k\neq j} = 1 - \sum_{i=1}^{C} X_{ik} \tag{6.90}$$

According to the stability criterion, $\lambda_k \leq 1$ for the phase k to be present, either as an incipient phase $\alpha_k = 0$ or as a fully developed phase $\alpha_k > 0$. Therefore, this criterion, together with Eqs. (6.89) and (6.90), means that, for the phase k to be present, the following conditions is fulfilled

$$\sum_{i=1}^{C} X_{ik} \geq 1 \tag{6.91}$$

and

$$\left(\frac{\partial Q}{\partial \alpha_k}\right)_{\alpha_j;k\neq j} \leq 0 \tag{6.92}$$

Otherwise, if $\lambda_k > 1$ then phase k is absent ($\alpha_k < 0$), and therefore, from the same Eqs. (6.89) and (6.90), one concludes that

$$\sum_{i=1}^{C} X_{ik} < 1 \tag{6.93}$$

together with

$$\left(\frac{\partial Q}{\partial \alpha_k}\right)_{\alpha_j;k\neq j} > 0 \tag{6.94}$$

Summarizing, the molar phase fractions α_j can be determined from minimizing the function as expressed by Eq. (6.83). On the condition that phase composition, and consequently the fugacity coefficients, remains constant. Based on the method proposed by Leibovici and Nichita [18], the minimization problem may be subjected to the following linear constraints

$$\alpha_k \epsilon (-\infty, +\infty) \tag{6.95}$$

$$E_i = \sum_{j=1}^{\sigma} \frac{\alpha_j}{\phi_{ij}} > 0 \tag{6.96}$$

$$\sum_{j=1}^{\sigma} \alpha_j = 1 \tag{6.97}$$

Once α_j has been calculated, the mole fractions x_{ij} have to be updated by the successive substitution method using Eq. (6.78), and hence, the fugacity coefficients ϕ_{ij} are also updated by using the considered equation of state. Using these values, the minimization problem given by Eq. (6.83) is repeated again in order to obtain new values of α_j, and so on, until convergence is reached in composition.

As in the two-phase case, the minimization problem can be expressed in terms of equilibrium constants. This implies that a given phase has to be selected as reference phase, from which the equilibrium constants are defined. Ideally, such a reference phase should be present in the system. However, it is not possible to know in advance what phases are present or not, because this is the answer that we precisely want to know from solving the phase equilibrium problem. To avoid such an issue, the equilibrium constants can be defined from using the whole system as reference phase, that is,

$$K_{ij} \equiv \frac{\phi_{ij}}{\phi_{it}} = E_i \phi_{ij} \tag{6.98}$$

From Eq. (6.78), the mole fractions are now determined using

$$x_{ij} = \frac{z_i}{K_{ij}} \tag{6.99}$$

and, from Eq. (6.76), we have that

$$\sum_{j=1}^{\sigma} \frac{\alpha_j}{K_{ij}} = 1 \tag{6.100}$$

The definition given by Eq. (6.98) does not imply that ϕ_{it} has to be calculated during the iterative calculations using the chosen equation of state as in the case of ϕ_{ij}, because $\phi_{it} = E_i^{-1}$ according to Eq. (6.76). In Fig. 6.6, the proposed algorithm is depicted. Similar to the vapor-liquid equilibrium algorithm from Fig. 6.5, this algorithm can be improved by incorporating strategies of phase identification such as choosing the phase with the lowest residual free Gibbs energy when several solutions are available for the used equation of state. As illustrated later, this algorithm can be

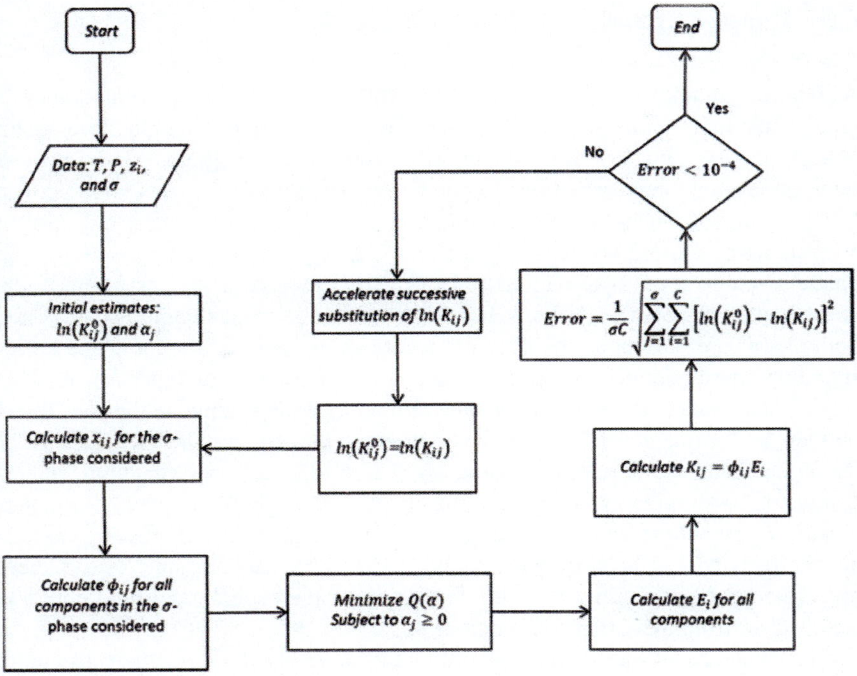

Fig. 6.6 Flowchart of the multiphase equilibrium algorithm

used in performing multiphase equilibrium calculations such as the asphaltene onset and water-heavy petroleum fluids as will be illustrated later.

As any other non-linear system of equations, initial estimates are required in order to solve the multiphase equilibrium algorithm numerically. According to Fig. 6.6, initial estimates have to be given for K_{ij} and α_j. For the former, this means that the composition for the considered phases have to be assumed so that the corresponding fugacity coefficients can be calculated, and hence, K_{ij} through Eq. (6.98). For the asphaltene onset and the presence of water shown later, it is considered that three phases may be present at most: vapor, and two liquid phases. It is suggested to assume a mole fraction equal to 0.99 for one chosen component, splitting out the residual 0.01 into the remaining components regardless of the considered phase. For the vapor phase, the lightest component is the chosen one, such as the solvent in calculating the asphaltene onset. On the other hand, the water or the lightest pseudocomponent from the distribution function are the chosen one for the first liquid phase. Finally, the heaviest pseudocomponent from the distribution function is the chosen one for the second liquid phase. More robust initialization schemes are provided elsewhere [25].

6.5 Phase Envelopes of Reservoir Fluids

A phase diagram for a pure component or mixture is a graphical representation of its phase behavior in terms of intensive properties such as pressure, temperature, molar volume, composition, specific enthalpy or entropy. These diagrams can be two- or three-dimensional representations, depending on the number of selected properties. In what follows, a phase envelope refers to a diagram where the phase behavior of a given petroleum fluid is depicted as a function of pressure and temperature. The phase envelope consists of regions separated from each other by curved lines, the phase boundaries between regions. For the presence of vapor and liquid phases only, both phases are present at equilibrium conditions in the inner region of the corresponding phase boundary, while just one phase is present in the outer region. This single phase can be liquid, vapor, or supercritical fluid, depending on the considered pressure and temperature. At low temperatures and pressures on the left-hand side of the phase boundary, liquids are present, whereas vapours are present at high temperatures and low pressures on the right-hand side of the phase boundary. At higher conditions than those on the critical point, supercritical fluids are present. There is not sharp boundaries between these single phases on the outer region; instead, there are transition regions. Therefore, the liquid, vapor and supercritical fluid concepts are clear far from these transition regions only.

The phase boundary between the inner and outer regions belongs to the inner region; therefore, vapor and liquid at equilibrium conditions are present on it, the so-called incipient or saturated points. The term incipient indicates that an incipient amount of either liquid or vapor is present. On the other hand, the term saturated indicates that either liquid or vapor cannot exist by itself; the presence of the other phase is required for the system to be thermodynamically stable. In turn, this boundary of saturated points is split out into two branches by the critical point. The right branch corresponds to the dew points, while the left branch corresponds to the bubble points. The dew and bubble terms indicate that liquid or vapor are the incipient phases, respectively. The highest pressure on the boundary of saturated points, where vapor and liquid can coexist at equilibrium, is called the cricondenbar, and the highest temperature is called the cricondentherm. In the context of the vapor-liquid equilibriums explained before, the dew points correspond to $\alpha_V = 1$, and the bubble points correspond to $\alpha_V = 0$. Hence, the vapor-liquid equilibrium algorithm from Fig. 6.5 can be used for constructing vapor-liquid phase envelopes. The general procedure consists of changing either temperature or pressure as input datum by either $\alpha_V = 1$ or $\alpha_V = 0$, depending on whether a dew or bubble point is intend to be determined. The Rachford-Rice equation is solved in order to calculate the remaining unknown variable, either temperature or pressure. There are most efficient algorithms that have been developed in order to construct phase vapor-liquid envelopes [27, 28]. For instance, in the method suggested by Michelsen and Mollerup [1], the set of algebraic Eqs. (6.72) and (6.73) is transformed into a set of ordinary differential equations so that the phase envelope can be constructed sequentially, starting from low pressures.

Table 6.2 Composition analysis of reservoir Petroleum fluids. *Data from Carreón-Calderón et al. [26]

°API	7.1*	9.9	10.3*	13.7*	22
% mol					
N_2	2.451	2.14	0.538	0.320	0.26
CO_2	0.011	6.38	7.373	2.400	1.64
H_2S	0.000	8.28	7.078	2.010	1.32
CH_4	21.593	9.42	7.461	25.35	29.58
C_2H_6	1.592	3.53	2.905	5.810	8.63
C_3H_8	0.220	4.64	3.905	4.690	6.37
i-C_4H_{10}	0.030	0.56	0.490	0.620	1.18
n-C_4H_{10}	0.060	2.28	1.991	2.390	3.07
i-C_5H_{12}	0.023	0.82	0.897	0.980	1.60
n-C_5H_{12}	0.029	1.17	1.358	1.550	2.17
C_6	0.0715	2.35	3.040	2.400	2.89
C7+	73.276	58.43	62.964	51.480	41.29

Table 6.3 Molecular Weight and Specific Gravity of Carbon Number Fractions corresponding to Table 6.2. *Data from Carreón-Calderón et al. [26]

Reservoir Petroleum Fluid	Fraction	M	S
7.1 °API*	C6	85.216	0.7056
	C7+	607.22	1.0246
9.9 °API	C6	85.589	0.6779
	C7+	593.5	1.0035
10.3 °API*	C6	85.837	0.6787
	C7+	513.33	1.0088
13.7 °API*	C6	86.2	0.6815
	C7+	412.30	0.9860
22 °API	C6	84.0	0.69
	C7+	326.0	0.9306

This algorithm is recommended because it can be used irrespective of the considered equation of state, as will be illustrated later.

For constructing phase envelopes of reservoir fluids, the needed data are the compositional analysis and two properties of the carbon number fractions, the molecular weight and the specific gravity, typically. Compositional analysis is commonly extended up to C_{20} or C_{30} fraction [29], this last one being the plus fraction. In Tables 6.2 and 6.3, the compositional analysis for five petroleum fluids from extra-heavy to medium are shown, where the analysis is extended up to C_{7+} fraction only. Although a more extended compositional analysis is always desirable, the information shown in these tables is usually enough for estimating phase envelopes and related properties. From Tables 6.4, 6.5, 6.6, 6.7 and 6.8, the corresponding ther-

Table 6.4 Characterization of unidentifiable fractions of 7.1 °API Reservoir Fluid using Exponential Distribution Function (EDF) and Gibbs free energy minimization (GFEM) (PR and MG-GCVOL)

i	x	M	S	HCR	Tb[K]	Tc[K]	Pc[MPa]	Vc[m^3/mol]	ω	CH3	CH_2(ch)	CH(ch)	C(ch)	ACH	AC	CH_2(cy)	CH(cy)	C(cy)
C6																		
1	0.0715	85.22	0.7056	2.17	360.08	550.87	3.2105	3.60E-04	0.2097	2.00	3.52	0.00	0.00	0.00	0.48	0.00	0.00	0.00
C7+																		
1	0.7328	607.22	1.0246	1.43	885.37	1154.63	0.9578	2.18E-03	0.3835	9.58	0.00	0.00	0.00	0.00	17.63	17.93	0.00	0.00
1	0.4693	316.05	0.9358	1.64	709.24	959.60	1.5336	1.13E-03	0.4404	2.83	0.00	0.00	0.00	0.00	5.60	14.70	0.00	0.00
2	0.1752	846.49	1.0396	1.40	838.49	1034.30	0.7899	3.11E-03	0.6658	12.51	0.00	0.00	0.00	10.71	0.00	0.00	39.87	0.00
3	0.0883	1680.31	1.1143	1.22	998.12	1208.55	0.6660	5.94E-03	0.6948	14.11	0.00	0.00	0.00	39.87	0.00	0.00	72.90	0.00
1	0.3199	231.81	0.8994	1.72	628.20	869.87	1.9664	8.28E-04	0.4396	1.70	0.00	0.00	0.00	0.00	3.19	11.98	0.00	0.00
2	0.1494	496.45	0.9805	1.53	825.08	1086.27	1.0920	1.77E-03	0.4070	5.92	0.00	0.00	0.00	0.00	11.49	19.21	0.00	0.00
3	0.1013	731.94	1.0210	1.44	890.61	1147.14	0.9231	2.48E-03	0.4398	0.00	0.00	0.00	0.00	8.87	10.80	34.70	0.00	0.00
4	0.0739	1003.56	1.0545	1.36	877.67	1076.64	0.7459	3.66E-03	0.6686	13.56	0.00	0.00	0.00	14.79	0.00	0.00	46.63	0.00
5	0.0539	1375.45	1.0886	1.28	951.01	1156.67	0.6897	4.93E-03	0.6819	14.59	0.00	0.00	0.00	27.17	0.00	0.00	61.63	0.00
6	0.0344	2158.80	1.1391	1.16	1056.48	1272.07	0.6439	7.53E-03	0.7214	13.48	0.00	0.00	0.00	59.87	0.00	0.00	90.39	0.00
1	0.2066	179.52	0.8691	1.79	561.73	796.04	2.4019	6.44E-04	0.4138	1.17	0.00	0.00	0.00	0.00	1.93	9.89	0.00	0.00
2	0.1134	327.10	0.9358	1.64	716.89	967.55	1.4914	1.16E-03	0.4396	2.93	0.00	0.00	0.00	0.00	5.80	15.21	0.00	0.00
3	0.0831	445.96	0.9683	1.56	797.00	1055.43	1.1776	1.59E-03	0.4171	4.92	0.00	0.00	0.00	0.00	9.64	18.27	0.00	0.00
4	0.0662	559.83	0.9920	1.51	855.84	1119.79	1.0087	2.00E-03	0.3960	7.19	0.00	0.00	0.00	0.00	13.79	20.39	0.00	0.00
5	0.0549	675.71	1.0116	1.46	904.59	1172.98	0.9019	2.41E-03	0.3803	9.84	0.00	0.00	0.00	0.00	18.41	21.86	0.00	0.00

(continued)

Table 6.4 (continued)

i	x	M	S	HCR	Tb[K]	Tc[K]	Pc[MPa]	Vc[m^3/mol]	ω	CH3	CH_2(ch)	CH(ch)	C(ch)	ACH	AC	CH_2(cy)	CH(cy)	C(cy)
6	0.0464	798.38	1.0292	1.42	824.59	1018.74	0.8054	2.95E-03	0.6678	12.49	0.00	0.00	0.00	8.96	0.00	0.00	37.93	0.00
7	0.0397	932.85	1.0456	1.38	860.63	1057.97	0.7626	3.42E-03	0.6680	13.30	0.00	0.00	0.00	12.59	0.00	0.00	43.71	0.00
8	0.0341	1085.87	1.0619	1.34	895.88	1096.38	0.7290	3.94E-03	0.6712	13.99	0.00	0.00	0.00	17.15	0.00	0.00	50.10	0.00
9	0.0292	1268.44	1.0787	1.31	932.03	1135.82	0.7015	4.57E-03	0.6778	14.53	0.00	0.00	0.00	23.16	0.00	0.00	57.49	0.00
10	0.0247	1502.18	1.0972	1.26	971.46	1178.90	0.6779	5.36E-03	0.6886	14.83	0.00	0.00	0.00	31.64	0.00	0.00	66.62	0.00
11	0.0201	1842.79	1.1200	1.21	1019.24	1231.16	0.6563	6.49E-03	0.7060	14.57	0.00	0.00	0.00	45.39	0.00	0.00	79.33	0.00
12	0.0142	2605.62	1.1596	1.12	1100.57	1320.28	0.6322	8.99E-03	0.7438	11.59	0.00	0.00	0.00	81.19	0.00	0.00	105.57	0.00

Table 6.5 Characterization of unidentifiable fractions from 9.9 °API Reservoir Fluid using Exponential Distribution Function (EDF) and Gibbs free energy minimization (GFEM) (PR and MG-GCVOL)

i	x	M	S	HCR	Tb[K]	Tc[K]	Pc[MPa]	Vc[m3/mol]	ω	CH3	CH2(ch)	CH(ch)	C(ch)	ACH	AC	CH2(cy)	CH(cy)	C(cy)
C6																		
1	0.0235	85.59	0.6779	2.24	359.19	548.36	3.1596	3.64E-04	0.2116	2.28	3.30	0.00	0.00	0.00	0.42	0.00	0.00	0.00
C7+																		
1	0.5843	593.50	1.0035	1.48	872.86	1138.97	0.9724	2.12E-03	0.3896	8.21	0.00	0.00	0.00	0.00	15.52	20.22	0.00	0.00
1	0.3734	309.57	0.9173	1.68	698.45	945.62	1.5546	1.10E-03	0.4440	2.49	0.00	0.00	0.00	0.00	4.85	15.25	0.00	0.00
2	0.1402	824.35	1.0179	1.45	830.79	1023.99	0.7901	3.06E-03	0.6735	13.68	0.00	0.00	0.00	8.02	0.00	0.00	39.51	0.00
3	0.0707	1635.49	1.0905	1.28	989.73	1197.12	0.6654	5.85E-03	0.7051	17.08	0.00	0.00	0.00	32.78	0.00	0.00	73.12	0.00
1	0.2540	227.57	0.8820	1.76	618.54	857.11	1.9862	813.36	0.4406	1.55	0.00	0.00	0.00	0.00	2.73	12.23	0.00	0.00
2	0.1194	483.97	0.9605	1.58	812.48	1070.42	1.1107	1721.38	0.4132	5.07	0.00	0.00	0.00	0.00	10.00	20.51	0.00	0.00
3	0.0811	712.91	0.9999	1.49	880.89	1135.19	0.9296	2422.94	0.4412	0.00	0.00	0.00	0.00	7.56	9.71	35.50	0.00	0.00
4	0.0592	977.09	1.0324	1.41	869.82	1066.08	0.7459	3599.34	0.6770	15.02	0.00	0.00	0.00	11.41	0.00	0.00	46.30	0.00
5	0.0432	1338.86	1.0656	1.34	942.84	1145.59	0.6892	4851.33	0.6914	16.83	0.00	0.00	0.00	21.88	0.00	0.00	61.53	0.00
6	0.0275	2100.97	1.1147	1.22	1047.88	1260.31	0.6433	7422.96	0.7326	17.57	0.00	0.00	0.00	49.99	0.00	0.00	91.10	0.00
1	0.1635	176.74	0.8527	1.83	553.33	784.67	2.4160	6.35E-04	0.4130	1.12	0.00	0.00	0.00	0.00	1.64	10.00	0.00	0.00
2	0.0905	319.43	0.9171	1.68	705.35	952.75	1.5154	1.14E-03	0.4435	2.57	0.00	0.00	0.00	0.00	4.99	15.75	0.00	0.00
3	0.0665	434.87	0.9487	1.61	784.64	1039.82	1.1980	1.55E-03	0.4229	4.23	0.00	0.00	0.00	0.00	8.36	19.31	0.00	0.00
4	0.0530	545.55	0.9717	1.55	843.02	1103.72	1.0255	1.94E-03	0.4024	6.15	0.00	0.00	0.00	0.00	12.03	22.01	0.00	0.00
5	0.0439	658.22	0.9908	1.51	891.45	1156.62	0.9158	2.35E-03	0.3866	8.39	0.00	0.00	0.00	0.00	16.11	24.13	0.00	0.00

(continued)

Table 6.5 (continued)

i	x	M	S	HCR	Tb[K]	Tc[K]	Pc[MPa]	Vc[m3/mol]	ω	CH3	CH2(ch)	CH(ch)	C(ch)	ACH	AC	CH2(cy)	CH(cy)	C(cy)
6	0.0372	777.52	1.0078	1.47	816.99	1008.60	0.8060	2.90E-03	0.6753	13.55	0.00	0.00	0.00	6.55	0.00	0.00	37.53	0.00
7	0.0318	908.31	1.0239	1.43	852.86	1047.55	0.7628	3.36E-03	0.6760	14.62	0.00	0.00	0.00	9.55	0.00	0.00	43.34	0.00
8	0.0273	1057.15	1.0396	1.40	887.95	1085.70	0.7289	3.88E-03	0.6798	15.61	0.00	0.00	0.00	13.38	0.00	0.00	49.79	0.00
9	0.0234	1234.75	1.0560	1.36	923.94	1124.89	0.7011	4.50E-03	0.6870	16.53	0.00	0.00	0.00	18.46	0.00	0.00	57.30	0.00
10	0.0198	1462.13	1.0740	1.32	963.21	1167.70	0.6774	5.28E-03	0.6984	17.34	0.00	0.00	0.00	25.68	0.00	0.00	66.61	0.00
11	0.0161	1793.51	1.0961	1.26	1010.80	1219.65	0.6558	6.40E-03	0.7166	17.87	0.00	0.00	0.00	37.48	0.00	0.00	79.65	0.00
12	0.0114	2535.67	1.1347	1.17	1091.81	1308.27	0.6316	8.86E-03	0.7558	16.81	0.00	0.00	0.00	68.48	0.00	0.00	106.88	0.00

Table 6.6 Characterization of unidentifiable fractions from 10.3 °API Reservoir Fluid using Exponential Distribution Function (EDF) and Gibbs free energy minimization (GFEM) (PR and MG-GCVOL)

i	x	M	S	HCR	Tb[K]	Tc[K]	Pc[MPa]	Vc[m3/mol]	ω	CH3	CH2(ch)	CH(ch)	C(ch)	ACH	AC	CH2(cy)	CH(cy)	C(cy)
C6																		
1	0.0304	85.84	0.6787	2.24	359.87	549.08	3.1541	3.65E-04	0.2129	2.28	3.31	0.00	0.00	0.00	0.43	0.00	0.00	0.00
C7+																		
1	0.6296	513.33	1.0088	1.47	842.43	1108.03	1.0667	1.84E-03	0.3986	7.34	0.00	0.00	0.00	0.00	13.79	16.92	0.00	0.00
1	0.3963	271.86	0.9229	1.67	671.38	918.33	1.7319	9.69E-04	0.4429	2.26	0.00	0.00	0.00	0.00	4.42	13.18	0.00	0.00
2	0.1549	695.39	1.0226	1.44	914.82	1184.98	0.8875	2.49E-03	0.3764	10.84	0.00	0.00	0.00	0.00	19.99	20.84	0.00	0.00
3	0.0784	1373.94	1.0961	1.26	951.50	1158.02	0.6915	4.90E-03	0.6777	13.69	0.00	0.00	0.00	28.71	0.00	0.00	61.02	0.00
1	0.2654	202.99	0.8878	1.75	594.91	833.47	2.1871	7.26E-04	0.4289	1.41	0.00	0.00	0.00	0.00	2.55	10.77	0.00	0.00
2	0.1309	411.49	0.9647	1.57	777.85	1035.03	1.2496	1.47E-03	0.4232	4.43	0.00	0.00	0.00	0.00	8.71	17.12	0.00	0.00
3	0.0895	602.14	1.0047	1.48	876.49	1142.89	0.9639	2.15E-03	0.3884	8.39	0.00	0.00	0.00	0.00	15.84	20.37	0.00	0.00
4	0.0655	822.82	1.0376	1.40	832.01	1027.36	0.7985	3.02E-03	0.6655	12.29	0.00	0.00	0.00	10.19	0.00	0.00	38.82	0.00
5	0.0479	1125.39	1.0712	1.32	904.70	1106.59	0.7237	4.07E-03	0.6697	13.61	0.00	0.00	0.00	19.31	0.00	0.00	51.42	0.00
6	0.0305	1763.41	1.1209	1.21	1009.54	1221.19	0.6614	6.21E-03	0.6983	13.80	0.00	0.00	0.00	43.70	0.00	0.00	75.81	0.00
1	0.1675	160.80	0.8590	1.82	534.16	765.76	2.6120	5.78E-04	0.3944	1.03	0.00	0.00	0.00	0.00	1.58	9.01	0.00	0.00
2	0.0979	275.19	0.9211	1.67	673.50	920.26	1.7140	9.81E-04	0.4434	2.26	0.00	0.00	0.00	0.00	4.42	13.41	0.00	0.00
3	0.0727	370.57	0.9529	1.60	750.48	1004.92	1.3532	1.32E-03	0.4315	3.70	0.00	0.00	0.00	0.00	7.31	16.19	0.00	0.00
4	0.0582	462.56	0.9762	1.54	807.86	1067.81	1.1473	1.65E-03	0.4129	5.37	0.00	0.00	0.00	0.00	10.46	18.27	0.00	0.00
5	0.0484	556.45	0.9956	1.50	855.74	1120.15	1.0125	1.99E-03	0.3956	7.32	0.00	0.00	0.00	0.00	13.97	19.87	0.00	0.00

(continued)

Table 6.6 (continued)

i	x	M	S	HCR	Tb[K]	Tc[K]	Pc[MPa]	Vc[m3/mol]	ω	CH3	CH2(ch)	CH(ch)	C(ch)	ACH	AC	CH2(cy)	CH(cy)	C(cy)
6	0.0411	656.02	1.0128	1.46	898.43	1166.73	0.9167	2.34E-03	0.3816	9.62	0.00	0.00	0.00	0.00	17.98	21.06	0.00	0.00
7	0.0352	765.28	1.0291	1.42	901.93	1159.57	0.9031	2.59E-03	0.4375	0.00	0.00	0.00	0.00	9.74	11.61	35.57	0.00	0.00
8	0.0303	889.71	1.0451	1.38	850.05	1046.91	0.7763	3.26E-03	0.6657	12.73	0.00	0.00	0.00	11.94	0.00	0.00	41.71	0.00
9	0.0259	1038.25	1.0616	1.35	885.88	1085.96	0.7396	3.77E-03	0.6678	13.40	0.00	0.00	0.00	16.37	0.00	0.00	47.91	0.00
10	0.0219	1228.49	1.0799	1.30	925.02	1128.68	0.7078	4.42E-03	0.6738	13.95	0.00	0.00	0.00	22.65	0.00	0.00	55.60	0.00
11	0.0179	1505.84	1.1023	1.25	972.51	1180.57	0.6785	5.36E-03	0.6860	14.21	0.00	0.00	0.00	32.90	0.00	0.00	66.36	0.00
12	0.0127	2127.27	1.1413	1.16	1053.43	1269.16	0.6454	7.41E-03	0.7173	12.88	0.00	0.00	0.00	59.76	0.00	0.00	88.77	0.00

Table 6.7 Characterization of unidentifiable fractions from 13.7 °API Reservoir Fluid using Exponential Distribution Function (EDF) and Gibbs free energy minimization (GFEM) (PR and MG-GCVOL)

i	x	M	S	HCR	Tb[K]	Tc[K]	Pc[MPa]	Vc[m³/mol]	ω	CH3	CH_2(ch)	CH(ch)	C(ch)	ACH	AC	CH_2(cy)	CH(cy)	C(cy)
C6																		
1	0.0240	86.20	0.6815	2.23	360.92	550.28	3.1485	3.67E-04	0.2149	2.26	3.36	0.00	0.00	0.00	0.43	0.00	0.00	0.00
C7+																		
1	0.5148	412.30	0.9860	1.52	785.67	1046.07	1.2483	1.47E-03	0.4190	5.10	0.00	0.00	0.00	0.00	9.84	15.51	0.00	0.00
1	0.3147	224.79	0.9045	1.71	623.00	865.25	2.0199	8.03E-04	0.4367	1.69	0.00	0.00	0.00	0.00	3.21	11.47	0.00	0.00
2	0.1324	534.37	0.9979	1.49	847.55	1111.94	1.0392	1.91E-03	0.3980	7.13	0.00	0.00	0.00	0.00	13.58	18.82	0.00	0.00
3	0.0677	1045.63	1.0696	1.33	888.20	1089.26	0.7403	3.78E-03	0.6646	12.78	0.00	0.00	0.00	17.70	0.00	0.00	47.85	0.00
1	0.2048	172.70	0.8715	1.79	553.87	788.25	2.4799	6.20E-04	0.4070	1.14	0.00	0.00	0.00	0.00	1.90	9.47	0.00	0.00
2	0.1099	321.89	0.9421	1.62	715.44	966.96	1.5122	1.15E-03	0.4389	3.00	0.00	0.00	0.00	0.00	5.94	14.64	0.00	0.00
3	0.0762	464.07	0.9809	1.53	810.22	1070.90	1.1447	1.65E-03	0.4117	5.55	0.00	0.00	0.00	0.00	10.77	17.91	0.00	0.00
4	0.0562	629.76	1.0131	1.46	889.44	1157.42	0.9383	2.25E-03	0.3838	9.25	0.00	0.00	0.00	0.00	17.29	20.18	0.00	0.00
5	0.0412	857.63	1.0459	1.38	841.95	1038.59	0.7882	3.14E-03	0.6637	12.21	0.00	0.00	0.00	11.61	0.00	0.00	40.17	0.00
6	0.0264	1339.19	1.0944	1.27	945.63	1151.73	0.6955	4.78E-03	0.6756	13.54	0.00	0.00	0.00	27.63	0.00	0.00	59.60	0.00
1	0.1249	141.57	0.8453	1.85	501.81	729.97	2.8663	5.11E-04	0.3659	0.90	0.00	0.00	0.00	0.00	1.22	8.09	0.00	0.00
2	0.0799	221.38	0.9003	1.72	618.25	859.68	2.0434	7.91E-04	0.4360	1.63	0.00	0.00	0.00	0.00	3.06	11.41	0.00	0.00
3	0.0607	291.32	0.9308	1.65	689.41	938.24	1.6381	1.04E-03	0.4422	2.53	0.00	0.00	0.00	0.00	5.00	13.77	0.00	0.00
4	0.0492	359.62	0.9534	1.60	743.96	998.21	1.3855	1.28E-03	0.4329	3.60	0.00	0.00	0.00	0.00	7.12	15.68	0.00	0.00
5	0.0412	429.74	0.9722	1.55	790.10	1048.80	1.2102	1.53E-03	0.4188	4.86	0.00	0.00	0.00	0.00	9.50	17.29	0.00	0.00

(continued)

Table 6.7 (continued)

i	x	M	S	HCR	Tb[K]	Tc[K]	Pc[MPa]	Vc[m^3/mol]	ω	CH3	CH_2(ch)	CH(ch)	C(ch)	ACH	AC	CH_2(cy)	CH(cy)	C(cy)
6	0.0351	504.36	0.9891	1.51	831.60	1094.21	1.0803	1.80E-03	0.4040	6.36	0.00	0.00	0.00	0.00	12.23	18.66	0.00	0.00
7	0.0302	586.42	1.0050	1.48	870.71	1136.92	0.9795	2.09E-03	0.3900	8.19	0.00	0.00	0.00	0.00	15.45	19.80	0.00	0.00
8	0.0260	680.02	1.0206	1.44	909.16	1178.85	0.8983	2.43E-03	0.3779	10.47	0.00	0.00	0.00	0.00	19.36	20.68	0.00	0.00
9	0.0223	791.88	1.0368	1.40	910.85	1169.46	0.8891	2.67E-03	0.4359	0.00	0.00	0.00	0.00	10.56	12.33	36.10	0.00	0.00
10	0.0189	935.28	1.0546	1.36	862.00	1060.37	0.7650	3.41E-03	0.6642	12.63	0.00	0.00	0.00	13.80	0.00	0.00	43.46	0.00
11	0.0155	1144.49	1.0765	1.31	908.94	1111.58	0.7214	4.13E-03	0.6687	13.33	0.00	0.00	0.00	20.52	0.00	0.00	52.00	0.00
12	0.0110	1613.73	1.1145	1.22	989.14	1199.25	0.6712	5.70E-03	0.6887	13.52	0.00	0.00	0.00	38.34	0.00	0.00	69.99	0.00

Table 6.8 Characterization of unidentifiable fractions from 22.0 °API Reservoir Fluid using Exponential Distribution Function (EDF) and Gibbs free energy minimization (GFEM) (PR and MG-GCVOL)

i	x	M	S	HCR	Tb[K]	Tc[K]	Pc[MPa]	Vc[m^3/mol]	ω	CH3	CH_2(ch)	CH(ch)	C(ch)	ACH	AC	CH_2(cy)	CH(cy)	C(cy)
C6																		
1	0.0289	84.00	0.6900	2.21	355.63	545.18	3.2206	3.57E-04	0.2037	2.12	3.35	0.00	0.00	0.00	0.44	0.00	0.00	0.00
C7+																		
1	0.4129	326.00	0.9306	1.65	714.38	964.16	1.4943	1.16E-03	0.4406	2.83	0.00	0.00	0.00	0.00	5.59	15.42	0.00	0.00
1	0.2421	185.36	0.8579	1.82	565.48	798.15	2.3300	6.66E-04	0.4202	1.19	0.00	0.00	0.00	0.00	1.80	10.40	0.00	0.00
2	0.1124	399.29	0.9397	1.63	762.57	1015.57	1.2757	1.42E-03	0.4301	3.67	0.00	0.00	0.00	0.00	7.26	18.32	0.00	0.00
3	0.0585	767.37	1.0061	1.47	813.93	1005.25	0.8099	2.86E-03	0.6755	13.49	0.00	0.00	0.00	6.29	0.00	0.00	37.08	0.00
1	0.1516	147.94	0.8294	1.89	507.09	732.89	2.7504	5.35E-04	0.3783	0.95	0.00	0.00	0.00	0.00	1.08	8.61	0.00	0.00
2	0.0904	248.12	0.8889	1.75	639.95	880.45	1.8526	8.86E-04	0.4451	1.73	0.00	0.00	0.00	0.00	3.14	13.14	0.00	0.00
3	0.0643	348.64	0.9244	1.66	727.23	976.59	1.4145	1.24E-03	0.4396	2.92	0.00	0.00	0.00	0.00	5.73	16.82	0.00	0.00
4	0.0480	467.16	0.9541	1.60	802.44	1059.09	1.1382	1.66E-03	0.4170	4.70	0.00	0.00	0.00	0.00	9.29	20.31	0.00	0.00
5	0.0356	631.04	0.9846	1.52	879.90	1143.79	0.9381	2.25E-03	0.3903	7.73	0.00	0.00	0.00	0.00	14.94	23.91	0.00	0.00
6	0.0229	978.85	1.0296	1.42	869.98	1065.94	0.7445	3.61E-03	0.6785	15.28	0.00	0.00	0.00	11.05	0.00	0.00	46.49	0.00
1	0.0887	126.41	0.8080	1.94	466.40	687.19	3.0626	4.60E-04	0.3364	0.86	0.00	0.00	0.00	0.00	0.72	7.47	0.00	0.00
2	0.0629	178.31	0.8518	1.83	555.04	786.32	2.3980	6.41E-04	0.4146	1.13	0.00	0.00	0.00	0.00	1.64	10.10	0.00	0.00
3	0.0495	226.45	0.8787	1.77	616.40	854.36	1.9921	8.10E-04	0.4407	1.52	0.00	0.00	0.00	0.00	2.64	12.25	0.00	0.00
4	0.0409	274.37	0.8992	1.72	665.64	908.74	1.7094	9.78E-04	0.4468	2.01	0.00	0.00	0.00	0.00	3.76	14.19	0.00	0.00
5	0.0346	324.06	0.9165	1.68	708.34	955.77	1.4977	1.15E-03	0.4433	2.59	0.00	0.00	0.00	0.00	5.04	16.00	0.00	0.00

(continued)

Table 6.8 (continued)

i	x	M	S	HCR	Tb[K]	Tc[K]	Pc[MPa]	Vc[m^3/mol]	ω	CH3	CH_2(ch)	CH(ch)	C(ch)	ACH	AC	CH_2(cy)	CH(cy)	C(cy)
6	0.0297	377.25	0.9321	1.65	747.36	998.65	1.3312	1.34E-03	0.4346	3.31	0.00	0.00	0.00	0.00	6.53	17.76	0.00	0.00
7	0.0257	435.97	0.9468	1.61	784.54	1039.43	1.1955	1.55E-03	0.4232	4.18	0.00	0.00	0.00	0.00	8.29	19.50	0.00	0.00
8	0.0223	503.14	0.9612	1.58	821.39	1079.79	1.0815	1.79E-03	0.4103	5.30	0.00	0.00	0.00	0.00	10.44	21.25	0.00	0.00
9	0.0192	583.59	0.9763	1.54	859.60	1121.57	0.9829	2.08E-03	0.3970	6.77	0.00	0.00	0.00	0.00	13.20	23.05	0.00	0.00
10	0.0163	686.89	0.9928	1.51	871.26	1124.53	0.9473	2.34E-03	0.4435	0.00	0.00	0.00	0.00	6.96	9.08	34.74	0.00	0.00
11	0.0134	837.81	1.0131	1.46	834.02	1026.84	0.7832	3.12E-03	0.6763	14.24	0.00	0.00	0.00	7.62	0.00	0.00	40.29	0.00
12	0.0095	1176.99	1.0484	1.38	912.61	1112.27	0.7084	4.30E-03	0.6858	16.51	0.00	0.00	0.00	16.33	0.00	0.00	55.01	0.00

modynamics characterization of the carbon number fractions is shown. The basic properties of pseudocomponents: mole fraction x, molecular weight M, and specific gravity S are calculated using the continuous exponential distribution as explained in the Chap. 5.

In turn, M and S from pseudocomponents are used to estimate the corresponding parameters entering into equations of state by means of either correlations or Gibbs energy minimizations, as explained in the Chaps. 3 and 4. As discussed in the Chap. 5, a strategy of characterization involving generalized correlations implies the adjustment of the parameters entering into the equation of state, especially the binary interaction parameters. The explanation of the mathematical strategies of such adjustments is beyond of the scope the current manuscript; hence, the reader is referred to specialized manuscripts [33, 34]. On the other hand, the use of Gibbs energy minimizations implies an adjustment strategy based on the search for the most suitable combination of equation of state and mixing rule. In this case, it has been illustrated that phase envelopes can properly predicted by using M, and S as constraints only [26, 35]. However, a third constraint may be necessary into the minimization process in some applications. For instance, in the case of the estimation of solubility of light components into bitumen, the molar ratio of hydrogen to carbon, the hydrogen-carbon ratio (*HCR*), is suggested as the third constraint [36, 37]. This constraint leads to pseudocomponents, whose pseudostructures represent their chemical nature better. A higher *HCR* gives rise to chemical pseudostructures where paraffinic functional groups prevail over naphthenic and aromatic ones. On the contrary, a low *HCR* gives rise to chemical pseudostructures where naphthenic and aromatic functional groups prevail over paraffinic ones. This can be observed by comparing pseudostructures of C_6 and C_{7+} fractions from Tables 6.4, 6.5, 6.6, 6.7 and 6.8. Hence, *HCR* should be always included as a third constraint in characterization processes using Gibbs energy minimizations. Most often *HCR* is not available experimentally, especially in compositional analysis for reservoir petroleum fluids. An option to estimate *HCR* for C_+ fractions, whole petroleum fluids and refining cuts is shown in Fig. 6.7 [38], where one can observed that experimental information agrees well with a linear function of S. In tables from Tables 6.4, 6.5, 6.6, 6.7 and 6.8, the *HCR* values are estimated using the correlation shown in this figure. The corresponding characterization is performed by Gibbs energy minimizations setting up M, S, and *HCR* as constraints as explained in the Chap. 3. The PR equation of state [30], the MG[31] and GCVOL[32] group contribution methods are used in such minimization processes.

Figure 6.8 illustrates the comparison between normal boiling temperatures from the Lee-Kesler and Soreide correlations, and those from Gibbs energy minimizations from Tables 6.4, 6.5, 6.6, 6.7 and 6.8. In this type of graphs, a good agreement is found between two sets of values, when a straight line is obtained with small intercept and slope about one. Thus, we have a good agreement between calculated values from the Soreide correlation and those values from minimizations. In contrast, values differ from those obtained using the Lee-Kesler correlation. Most important, the calculated values from the three approaches agree well with each other for temperatures lesser than 800 K approximately. Above such a normal boiling temperature, one

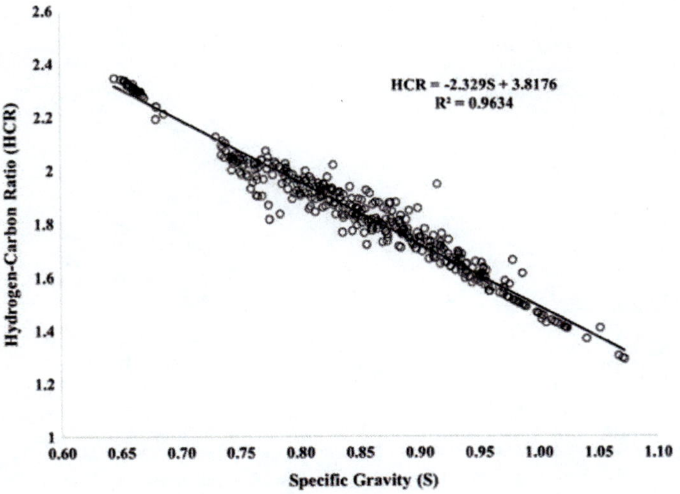

Fig. 6.7 Molar Ratio of Hydrogen to Carbon of Petroleum Fractions [38]

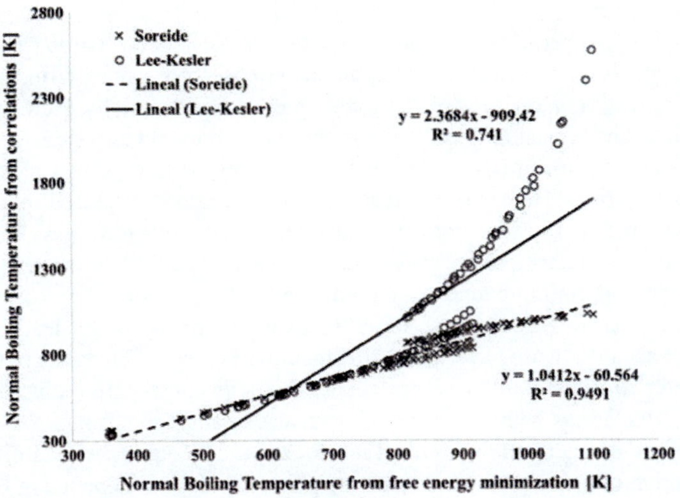

Fig. 6.8 Comparison between calculated values of normal boiling temperatures from free energy minimization and correlations

might expect an increasing uncertainty in predicting properties of the corresponding pseudocomponents.

Similarly, Fig. 6.9 illustrates the comparison of the critical temperatures, but, in this case, those corresponding to the Lee-Kesler correlation are calculated using two values of normal boiling temperatures as input datum. On the one hand, temperatures calculated from minimizations are used, which are indicated by the symbol (*). On

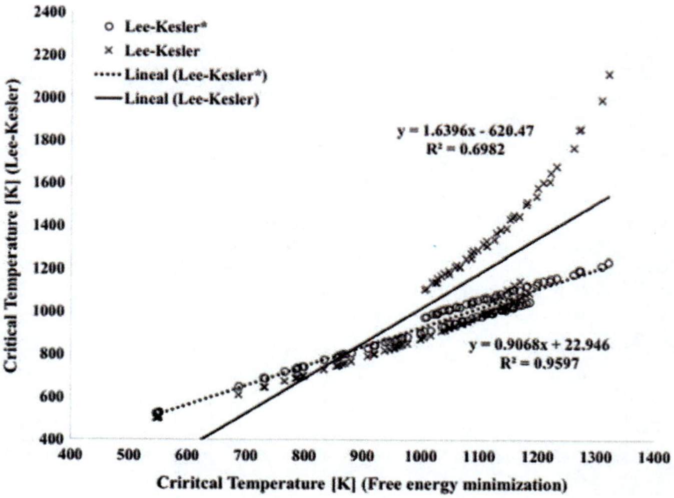

Fig. 6.9 Comparison between calculated values of critical temperatures from free energy minimization and the Lee-Kesler correlation

the other hand, temperature calculated from the Lee-Kesler correlation from Fig. 6.8 are used, which are indicated by no symbol. As observed, values differ from each other for critical temperatures higher than 1000 K, approximately. Here again, one might expect an increasing uncertainty above such a critical temperature.

As in the case of critical temperatures, the corresponding critical pressures are shown in Fig. 6.10. Here, the calculated values are more dispersed, and a sharp trend is not present, showing less agreement between the sets of values. What draws attention is that the critical pressure vanishes for the heaviest pseudocomponents, when the normal boiling temperature predicted by the corresponding correlation of Lee-Kesler is used. Indeed, these are unrealistic outcomes, which lead in turn to unrealistic acentric factors as well, as illustrated in Fig. 6.11. Thus, the heaviest the considered petroleum fluid, it is more likely that adjustments are required in order to match calculations with experimental data, as explained in Chap. 5.

Using the characterization given in Table 6.7 and the PR equation of state, the effect of the number of pseudocomponents on the phase envelope is depicted in Figs. 6.12 and 6.13. In the former case, the MHV1 mixing rule ($q_1 = -0.519$) is used together with the UNIFAC solution model, whereas, in the latter case, the classical mixing rule of van der Waals is used together with the Tsonopoulos expression ($k_0 = 2$ and $k_1 = 1$) for estimating the binary interaction parameters k_{ij}. Once the C_{7+} fraction is split out into pseudocomponents, a little impact is observed on the phase envelop regardless of the used mixing rule, especially on the bubble points region. Therefore, the number of pseudocomponents to be used can be determined from other considerations, such as the required time for performing reservoir simulations, a detailed compositional description for separation processes, etc.

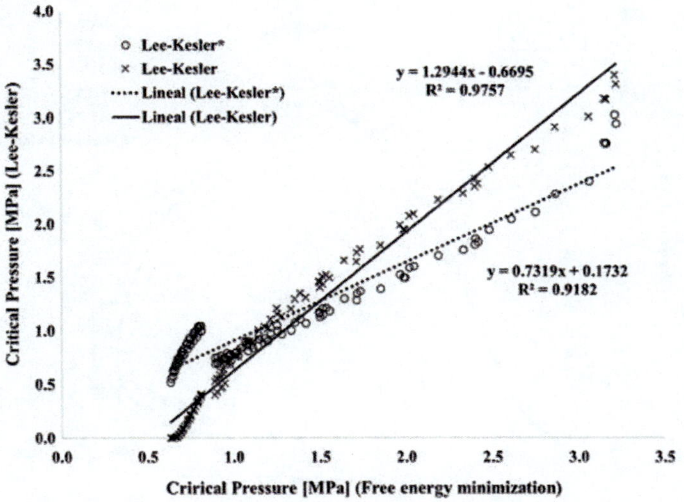

Fig. 6.10 Comparison between calculated values of critical pressures from free energy minimization and the Lee-Kesler correlation

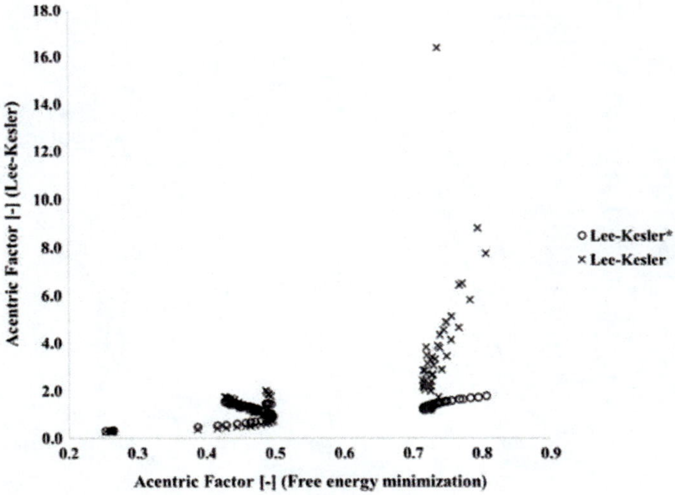

Fig. 6.11 Comparison between calculated values of acentric factors from free energy minimization and the Lee-Kesler correlation

In addition to the number of pseudocomponents, it is necessary to decide which equation of state and mixing rule is suitable. For the considered reservoir fluids, phase envelopes are depicted from Figs. 6.14, 6.15, 6.16, 6.17 and 6.18, where the mixing rules from the Chap. 3 are used together with the PR equation of state. In these figures, the Predicative Cubic Equation of State (PCES) refers to the MHV1

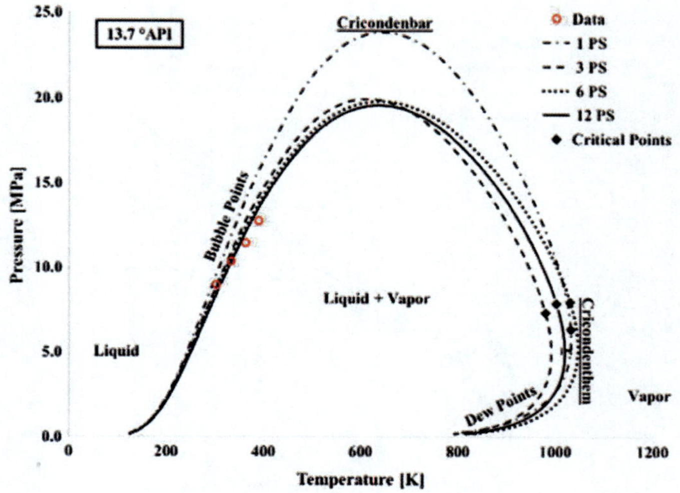

Fig. 6.12 Effect of the number of pseudocomponents on the Phase Envelope using the PR equation of state, the MHV1 mixing rule ($q_1 = -0.519$) and the UNIFAC solution model. Characterization from Table 6.7

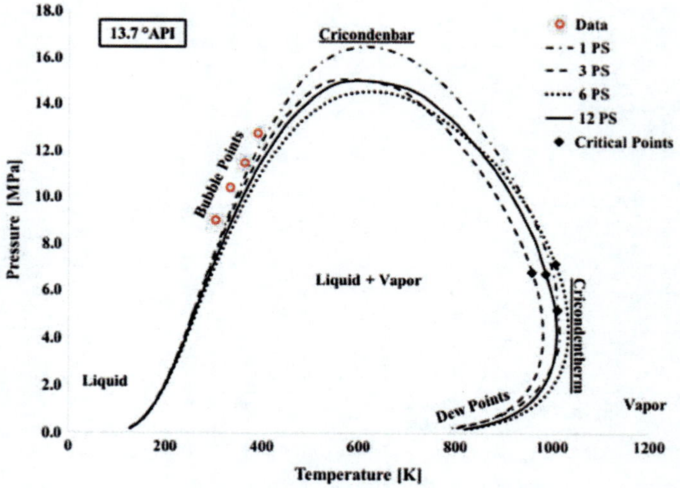

Fig. 6.13 Effect of the number of pseudocomponents on the phase envelope using the PR equation of state and the vdW mixing rule together with k_{ij} from the Tsonopoulos correlation ($k_0 = 2$ and $k_1 = 1$). Characterization from Table 6.7

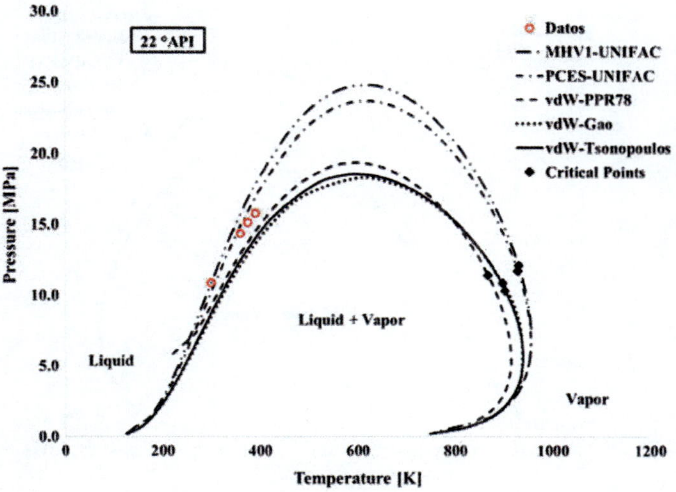

Fig. 6.14 Effect of the mixing rule on the phase envelope (22.0 °API) using the PR equation of state with 6 pseudocomponents. Characterization from Table 6.8

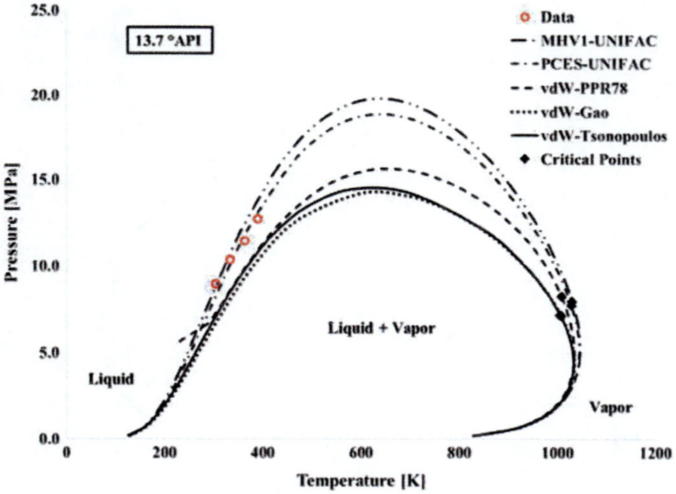

Fig. 6.15 Effect of the mixing rule on the phase envelope (13.7 °API) using the PR equation of state with 6 pseudocomponents. Characterization from Table 6.7

mixing rule in combination with the parameter q_1 from experimental data of liquid molar volumes, as explained in the Chap. 3. In the case of the PR equation of state, we have that $q_1 = -0.531$.

The outcomes show that predictions may be classified in two sets. On the one hand, we have those obtained from the MHV1 mixing rule either with $q_1 = -0.519$

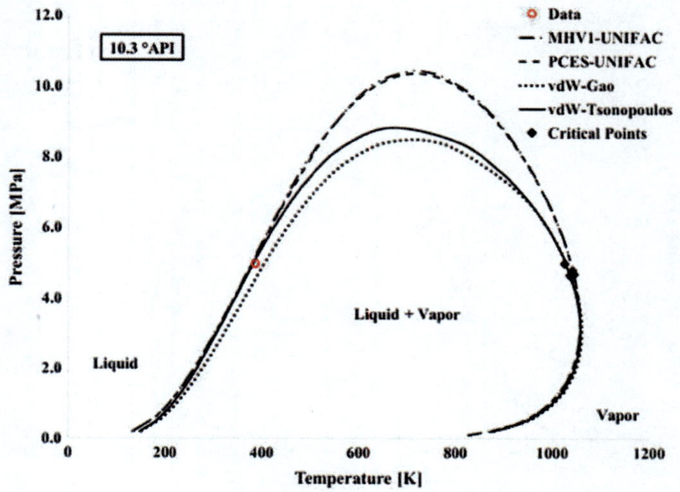

Fig. 6.16 Effect of the mixing rule on the phase envelope (10.3 °API) using the PR equation of state with 6 pseudocomponents. Characterization from Table 6.6

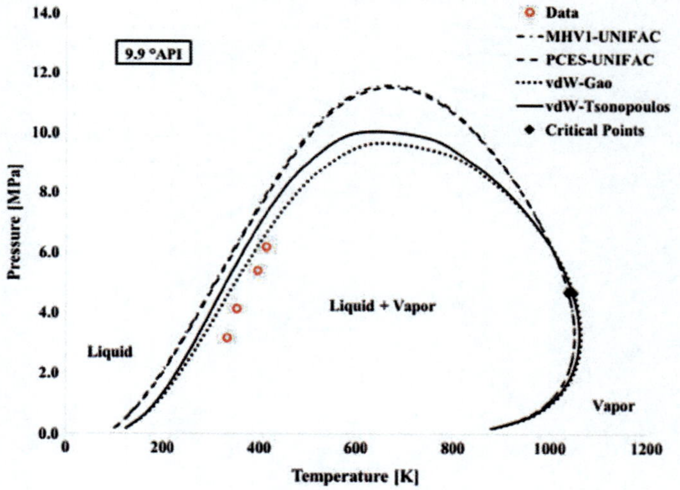

Fig. 6.17 Effect of the mixing rule on the phase envelope (9.9 °API) using the PR equation of state with 6 pseudocomponents. Characterization from Table 6.5

or $q_1 = -0.531$ (PCES). On the other hand, we have those obtained from the vdW mixing rule with the Tsonopoulos, Gao, and PPR78 expressions for k_{ij}. In the specific case of the PPR78 expression, unrealistic outcomes are found for the heaviest petroleum fluids. For this reason, this option does not appear in the corresponding figures. As seen, the former set predicts higher cricondenbar pressures than those

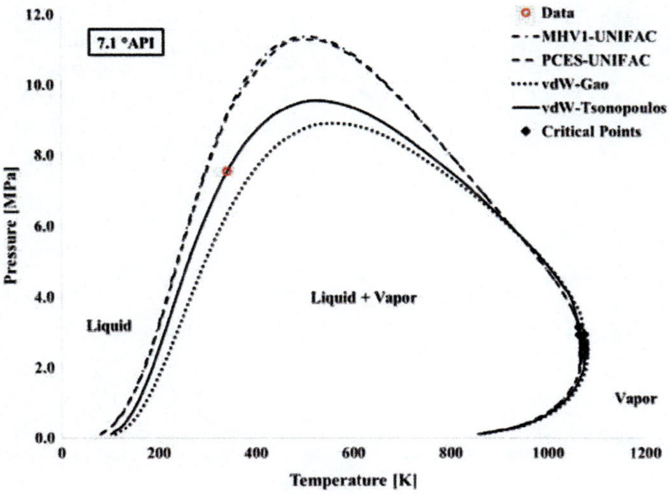

Fig. 6.18 Effect of the mixing rule on the phase envelope (7.1 °API) using the PR equation of state with 6 pseudocomponents. Characterization from Table 6.4

predicted using the latter set. From the data available usually (saturation pressure at the reservoir temperature), it is not possible to establish which the best option is, as can be observed from the same figures. Actually, except for the case in Fig. 6.17, all options can be adjusted to achieve a better match by changing the C_{7+} fraction properties such as the molecular weight.

The suggested strategic consists of searching for the best combination of pseudocomponents, cubic equation of state, and mixing rule, as illustrated in Fig. 6.19, where neither the equation-of-state parameters nor the C_{7+} fraction properties have been modified. This strategy intends to preserve the predictive capabilities of models, as can be observed in Fig. 6.20. In this figure, it is illustrated the effect of nitrogen injection in the original reservoir fluid. From a qualitative point of view, a better description is obtained from using the PCES mixing rule than from using the vdW-Tsonopoulos mixing rule. The former takes into account nitrogen-hydrocarbon interactions through the UNIFAC solution model, while the latter represents well binary interaction for nonpolar components, especially for hydrocarbons.

A major disadvantage of this strategy is that the adjustment of the molecular weight from the C_{7+} fraction, as well as its basic properties in general, has little impact on calculations, as can be observed in the same Fig. 6.20. Here, a change in the molecular weight of 20% with respect to the original value has only a minor influence on the outcomes. In other words, considering that the equation-of-state and mixing-rule parameters are held fixed and equal to their original values, there is little margin for regression to experimental data.

The other property related to the phase equilibrium calculations is the density, especially the liquid density. In Fig. 6.21, it is illustrated the liquid density from the

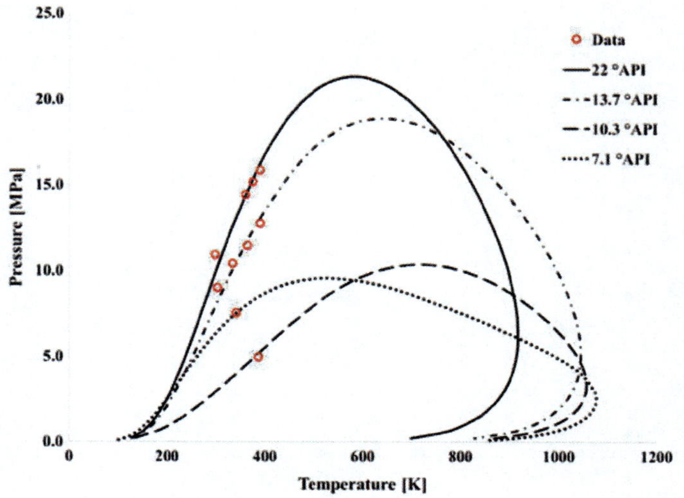

Fig. 6.19 Phase envelopes of reservoir petroleum fluids using the best combination of pseudocomponents, cubic equation of state, and mixing rule

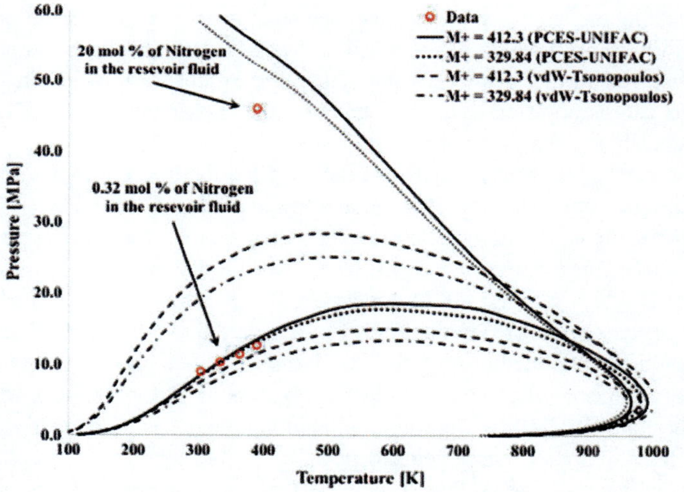

Fig. 6.20 Effect of nitrogen injection in the original reservoir fluid (13.7 °API) using the PR equation of state and with 6 pseudocomponents

isothermal differential vaporization experiment, which is also known as the differential liberation or the differential depletion experiment. This experiment search for emulating changes that occur in the reservoir petroleum fluid during its production from the reservoir to the surface. The experiment starts from a pressure above the saturation one, where a single phase is present. Then, the pressure is reduced

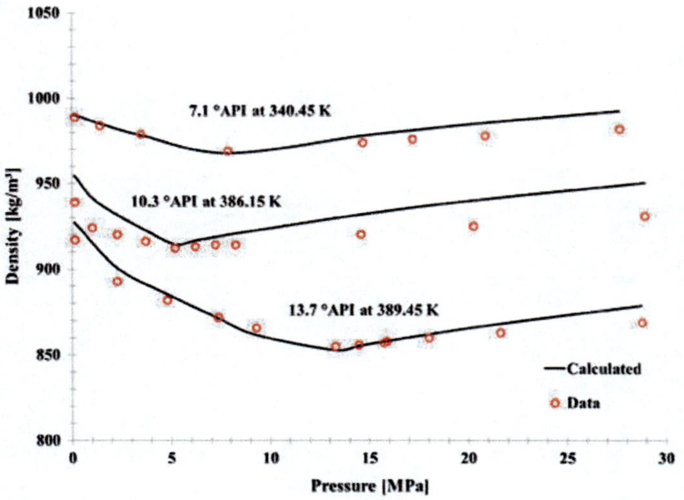

Fig. 6.21 Isothermal differential vaporization corresponding to Fig. 6.19. Volume Translation from the Peneloux correlation and GCVOL group contribution for identifiable and unidentifiable components, respectively [26]

in steps up to atmospheric conditions, keeping constant the temperature, which corresponds usually to the reservoir one. Above the saturation pressure, the density decreases because the fluid swells. On the other hand, below the saturation pressure, light components are released from the fluid, with which the fluid becomes heavier, increasing its density. As in the case of phase envelopes, the C_{7+} fraction molecular weight can be adjusted to match the experiments. If this adjustment is not enough, the C_{7+} fraction specific gravity can be adjusted by up to $\pm 1\%$.

For some cases, it is not possible to represent the experiment data as explained before, for example, the phase envelope in Fig. 6.17. Under these circumstances, the regression of the equation-of-state parameters to measured PVT data is an option. In practice, any equation-of-state parameter has been suggested to be modified for regression purposes; however, the adjustment of the binary interaction parameters k_{ij} is the most usual and suggested option, because they are the most evident deficiency of the equations of state, either the CEoS or the PC-SAFT equations of state. With the former, the k_{ij} parameters can be estimated through either the Tsonopoulos or Gao expressions, being k_0 and k_1 the only adjustable parameters for the whole mixture. With the latter, empirical expressions as a function of temperature are commonly used. For instance, the adjustment corresponding to the phase envelope of Fig. 6.17 is illustrated in Fig. 6.22. Here, a generalized four-parameter expression [39], which is also a function of the molecular weight, is used. The PC-SAFT parameters, which are shown in Table 6.9, are estimated from the minimization procedure of the Gibbs free energy using the functional groups from Burgess et at. [40], as explained in the Chap. 4. The outcomes are weak function of the number of pseudocomponents as before; therefore, the smallest number is recommended for regression purposes.

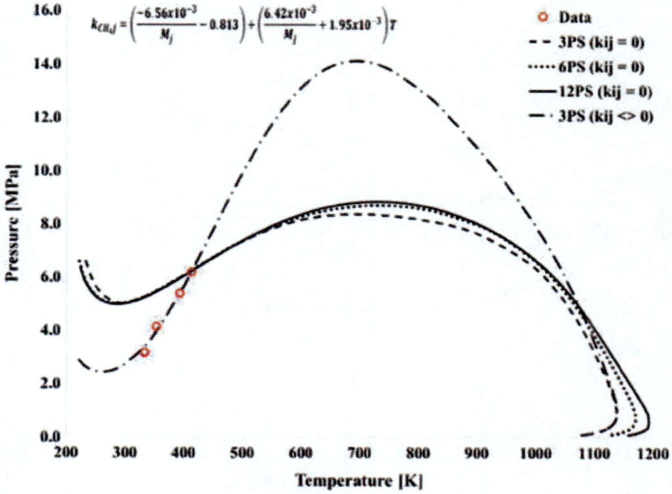

Fig. 6.22 Regression to the Phase Envelope of 9.9 °API Reservoir Petroleum Fluid using the sPC-SAFT equation of state with 3 pseudocomponents

Besides, the regression is performed for the k_{ij} parameters corresponding to the lightest component (methane) and the remaining ones.

6.6 Solvent-Heavy Oil

6.6.1 Gases and Light Hydrocarbons

One of the main characteristics of heavy petroleum fluids, concerning their thermophysical properties, is their high viscosity. In contrast to lighter petroleum fluids, it prevents a suitable mobility throughout production system from reservoirs to processing facilities. To reduce viscosity sufficiently for heavy petroleum to flow conveniently, the injection of solvents into it is a technology commonly used. For instance, to maximize the petroleum recovery from reservoirs, the process of injecting a solvent via a mixing process (mass transfer) is known as vapor extraction (VAPEX) [43]. Warm VAPEX refers to VAPEX process where the solvent is heated so that it condenses in the reservoir (in situ), taking advantage of both mass and heat transfer in order to reduce the petroleum viscosity. Similarly, Hybrid VAPEX process combines the solvent with steam. These techniques belong to the so-called enhanced oil recovery techniques (EOR) [44].

The key physical parameter in such EOR processes is the solvent solubility in the heavy petroleum fluid, which may be defined as the maximum amount of solvent that can be accepted into a phase liquid at a given temperature and pressure. As illustration,

Table 6.9 Characterization of unidentifiable fractions from 9.9 °API Reservoir Fluid using EDF and GFEM

9.9 °API							
i	x	M	S	HCR	m	σ [Å]	ε/k [K]
C6							
1	0.0235	85.59	0.6779	2.24	3.2550	3.6870	237.59
C7+							
1	0.5843	593.50	1.0035	1.48	14.4313	3.9403	291.76
1	0.3734	309.57	0.9173	1.68	8.5382	3.8465	270.93
2	0.1402	824.35	1.0179	1.45	19.9835	3.9372	293.69
3	0.0707	1635.49	1.0905	1.28	37.4672	3.9484	307.63
1	0.2540	227.57	0.8820	1.76	6.3602	3.8579	267.84
2	0.1194	483.97	0.9605	1.58	13.0562	3.8395	274.99
3	0.0811	712.91	0.9999	1.49	17.5477	3.9334	290.37
4	0.0592	977.09	1.0324	1.41	23.4371	3.9391	296.30
5	0.0432	1338.86	1.0656	1.34	31.2840	3.9445	302.65
6	0.0275	2100.97	1.1147	1.22	47.2965	3.9508	312.48
1	0.1635	176.74	0.8527	1.83	4.9381	3.8792	265.84
2	0.0905	319.43	0.9171	1.68	8.8423	3.8436	270.65
3	0.0665	434.87	0.9487	1.61	11.8304	3.8394	273.71
4	0.0530	545.55	0.9717	1.55	14.6102	3.8392	276.16
5	0.0439	658.22	0.9908	1.51	17.3764	3.8404	278.31
6	0.0372	777.52	1.0078	1.47	19.0252	3.9345	291.76
7	0.0318	908.31	1.0239	1.43	21.9434	3.9374	294.68
8	0.0273	1057.15	1.0396	1.40	25.2197	3.9401	297.63
9	0.0234	1234.75	1.0560	1.36	29.0778	3.9428	300.75
10	0.0198	1462.13	1.0740	1.32	33.9503	3.9456	304.27
11	0.0161	1793.51	1.0961	1.26	40.9440	3.9485	308.68
12	0.0114	2535.67	1.1347	1.17	56.2738	3.9525	316.59

Fig. 6.23 and Fig. 6.24 show experimental solubility of gases and light alkanes in some heavy petroleum fluids as a function of pressure. In the former case, the carbon dioxide solubility is more than twice as large as the hydrogen and nitrogen solubility, which, in turn, are comparable to each other at a given pressure. In the latter case, we observe that the smaller the light alkane, the smaller the solubility. In general, for a specific solvent, solubility increases with increasing pressure and decreasing temperature. However, in contrast to the behavior shown by other solvents, opposite experimental evidence has been found for the solubility of nitrogen and hydrogen in heavy petroleum fluids. For instance, it has been found that the solubility of nitrogen [46, 47] and hydrogen [45, 49] increase with increasing temperature, at constant pressure. Besides, in the case of the hydrogen solubility, it has been found that its

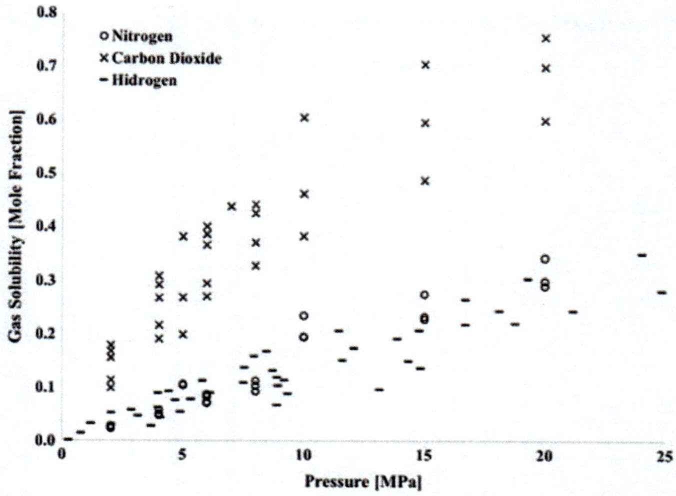

Fig. 6.23 Experimental Solubility of gases in Heavy Oils in the interval from 323.15 K to 573.15 K [45–47]

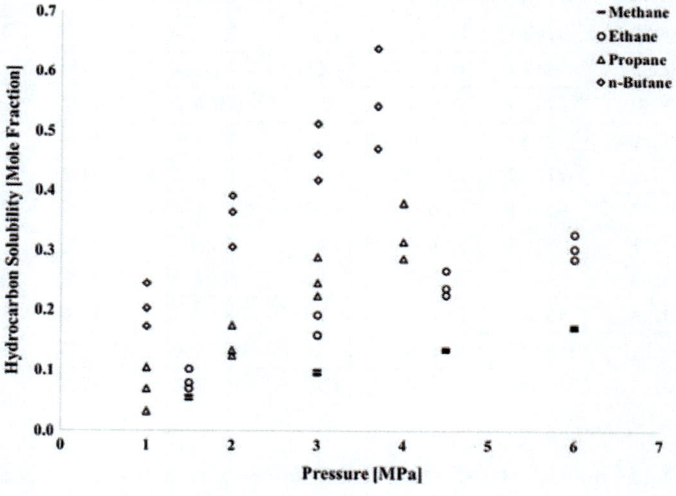

Fig. 6.24 Experimental Solubility of light hydrocarbons in Bitumen at 473.15 K, 503.15 K, and 533.15K [48]

solubility approaches the solubility found in lighter hydrocarbons as temperature increases [50]. This might explain why binary interactions parameters in the PC-SAFT equation of state between hydrogen and long-chain alkanes can be used in order to predict the hydrogen solubility in heavy petroleum fractions [51].

6.6.2 *The sPC-SAFT Equation of State*

The solvent solubility has been estimated using both CEoS and sPC-SAFT equations of state. When the PC-SAFT equation of state is intended to be used, we can make use of Generalized Pseudocomponents (GP) such as the 37 pseudocomponents suggested by Saajanlehto and Alopaeus [52]. The main advantage of the use of a set of GP is that the pseudocomponent properties are held constant regardless of the petroleum fluid. Thus, the number of pseudocomponents will remain fixed, if more than one petroleum fluid is involved in calculations. On the other hand, the main disadvantage is that a single set of GP might not be suitable for prediction purposes in all considered system, as will be illustrated later.

As explained in Chap. 4, it is advisable to reduce the number of pseudocomponents in order to decrease the time spent in calculations. Table 6.10 shows the set of GP as obtained in Chap. 4 from a lumping process of the pseudocomponents of Saajanlehto and Alopaeus [52]. To calculate the mole fraction corresponding to each pseudocomponent from Table 6.10, one iteratively searches for a weight fraction (w_c) so that the calculated either T_b or M of a given pseudocomponent, depending on the available experimental distribution, coincides with the value from Table 6.10. According to the generalized distribution function (GDF) from Chap. 5, T_b or M of a given pseudocomponent is determined from

Table 6.10 Generalized Pseudocomponents (GP) for the PC-SAFT equation of state. Mole fractions from GDF using data from Athabasca Bitumen [45], MacKay River Bitumen [53], and Xinjian Extra Heavy Oil [47]

Tb[K]	M	S	HCR	m	σ [Å]	ε/k [K]	Athabasca	MacKay River	Xinjian
473	144	0.85	1.84	4.0354	3.8747	282.21	0.0183	0.0222	0.0667
523	172	0.87	1.79	4.6290	3.9203	289.36	0.0956	0.0702	0.1321
573	204	0.89	1.74	5.2798	3.9586	295.76	0.0808	0.0831	0.0710
623	240	0.91	1.70	5.9928	3.9917	301.60	0.1260	0.1117	0.1112
673	280	0.93	1.66	6.7738	4.0208	307.00	0.0936	0.1062	0.0642
723	324	0.94	1.62	7.6292	4.0468	312.04	0.1178	0.1163	0.0907
773	373	0.96	1.58	8.5661	4.0704	316.81	0.0826	0.0996	0.0572
823	428	0.97	1.55	9.5921	4.0921	321.37	0.0949	0.0979	0.0764
873	489	0.99	1.52	10.7156	4.1123	325.78	0.0632	0.0774	0.0492
923	557	1.00	1.48	11.9458	4.1314	330.09	0.0693	0.0706	0.0627
973	633	1.02	1.45	13.2924	4.1495	334.35	0.0435	0.0517	0.0417
1023	719	1.03	1.42	14.8121	4.1662	338.21	0.0468	0.0446	0.0513
>1023	2250	1.23	0.95	40.0000	4.2000	370.00	0.0676	0.0485	0.1256

$$\langle P_c \rangle = P_0 \left\{ 1 + \frac{1}{w_c} \left(\frac{A}{B} \right)^{\frac{1}{B}} \left[\Gamma \left(1 + \frac{1}{B}, q_{i-1} \right) - \Gamma \left(1 + \frac{1}{B}, q_i(w_c) \right) \right] \right\} \tag{6.101}$$

where P_0, A, and B are specific for the petroleum fluid under study, and for each property. For instance, from Chap. 5, we find that $T_0 = 450.04$, $A_T = 2.254$, and $B_T = 1.888$ using data from MacKay River Bitumen[53]. Similarly, we find that $M_0 = 135.24$, $A_M = 7.159$, and $B_M = 1.042$ using data from Xinjian Extra Heavy Oil [47]. Once weights fractions have been calculated for each pseudocomponent $T_b \leq 1023$, then the weight fraction for the asphaltene pseudocomponent $T_b > 1023$ is obtained from the mass balance constraint as follows:

$$w_{>1023} = 1 - \sum_{j=1}^{12} w_j \tag{6.102}$$

Finally, the corresponding mole fractions are calculated using

$$\langle x_c \rangle = \frac{\frac{w_c}{M_c}}{\sum_{j=1}^{13} \frac{w_j}{\langle M_j \rangle}} \tag{6.103}$$

In the Table 6.10, mole fractions for three heavy petroleum fluids are shown.

Figures 6.25 and 6.26 illustrate a comparison between experimental [45–48] and calculated solubility of gases and light hydrocarbons using GP from Table 6.10. The solubility calculations, where the sPC-SAFT equation of state [42] with $k_{ij} = 0$

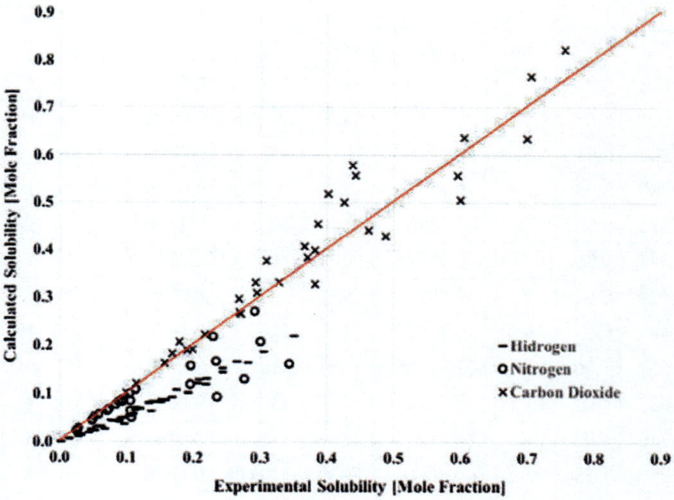

Fig. 6.25 Comparison of experimental [45–47] and calculated solubility of gases in Heavy Oils using GP from Table 6.10 and the sPC-SAFT equation of state $(k_{ij} = 0)$

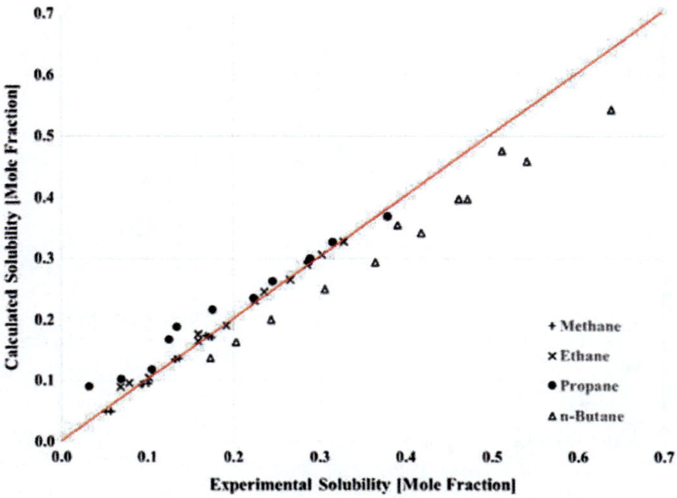

Fig. 6.26 Comparison of experimental [48] and calculated solubility of light hydrocarbons in Bitumen using GP from Table 6.10 and the sPC-SAFT equation of state $(k_{ij} = 0)$

is used, are performed assuming vapor-liquid systems according to Fig. 6.5. To assess the prediction capabilities of the model (thermodynamic characterization and equation of state), the binary interaction parameters are set equal to zero, that is, $k_{ij} = 0$. In this sort of diagrams, if experimental and calculated values coincide with each other (an exact prediction), then all points lie on the straight line with slope and intercept equal to one and zero, respectively (red line). As observed, better predictions are obtained for light hydrocarbons than for gases. This is an expected result because it has been illustrated that $k_{ij} \neq 0$ is required for asymmetric systems composed of heavy hydrocarbons and gases [54]. From a quantitative point of view, the errors can be calculated as the Absolute Average Deviation percent (AAD %), which is defined as

$$AAD\% = \frac{100}{ND} \sum_{k=1}^{ND} \left| \frac{x_k^{Exp} - x_k^{Cal}}{x_k^{Exp}} \right| \tag{6.104}$$

where x stands for the solvent solubility expressed as mole fraction, and *ND* for the number of data. The superscripts *Exp* and *Cal* indicate an experimental and calculated value, respectively. Thus, the smaller AAD%, the better the prediction. In Fig. 6.25, the AAD% are 45.4, 16.8, and 10.3 for hydrogen, nitrogen and carbon dioxide respectively, while in Fig. 6.26 the AAD% are 4.6, 6.2, 30.2 and 16.0 for methane, ethane, propane and n-butane, respectively. We may establish an AAD% smaller than 5% for calculations to be considered in a good agreement with experiments; hence, adjustment of k_{ij} is not suggested for methane in this case.

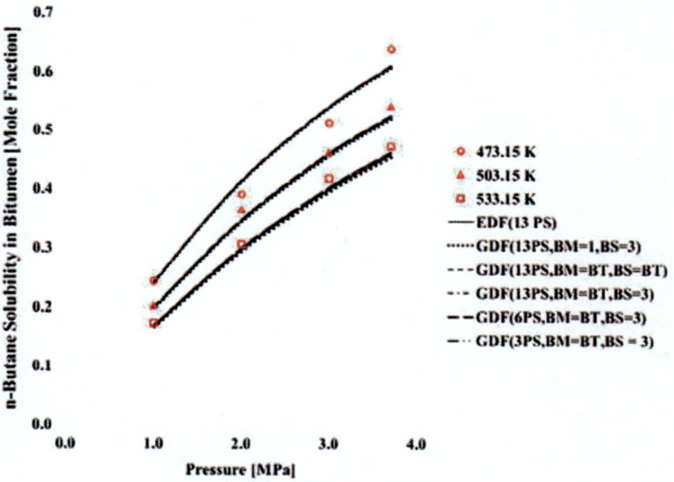

Fig. 6.27 Effect of the distribution function (GDF and EDF) on solubility. Characterization from GFEM, and phase equilibrium calculations using the sPC-SAFT equation of state $\left(k_{ij} = 0\right)$. Data from Haddadnia et al. [48]

On the other hand, specific pseudocomponents may be required for a given petroleum fluid. In this case, one has to select a proper distribution function and its corresponding parameters in accordance with the available experimental information. After, it is necessary to establish the number of pseudocomponents according to the considered application, so that the desired accuracy with experiments is not lost. Fortunately, likewise the phase envelopes, the solvent solubility is a weak function of all these variables for larger solvents than n-propane, especially when thermodynamic characterization is performed by the Gibbs free energy minimization (GFEM) of pseudocomponents. For instance, Fig. 6.27 shows the n-butane solubility in MacKay River Bitumen as function of GDF. The adjustment of GDF to the simulated Distillation is shown in Chap. 5; the calculated parameters are $T_0 = 435.04$, $A_T = 2.531$, and $B_T = 2.207$. To carry out GFEM, it is necessary that molecular weight and specific gravity of each proposed pseudocomponent are available. This means that, for the bitumen in Fig. 6.27, the distribution of such properties should be available as well. In this case, however, the bulk properties ($S = 1.0004$ and $M = 513$) are the only available data; therefore, the distributions functions have to be calculated as described in Chap. 5. These calculations need the knowledge of S_0 and M_0, that is, the initial values of the property distribution, which are unknown variables generally. Assuming that T_0 corresponds to the normal boiling temperature of an n-alkane (T_b), one may take M_0 to be equal to the molecular weight of such an n-alkane. Often, there is not an n-alkane component, whose T_b is equal to T_0 exactly. Thus, it is suggested to take the n-alkane component, whose T_b is the closest and largest to T_0. For this bitumen, this corresponds to the n-decane component with $T_b = 447.3$, and therefore, one may assumed that $M = M_0 = 142.27 \approx 143$. Finally, with M_0 and T_0, S_0 can be iteratively calculated by using the Lee-Kesler correlation. In Fig. 6.27, GDF is split out into three, six and thirteen pseudocomponents by using the distribution functions so calculated. Besides, GDF with $B_M = 1$ and the Exponential Distribution Function (EDF) are also included. The solubility calculations are performed in the same way as with GP. As observed, the selected distribution functions have little impact on the n-butane solubility, obtaining a maximum and minimum value of AAD equal to 4.1%, and 2.8%, respectively. Finally, it is worth highlighting that this example illustrates the prediction capability of the chemical thermodynamics for this sort of systems.

To compare the outcomes by using GP with those obtained by using specific pseudocomponents, the characterization of the three heavy petroleum fluids from Table 6.10 is now shown in Table 6.11, but using EDF with thirteen pseudocomponents as well. GFEM is carried out by using the PC-SAFT equation of state[41] with group contribution parameters from Burgess et at. [40], as described in Chap. 4. As before, the solubility calculations are performed according to Fig. 6.5 with the sPC-SAFT equation of state and $k_{ij} = 0$. For gases and light hydrocarbons, the outcomes are depicted in Figs. 6.28 and 6.29, respectively. The AAD % are 26.4, 27.2, and 25.6 for hydrogen, nitrogen and carbon dioxide, while the AAD% are 21.5, 29.2, 57.9 and 3.5 for methane, ethane, propane and n-butane, respectively. As observed, larger AAD are obtained by using specific pseudocomponents than by using GP for nitrogen, carbon dioxide, methane, ethane, and propane. Therefore, GP are suggested to estimate solubility of lighter components than n-butane with the exception of hydrogen. Accordingly, specific pseudocomponents should be used for hydrogen, n-butane and heavier solvents.

As mentioned before, predictions may require an adjustment of either petroleum-fluid properties, equation-of-state parameters, or both in order to improve the accuracy of the results. In the latter case, the binary interaction parameters are the most logical option because, they are the evident deficiency of equations of state in general. For the solvent solubility illustrated in the previous examples, a temperature-dependent expression is suggested again as illustrated in Figs. 6.30 and 6.31. In these figures, the AAD is about or less than 5%, except for the propane whose AAD is about 19%. For the methane and n-butane solubility, adjustments are not needed, as demonstrated before. Finally, the liquid density of the nitrogen-and carbon dioxide-bitumen systems are illustrated in Figs. 6.32 and 6.33 with AAD of 0.3 and 1.0%, respectively. These outcomes correspond to predictions, where no adjustments are performed in order to match experimental densities. One advantage of the specific and generalized pseudocomponents as used above is that the liquid-density calculations are a weak function of the adjustment procedure of the solvent solubility. In other words, liquid-density data are not required to be included in the adjustment procedure.

Table 6.11 Characterization of heavy petroleum fluids using EDF with 13 PS and GFEM

MacKay River (S = 1.0004 and M = 513) [53]							Athabasca Bitumen (S = 1.009 and M = 522) [45]							Xinjian Oil (S = 1.0352 and M = 574.5) [47]						
x	M	S	HCR	m	σ[Å]	ε/k[K]	x	M	S	HCR	m	σ[Å]	ε/k[K]	x	M	S	HCR	m	σ[Å]	ε/k[K]
0.2125	185.3	0.873	1.78	4.5323	3.9794	275.13	0.2046	196.3	0.886	1.75	4.7458	3.9842	278.74	0.2318	190.6	0.893	1.74	5.0770	3.8922	271.94
0.1470	268.0	0.915	1.69	6.6026	3.9443	279.52	0.1457	275.5	0.924	1.66	6.7332	3.9446	280.92	0.1489	296.7	0.943	1.62	7.8519	3.8647	275.61
0.1149	342.7	0.941	1.63	8.4221	3.9279	281.59	0.1153	348.4	0.950	1.61	8.4872	3.9296	282.94	0.1137	388.8	0.973	1.55	10.1570	3.8582	278.29
0.0947	415.8	0.961	1.58	10.1558	3.9197	283.47	0.0956	420.2	0.969	1.56	10.1714	3.9220	284.78	0.0925	477.7	0.995	1.50	12.3246	3.8561	280.52
0.0803	490.4	0.979	1.54	11.8888	3.9151	285.22	0.0813	493.6	0.986	1.52	11.8618	3.9175	286.51	0.0778	567.9	1.013	1.46	13.6199	3.9460	294.05
0.0692	568.9	0.994	1.50	13.0254	3.9851	295.60	0.0703	571.0	1.002	1.48	12.9554	3.9886	297.34	0.0667	662.2	1.030	1.42	15.7043	3.9477	296.97
0.0603	653.7	1.008	1.47	14.8150	3.9869	298.27	0.0613	654.8	1.016	1.45	14.7097	3.9902	300.02	0.0578	763.9	1.045	1.38	17.9219	3.9494	299.78
0.0527	747.8	1.023	1.44	16.7800	3.9886	300.94	0.0537	747.9	1.030	1.42	16.6384	3.9918	302.69	0.0504	876.7	1.060	1.35	20.3526	3.9509	302.57
0.0460	856.1	1.037	1.40	19.0120	3.9905	303.70	0.0470	855.0	1.045	1.38	18.8315	3.9933	305.47	0.0439	1006.1	1.075	1.31	23.1107	3.9524	305.44
0.0399	986.2	1.052	1.37	21.6639	3.9923	306.68	0.0408	983.8	1.060	1.35	21.4395	3.9950	308.46	0.0380	1161.5	1.091	1.28	26.3853	3.9538	308.52
0.0341	1153.7	1.069	1.33	25.0374	3.9943	310.09	0.0349	1149.7	1.076	1.31	24.7599	3.9967	311.90	0.0325	1361.4	1.109	1.24	30.5496	3.9552	312.04
0.0282	1399.0	1.090	1.28	29.9104	3.9966	314.43	0.0288	1392.8	1.098	1.26	29.5597	3.9986	316.28	0.0267	1653.9	1.131	1.18	36.5645	3.9566	316.50
0.0202	1952.4	1.127	1.19	40.6823	3.9995	322.35	0.0207	1941.5	1.135	1.17	40.1821	4.0010	324.27	0.0191	2313.0	1.170	1.09	49.8635	3.9577	324.59

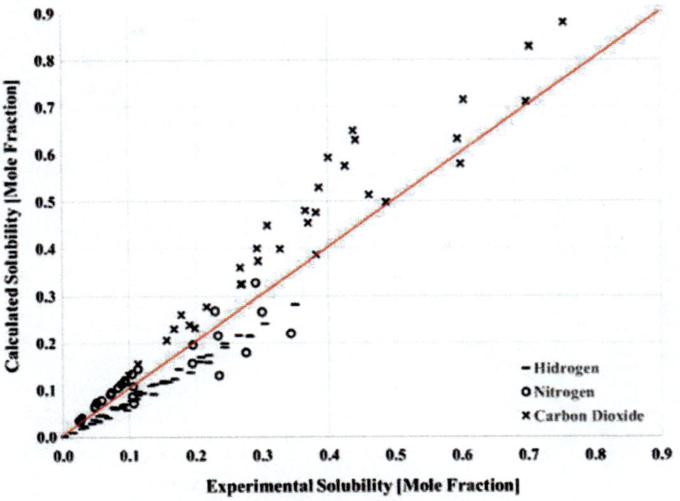

Fig. 6.28 Comparison of experimental [45–47] and calculated solubility of gases in Heavy Oils using Pseudocomponents from Table 6.11 and the sPC-SAFT equation of state $\left(k_{ij} = 0\right)$

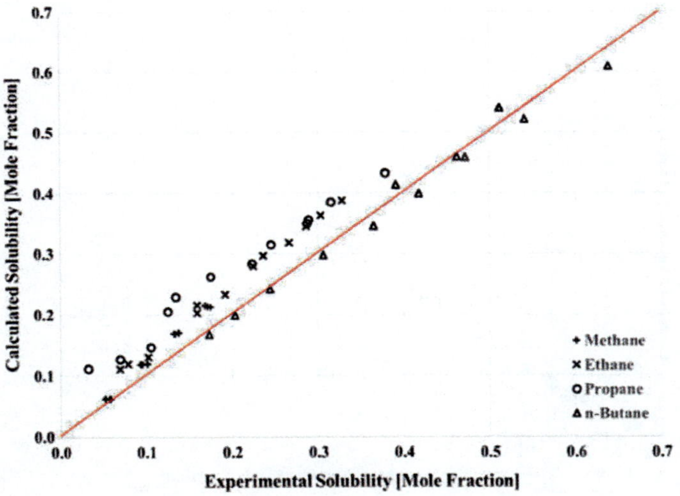

Fig. 6.29 Comparison of experimental [48] and calculated solubility of light hydrocarbons in Bitumen using Pseudocomponents from Table 6.11 and the sPC-SAFT equation of state $\left(k_{ij} = 0\right)$

6.6.3 Cubic Equations of State

The characterization procedure using correlations for estimating cubic-equation-of-state parameters are normally used [55]. As mentioned in Chap. 5, in addition to modify physical properties, binary interactions parameters from the classical mixing

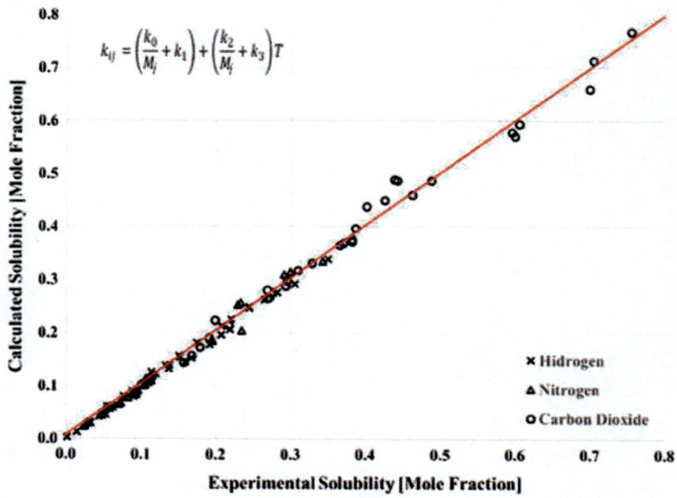

Fig. 6.30 Comparison between experimental [45–47] and calculated solubility of gases in Heavy Oils using the sPC-SAFT equation of state $\left(k_{ij} \neq 0\right)$

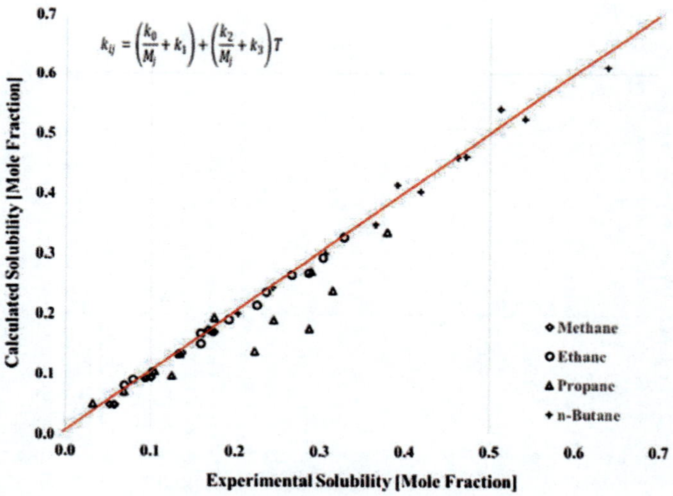

Fig. 6.31 Comparison between experimental [48] and calculated solubility of light hydrocarbons in Bitumen using the sPC-SAFT equation of state $\left(k_{ij} \neq 0\right)$

rule between solvent and pseudocomponents are usually optimized to fit the experimental saturation pressure. A generalized expression, as those shown in Chap. 3 with two adjustable parameters k_0 and k_1, is usually enough in order to obtain satisfactory outcomes at low temperatures[56]. For high temperatures, it has been illustrated that temperature-dependent binary interaction parameters are required generally [57].

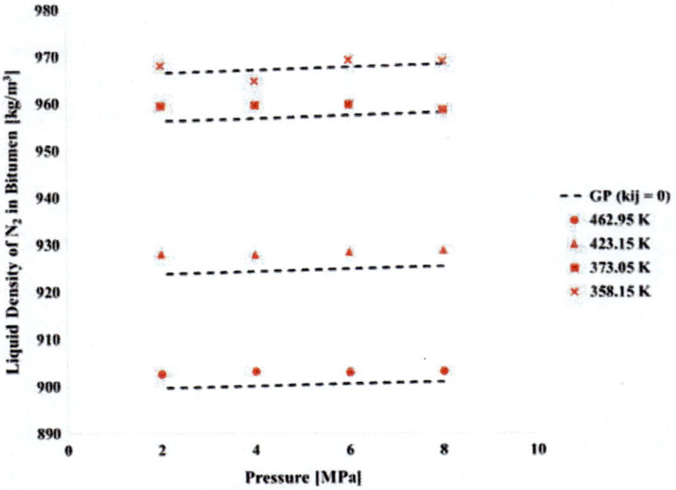

Fig. 6.32 Liquid Density of nitrogen-Bitumen mixtures. Data from Haddadnia et al. [46]

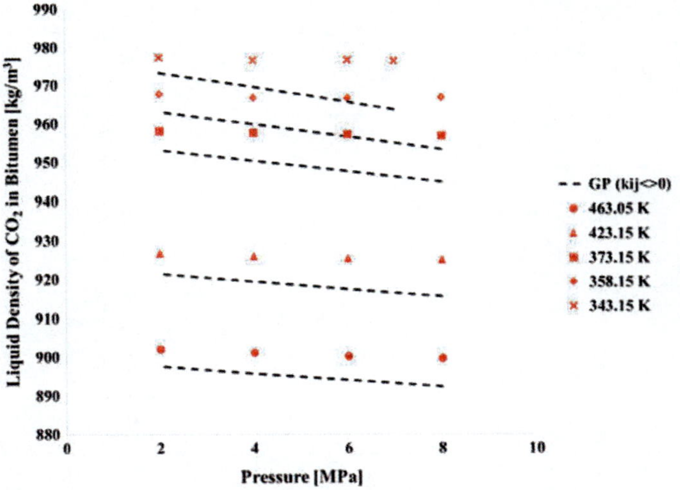

Fig. 6.33 Liquid Density of carbon dioxide-Bitumen mixtures. Data from Haddadnia et al. [46]

Alternatively, one may consider the use of mixing rules based on excess Gibbs energy models, for example, the MHV1 or PCES mixing rules used in calculating phase envelopes of reservoir fluids from Figs. 6.12, 6.13, 6.14, 6.15, 6.16, 6.17, 6.18, 6.19 and 6.20. As pointed out in Chap. 3, however, these mixing rules are not a suitable option here, because they give unsatisfactory results in predicting vapor-liquid equilibriums for asymmetric mixture such as those formed by solvent and heavy petroleum fractions. At the same time, it was shown that this disadvantage could be overcome by using the Universal Mixing Rule (UMR) in combination with the UNIFAC solution model or by using the so-called predictive Peng-Robinson equation of state (PPR78). For instance, Fig. 6.34 illustrates estimations of gas solubility of carbon dioxide, nitrogen, and methane in pure hydrocarbons as a function of

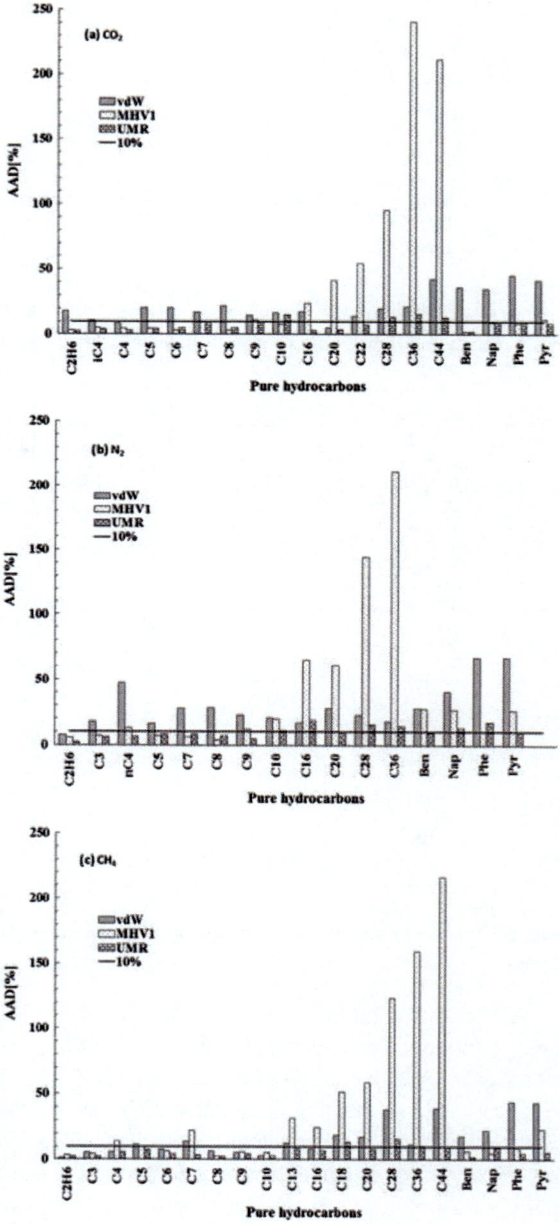

Fig. 6.34 Prediction of Gas Solubility in Pure Hydrocarbons (**a**) CO_2, (**b**) N_2 and (**c**) CH_4 [38]

Table 6.12 Interaction Parameters for the UNIFAC solution model in combination with the UMR mixing rule and the PR equation of state

Gas		α_{nm},	β_{nm}	χ_{nm}	α_{mn},	β_{mn}	χ_{mn}
CO_2	$-CH_2$	−38.67*	8.615E-01*	−1.791E-03*	919.80*	2.193E+02*	4.631E-03*
CH_4		68.14	−0.74	3.410E-05	−39.10	8.457E-01	−3.721E-04
N_2		3.75	0.33	−1.286E-03	271.12	−9.902E-01	1.568E-03
CO_2	ACH	−29.40*	0.00*	0.00*	−3.91*	0.00*	0.00*
CH_4		−42.11	138.92	0.00	180.01	−1.37	0.00
N_2		−19.21	−1.01	0.00	222.56	2.10	0.00

*Horstmann original values [58]

three mixing rules; the used interaction parameters for the UNIFAC solution model are shown in Table 6.12. As seen from this figure, the larger the asymmetry of the gas-hydrocarbon mixture, the larger the errors for the three gases when the MHV1 mixing rules is used, obtaining acceptable estimations for non-asymmetric mixtures and in the presence of aromatic components, except for mixtures involving nitrogen. Besides, it can be observed that the classical mixing rule with the Tsonopoulos correlation gives the largest errors generally. Thus, we have that, in this example, the UMR mixing rule appears as the suitable option for asymmetric mixtures in the presence of aromatic components, such as those present in mixtures of solvent and heavy petroleum fractions.

Figure 6.35 shows the comparison of gas solubility in different hydrocarbons including the pure components from Fig. 6.34, and petroleum fractions from Table 6.13. It is worth pointing out that the UNIFAC interactions parameters entering

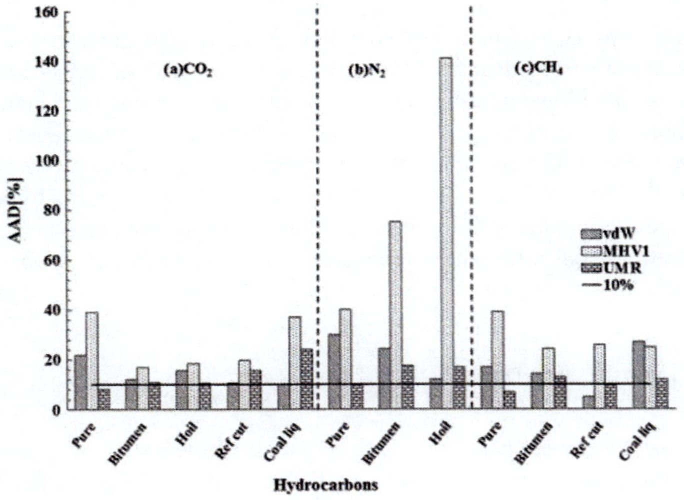

Fig. 6.35 Comparison of gas solubility in different hydrocarbons [38]

Table 6.13 Experimental Properties of Petroleum Fractions [38]

Petroleum fraction	MW	Density [kg/m^3]	HCR
Athabasca Bitumen	552	1007.0	
Mackay River Bitumen	513	1000.4	
Surmont Bitumen	539	1010.0	
Wabasca Bitumen	447	1007.0	
Peace River Bitumen	527	1018.1	
Cold Lake Bitumen	533	1060.9	
Heavy Oil	575	1035.2	
Heavy OilA	350	985.2	
Heavy Oil B	236	889.5	
Heavy Oil D	358	978.6	
Heavy Oil E	463	1011.0	
Heavy Oil G	373	931.9	
Heavy Oil H	458	1022.3	
Heavy Oil I	370	979.3	
EXXON A1	224	903.0	1.70
EXXON A5	248	930.0	1.65
EXXON B1	199	917.0	1.63
EXXON B4	232	917.0	1.59
CLPPA5(400–450)	154	965.0	1.29
CLPPA5(500–600)	182	984.0	1.38
WCLP CUT8	163	961.0	1.43
WCLP CUT7	185	984.0	1.38

into the MHV1 and UMR mixing rules are those values shown in Table 6.12. In other words, parameters from identifiable components were also used with pseudocomponents without additional adjustments. Once petroleum fractions were split out into pseudocomponents according to GDF using the experimental information available in the same Table 6.13, the cubic-equation-of-state parameters were determined by using GFEM. Figure 6.35 illustrates that predictions of gas solubility in petroleum fractions are comparable with those predictions of identifiable components as long as a suitable thermodynamic characterization is available for the petroleum fractions.

6.6.4 Asphaltenes Onset

Asphaltenes is the most heaviest and polarizable petroleum fraction; polynuclear aromatic rings, aliphatic rings and chains, heteroatoms (nitrogen, sulfur, etc.) and metals (vanadium, nickel, etc.) form them. From the practical point of view, asphaltenes are conceived as a set of components insoluble in low-molecular-weight alkanes such as n-pentane and n-heptane, and soluble in aromatic components such as benzene and toluene. However, the quantity and quality of asphaltene fraction depend not only on the used solvent for its separation from the petroleum fluid (the asphaltene onset), but also on the volume of the solvent. For instance, its molecular weight may lie in the interval between 1000 and 2500 [59–61]; however, a fixed value between 1700 and 1800 is typically taken in calculations. Actually, the asphaltene onset is nothing more than a saturation point like the vapor-liquid saturation points on the phase envelopes shown before. Therefore, the asphaltene onset indicates that the solvent-heavy petroleum system cannot exist by itself as a single liquid phase; the presence of the other liquid phase is required for the system to be thermodynamically stable. Accordingly, the asphaltene precipitation is treated as liquid-liquid (solvent-rich and asphaltene-rich phases) phase separation from calculations point of view.

When the asphaltene onset is intended to be predicted, the heavy petroleum fluid is experimentally characterized through a SARA-compositional analysis formed by Saturates, Aromatics, Resins, and Asphaltenes chemical families. Numerically, aromatics and resins, whose molecular weights are about 500 and 800 approximately, are usually combined into a single pseudocomponent because both have polar-polar interactions, yielding thus a three-pseudocomponent mixture. The characterization procedure is basically the same as before, because the asphaltene onset is a saturation point of the same solvent-heavy petroleum system as explained already. Therefore, there are two options for the asphaltene onset calculations to match experiments; that is, adjust the properties of the heavy petroleum fluid and its pseudocomponents, and/or adjust the equation-state-parameters such the binary interaction parameters. In case of the PC-SAFT equation of state, it is common that the non-associating properties of the asphaltene fraction are fitted to match the measured onset, keeping constant the equation-of-state parameters. If the vapor-liquid-liquid equilibriums, where liquid-liquid equilibrium corresponds to the asphaltene onset, are intend to be determined, the aromatic-resin properties are fitted to match the experimental values of the liquid density and bubble point at standard conditions. This can be performed by using the aromaticity parameter as explained in the Chap. 4. Once such an aromaticity is established, then the PC-SAFT parameters of the asphaltene fraction are adjusted to match the experimental onset [62–64].

On the other hand, the adjustment of the equation-of-state parameters is common with CEoS, especially temperature-dependent binary interaction parameters of the van der Waals mixing rule [56, 57]. When asphaltene precipitation is intend to be described by using the so-called the Cubic-Plus-Association equations of state (CPA), the corresponding association properties of asphaltene fractions are fitted to match the measured onset as well, keeping constant the equation-of-state parameters [65–67]. The CPA equations of state, as will be explained later, are a combination of CEoS and an association term, which accounts explicitly for effects of hydrogen bonding in either pure components or mixtures.

When the SARA-compositional analysis is not available for the considered petroleum fluid, one may make use of GFEM in order to estimate the asphaltene onset. For instance, the characterization of WC-B-B2 Bitumen, whose experimental

Table 6.14 Characterization of WC-B-B2 Bitumen using GDF ($M_0 = 115.73$, $B_M = 1$, and $B_S = 3$) and GFEM

PS	x	Tb[K]	M	S	HCR	m	σ[Å]	ε/k [K]
1	0.4232	569.62	182.5	0.933	1.645	4.552	3.921	279.80
2	0.2308	677.01	334.6	0.976	1.545	8.593	3.870	279.92
3	0.1470	750.84	525.5	1.004	1.480	12.696	3.945	292.33
4	0.0987	821.27	782.3	1.030	1.419	18.680	3.943	296.36
5	0.0656	901.96	1177.6	1.059	1.352	27.623	3.944	301.43
6	0.0347	1043.50	2225.0	1.108	1.237	50.421	3.949	310.96

and calculated Simulated Distillation [68] are depicted in Chap. 5, is shown in Table 6.14. In addition to the Simulated Distillation, the only bulk property available is the specific gravity ($S = 1.015$); hence, the corresponding M must be estimated. From the Simulated Distillation, we have that $T_b = 743.72$ K as the expected value. Using these values of S and T_b for the whole bitumen, we estimate M from both the Lee-Kessler and Soreide correlations, choosing the largest value as M. In this case, $M = 463.42$ from the Soreide correlation. Then, following the procedure explained before for the solvent solubility, the properties pseudocomponents are calculated. However, unlike the calculation of the solvent solubility, the asphaltene onset is a strong function the proposed pseudocomponents. It is suggested to choose a number of pseudocomponents so that the molecular weight of the heaviest two pseudocomponents is larger than the lower limit of 1000. This means that the unidentifiable petroleum fractions should be split out into five or six pseudocomponents at least. If this condition is not fulfilled, then the heaviest components are too light in order to predict a liquid-liquid equilibrium, as required on the asphaltene onset point; unless, the equation-of-state parameters or petroleum-fraction properties are modified.

As before, to split out the petroleum fraction into pseudocomponents by using a distribution function, the initial values of such distribution are needed. These initial values are estimated as before, but in this case, it is necessary to adjust M_0 in order to calculations match the experimental value of the asphaltene onset. Thus, for the example illustrate in Table 6.14, $T_0 = 413.68$, with which the n-Nonane component ($T_b = 447.3$) is chosen in order to start the iterative calculations. This means that M_0 is first set equal to 128.25. Thus, it is found that $M_0 = 115.73$ in order to calculations match the asphaltene onset for n-pentane-bitumen mixtures at 296.15 K, as can be seen from Fig. 6.36. During calculations, the corresponding S_0 is estimated using T_0 and M_0 through the Lee-Kessler correlation, while the non-associating parameters of the PC-SAFT equation of state are calculated through GFEM by using the group contribution parameters from Burgess et at. [40], as in the case of the calculations of the solvent solubility described before. The corresponding vapor-liquid-liquid equilibriums for n-pentane-bitumen mixtures are depicted from Fig. 6.36, 6.37, 6.38, 6.39 and 6.40. The experimental asphaltene onset and saturation pressures of vapor-liquid equilibriums, when they are available, are also included in these figures. It is worth pointing out that the experimental asphaltene onset at a single temperature

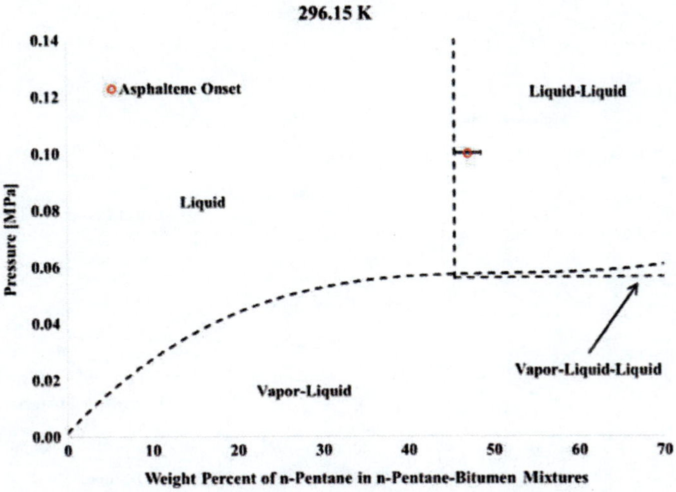

Fig. 6.36 Vapor-Liquid-Liquid Phase Diagram for n-Pentane-Bitumen at 296.15 K, using the sPC-SAFT equation of state $\left(k_{ij} = 0\right)$ and pseudocomponents from Table 6.14. Data from Agrawal et al. [57] and Johnston et al. [68]

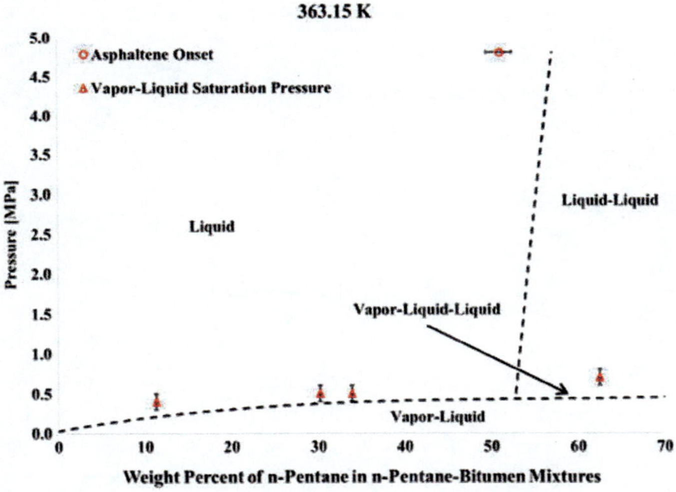

Fig. 6.37 Vapor-Liquid-Liquid Phase Diagram for n-Pentane-Bitumen at 363.15 K, using the sPC-SAFT equation of state $\left(k_{ij} = 0\right)$ and pseudocomponents from Table 6.14. Data from Agrawal et al. [57] and Johnston et al. [68]

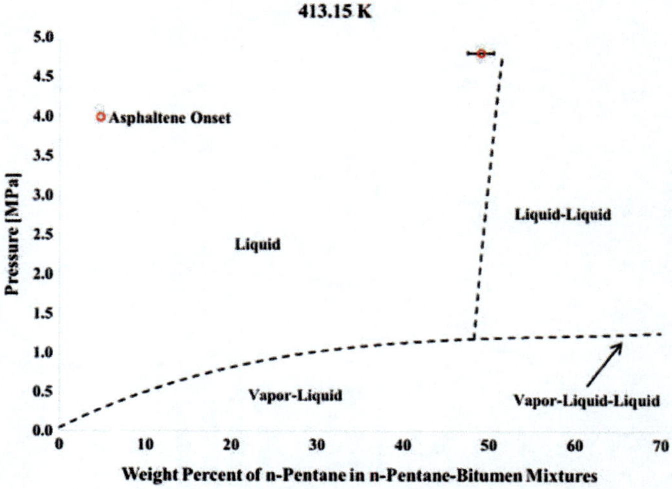

Fig. 6.38 Vapor-Liquid-Liquid Phase Diagram for n-Pentane-Bitumen at 413.15 K, using the sPC-SAFT equation of state $(k_{ij} = 0)$ and pseudocomponents from Table 6.14. Data from Agrawal et al. [57] and Johnston et al. [68]

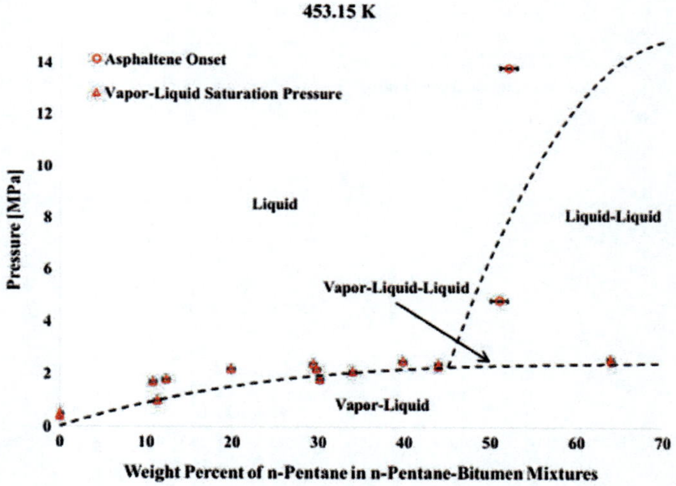

Fig. 6.39 Vapor-Liquid-Liquid Phase Diagram for n-Pentane-Bitumen at 453.15 K, using the sPC-SAFT equation of state $(k_{ij} = 0)$ and pseudocomponents from Table 6.14. Data from Agrawal et al. [57] and Johnston et al. [68]

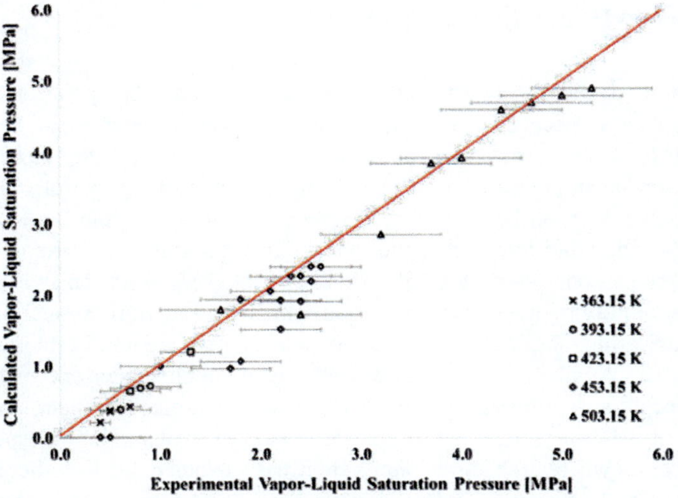

Fig. 6.40 Saturation pressures of n-pentane-bitumen mixtures, using the sPC-SAFT equation of state $\left(k_{ij}=0\right)$ and pseudocomponents from Table 6.14. Data from Agrawal et al. [57] and Johnston et al. [68]

(296.15 K) is the only experimental incipient point used for regression purposes. The remaining ones, asphaltene onsets and saturation pressures, are predictions of the proposed thermodynamic model. From a qualitative point of view, these results compares well with those obtained by using CEoS when k_{ij} are adjusted to minimize the error between calculations and experiments [68, 69]. This might indicate that the results could be improved by performing similar adjustments.

As discussed in the Chap. 3, one the major disadvantage of CEoS is that they are unable to handle asymmetric mixtures properly for phase equilibrium calculations. For saturation pressures, asphaltene onset and yield of asymmetric mixtures formed by n-pentane and bitumen, Johnston et al. [69] assessed several mixing rules, including EoS/G^E-type mixing rules such as the Huron-Vidal mixing rule. It was found that temperature- and composition-dependent k_{ij} are needed for the classical mixing rule of van der Waals in order to match calculations with experiments. Thus, the Universal Mixing Rule shown in Chap. 3 appears as a promising option for such asymmetric systems. However, its use has not been assessed for the asphaltene onset and yield predictions yet. Therefore, the PC-SAT equation of state seems to be the suggested option in such calculations for the time being.

From the shown calculations and other analysis [65, 66, 68, 69], it may be asserted that asphaltene precipitation is mostly caused by the compositional and pressure changes. In general, the asphaltene solubility increases as the carbon number of the n-alkane increases, while an increasing pressure leads to a higher solubility of the asphaltene; temperature does not play a significant role. A higher solubility means that more solvent is needed for the asphaltene onset point to appear.

6.7 Water-Heavy Petroleum Fluids

The solubility of water in hydrocarbons increases as hydrocarbon polarity increases, which is expected because of the polar nature of water. In general, the heavier the petroleum fluid, the larger the content of asphaltene, which is the more polarizable fraction, resulting thus in an increasing tendency of the petroleum fluid to be polarizable. Hence, the solubility of water in heavy petroleum fluids turns out to be higher than that found in lighter ones, where paraffinic and aromatic chemical families predominate instead of resins and asphaltene ones. This is the reason why the solubility of water has been usually neglected in conventional petroleum fluids, considering thus that both aqueous and oleic phases are immiscible each other. However, the solubility of water increases when temperature increases even for paraffinic hydrocarbons as illustrated in Fig. 6.41. Therefore, unlike light petroleum fluids, the solubility of water should not be ignored in heavy petroleum fluids, especially in EOR processes where high temperatures are usually handled. On the other hand, the solubility of hydrocarbons in water in water-heavy petroleum systems is generally ignored, because the main concern is about its effect on the petroleum properties, leading to unavailable experimental information.

The increase in the solubility of water means that all resultant thermophysical properties of the water-heavy petroleum system differ from those of the heavy petroleum fluid alone. For instance, it has been found that the dissolution of water in bitumen decreases density and viscosity of the oleic phase (bitumen-rich phase) under liquid-liquid equilibrium conditions [73]. As shown in Fig. 6.42, this viscosity reduction may be significant: as much as 21% at 470 K.

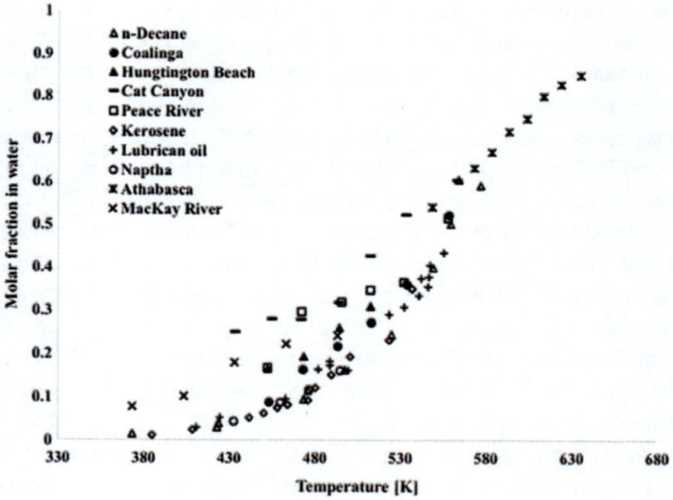

Fig. 6.41 Experimental solubility of water in heavy petroleum fluids [53, 70–72]

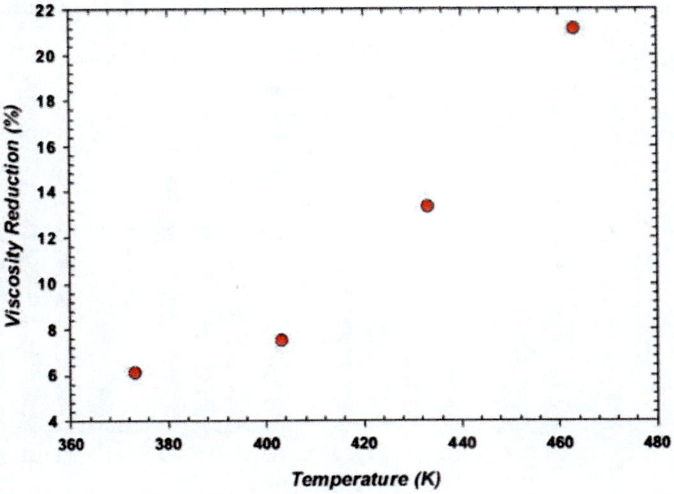

Fig. 6.42 Effect of the solubility of water in MacKay River bitumen on its viscosity [73]

On the other hand, the presence of water in solvent-heavy petroleum systems modify the solvent solubility in the heavy petroleum because of the dissolution of water in heavy petroleum- and solvent-rich phases, the oleic liquid and vapor phases respectively. In other words, these dissolutions modify the equilibrium between solvent and heavy petroleum fluid with respect to the equilibrium in the absence of water, and therefore, the corresponding solvent solubility. As before, variations in density and viscosity of the oleic phase are generally expected. In this case, however, variations arise from the effect of both water and solvent solubility in the heavy petroleum fluid. For instance, Zirrahi et al. [73] measured the n-alkane solubility and its effects on density and viscosity of the oleic phase in the presence and absence of water as illustrated in Fig. 6.43. As observed, the presence of water reduces the solvent solubility: the larger the solvent solubility, the larger the reduction of such a solubility. Thus, we have that the effect of the presence of water is more significant for propane and butane than for methane and ethane. Besides, their results show that the effect of water on solvent solubility is larger at higher temperatures because of the higher temperature, the larger the solubility of water in the oleic phase. Concerning pressure, they concluded that the effect of water on solvent solubility decreases as pressure increases. This finding may be explained by the fact that the solubility of n-alkanes, which are more volatile components than water, is a stronger function of pressure than the solubility of water. As mentioned before, changes of solvent solubility lead to changes of density and viscosity of the oleic phase. Thus, for the solubility of the n-butane illustrated in Figs. 6.43, 6.44 and 6.45 show the corresponding effects on density and viscosity as a function of temperature and pressure. As observed, the change of n-butane solubility because of the presence of water increases both density and viscosity of the oleic phase with respect to their values in the absence of water. The reason for this result is that the presence of water leads

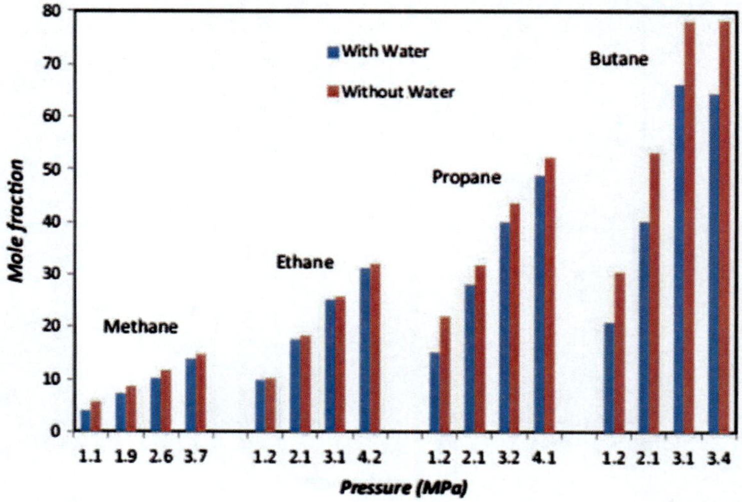

Fig. 6.43 Solubility of n-alkane solvents in MacKay River bitumen in the presence and absence of water at 423 K [73]

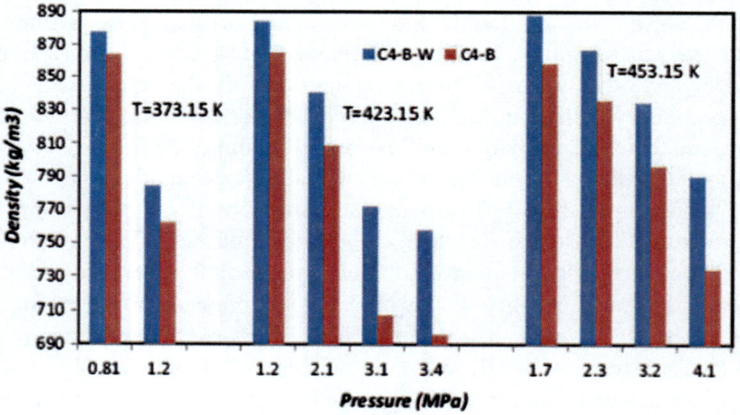

Fig. 6.44 Density of oleic phase saturated with butane in absence (C_4-B) and presence of water (C_4-B-W) at 373.15, 423.15 and 453.15 K [73]

to a more diluted oleic phase with respect to n-butane, resulting in a heavier oleic phase. In other words, it could be said that n-butane is somehow replaced by water, which is a denser component at the same temperature and pressure. The viscosity is a strong function of the density and therefore similar trends are obtained.

On the other hand, Chen and Yang analyzed water-solvent-heavy petroleum phase equilibriums for a wide variety of systems [74]. Their main findings can be summarized as follows:

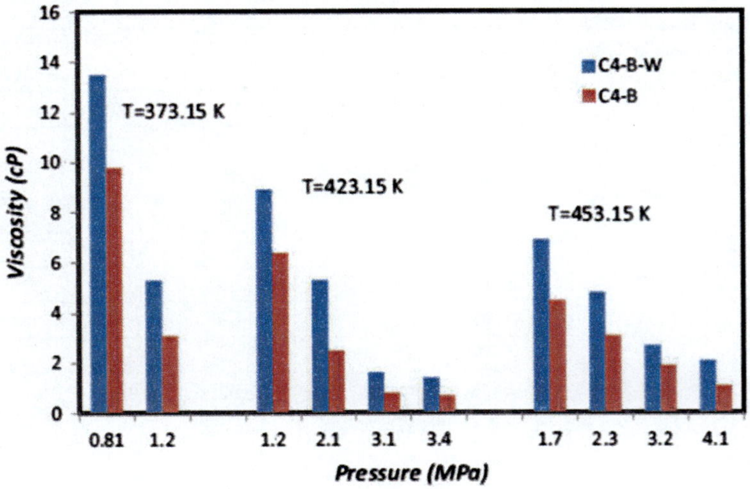

Fig. 6.45 Viscosity of oleic phase saturated with butane in absence (C_4-B) and presence of water (C_4-B-W) at 373.15, 423.15 and 453.15 K [73]

- The dissolution of water can modify the phase boundaries for solvent-heavy petroleum systems, affecting solvent solubility in heavy petroleum, especially at high temperatures. For instance, the dissolution of water enhances the carbon dioxide solubility, reducing alkane solubility, while it can significantly reduce the toluene solubility.
- The dissolution of water is the fundamental significance for solvent solubility in heavy petroleum on Vapor-Liquid equilibrium regions but no effects are shown on Vapor-Liquid-Aqueous equilibrium regions. The effects of pressure on solvent solubility in heavy petroleum are much larger on Vapor-Liquid-Aqueous equilibrium regions than on Vapor-Liquid equilibrium regions. Addition of solvents can enhance the solubility of water in heavy petroleum, while the enhancement is increased with carbon number of alkanes.

6.7.1 Compressibility and Fugacity Factors for Associating Systems

In the Chap. 4, it is shown that the PC-SAFT equation of state is derived from an general expression of the molar residual Helmholtz free energy $\left(\tilde{A}^R/RT\right)$, which is composed of two contributions: the hard-sphere-chain contribution $\left(\tilde{A}^{hsc}/RT\right)$ and the dispersion contribution $\left(\tilde{A}^{dsp}/RT\right)$. As have been shown by all the studied previously systems until now, these two contributions have proved to be enough for non-associating system, that is, for systems where hydrogen bonds are absent in the

Table 6.15 Bond Dipole Moments for selected pure components [76]

Component	Formula	Dipole [Debye]
Methane	CH_4	0.0
Carbon Dioxide	CO_2	0.0
Nitrogen	N_2	0.0
Hydrogen Sulfide	H_2S	0.9
n-Octane	C_8H_{18}	0.0
Cyclohexane	C_6H_{12}	0.3
Benzene	C_6H_6	0.0
Toluene	C_7H_8	0.4
1-Methylnaphthalene	$C_{11}H_{10}$	0.5
Diphenylmethane	$C_{13}H_{12}$	0.4
1,4-Diphenylbenzene	$C_{18}H_{14}$	0.7
Eicosane	$C_{20}H_{42}$	0.0
Water	H_2O	1.8

molecular interactions. Indeed, some of the forming components of petroleum fluids can present hydrogen bonding such as asphaltenes; consequently, some methods have treated the petroleum mixtures as associating systems [64, 66, 72]. Water, however, is a component with a substantial dipole as observed in Table 6.15. As illustrated in Fig. 6.46, this dipole leads to strong electrostatic interactions between oxygen and hydrogen atoms from different water molecules.

Irrespective of the used equation of state either the CEoS or sPC-SAFT, the association contribution due to the presence of hydrogen bonding is rather similar. Hence, we use the sPC-SAFT equation of state in order to introduce such a contribution, which is incorporated to $\tilde{A}^R/RT$ merely as one additional term, that is,

$$\frac{\tilde{A}^R}{RT} = \frac{\tilde{A}^{hsc}}{RT} + \frac{\tilde{A}^{dsp}}{RT} + \frac{\tilde{A}^{ass}}{RT} \tag{6.105}$$

where the molar Helmholtz free energy of association $\tilde{A}^{ass}/RT$ is [77–80]

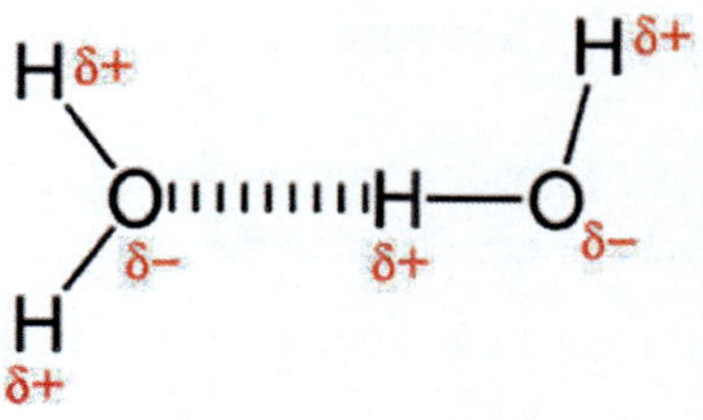

Fig. 6.46 Hydrogen Bond (IIIIIIII) between two water molecules (H_2O). The symbol δ indicates slightly electrostatic: positive (+) and negative (−)

$$\frac{\tilde{A}^{ass}}{RT} = \sum_{i=1}^{C} x_i \sum_{A_i} \left[ln(X_{A_i}) - \frac{X_{A_i}}{2} \right] + \frac{AS}{2} \tag{6.106}$$

AS stands for the number of association sites on each molecule, and $\sum_{A_i}$ represents the sum over all associating sites on the molecule. X_{A_i} corresponds to the mole fraction of molecules from a component i not bonded at site A, and x_i is the corresponding mole fraction. From the knowledge of X_{A_i}, one can estimate the monomer fraction X_1 of associating components either in a pure state or in a mixture. A monomer is a molecule of associating component, which does not undergo hydrogen bounding . To derive the association compressibility factor Z^{ass} corresponding to $\tilde{A}^{ass}/RT$, the following identity resulting from the fundamental equations of the thermodynamic is used.

$$\frac{p^{ass}}{RT} = -\left[\frac{\partial}{\partial \tilde{V}} \left(\frac{\tilde{A}^{ass}}{RT} \right) \right]_T \tag{6.107}$$

From which, it can be demonstrated that Z^{ass} is equal to [81]

$$Z^{ass} \equiv \frac{p^{ass}\tilde{V}}{RT} = -\frac{1}{2} \left\{ 1 - \tilde{V} \left[\frac{\partial ln(g)}{\partial \tilde{V}} \right]_{T,n} \right\} \sum_{i=1}^{C} x_i \sum_{A_i} (1 - X_{A_i}) \tag{6.108}$$

For the association fugacity coefficient ϕ_i^{ass} for the component i, we use another identity resulting from the fundamental equations, that is,

$$ln(\phi_i^{ass}) = -\left[\frac{\partial}{\partial n_i} \left(\frac{\tilde{A}^{ass}}{RT} \right) \right]_{T,P} \tag{6.109}$$

From which, it can also be demonstrated that [81]

$$ln(\phi_i^{ass}) = \sum_{A_j} ln(X_{A_j}) - \frac{1}{2} \left[\frac{\partial ln(g)}{\partial n_i} \right]_{T,\tilde{V},n_k} \sum_{j=1}^{C} x_j \sum_{A_j} (1 - X_{A_j}) \tag{6.110}$$

In the above expressions, g accounts for the hard sphere radial distribution function from the Chap. 4, that is,

$$g = g^{hs} = \frac{2 - \eta}{2(1 - \eta)^3} \tag{6.111}$$

For the sPC-SAFT equation of state, the reduced packing fraction η is

$$\eta = \frac{\pi}{6\tilde{V}} \tilde{m} d^3 \tag{6.112}$$

Thus, the needed derivatives for Z^{ass} and ϕ_i^{ass} in Eqs. (6.108) and (6.110) are then

$$\tilde{V}\left[\frac{\partial ln(g)}{\partial \tilde{V}}\right]_{T,n} = -\frac{5\eta - 2\eta^2}{(1-\eta)(2-\eta)} \tag{6.113}$$

$$\left[\frac{\partial ln(g)}{\partial n_i}\right]_{T,\tilde{V},n_k} = \left[\frac{\partial ln(g)}{\partial x_i}\right]_{T,\tilde{V},\underline{x}} = \frac{m_i d_i^3}{\tilde{m} d^3}\left[\frac{5\eta - 2\eta^2}{(1-\eta)(2-\eta)}\right] \tag{6.114}$$

As observed, except for X_{A_i}, the incorporation of the association contribution to the sPC-SAFT does not involve additional variables to those handled previously for non-associating mixtures. Therefore, the numerical strategies for solving phase equilibrium problems involving heavy petroleum fluids are not modified considerably, with exception of the initial estimates as explained in the multiphase equilibrium algorithm from Fig. 6.6. X_{A_i} is determined by solving iteratively the following expression

$$X_{A_i} = \left[1 + \frac{1}{\tilde{V}}\sum_{j=1}^{C} x_j \sum_{B_j} X_{B_j} \Delta^{A_i B_j}\right]^{-1} \tag{6.115}$$

For an associating pure component such a pure water, this expression reduces to

$$X_A = \left[1 + \frac{1}{\tilde{V}}\sum_{B} X_B \Delta^{AB}\right]^{-1} \tag{6.116}$$

$\tilde{V}$ is the molar volume of the system. $\Delta^{A_i B_j}$ is the association strength, which is the key property characterizing the molecules association due to hydrogen bonding. For the sPC-SAFT, $\Delta^{A_i B_j}$ is calculated from

$$\Delta^{A_i B_j} = N_A \sigma_{ij}^3 \kappa^{A_i B_j}\left(\frac{\pi}{6}\right)\left[e^{\frac{\varepsilon^{A_i B_j}}{kT}} - 1\right] g \tag{6.117}$$

N_A is the Avogadro number. The two parameters characterizing the hydrogen association are the cross-association energy $\varepsilon^{A_i B_j}$ and the cross-association volume $\kappa^{A_i B_j}$. When the original version of the PC-SAFT equation of state is intended to be used, the corresponding $\kappa^{A_i B_j}$ must be divided by $\pi/6$ for being used into the simplified version [82]. σ is the segment diameter parameter from the PC-SAFT equation of state; therefore, as shown in the Chap. 4, the cross parameter σ_{ij} is determined from

$$\sigma_{ij} = \frac{\sigma_i + \sigma_j}{2} \tag{6.118}$$

The cross-association energy is similarly estimated

$$\varepsilon^{A_iB_j} = \frac{\varepsilon^{A_iB_i} + \varepsilon^{A_jB_j}}{2} \tag{6.119}$$

However, concerning the cross-association volume, there have been suggested at least two options [82, 83]:

$$\kappa^{A_iB_j} = \sqrt{\kappa^{A_iB_i}\kappa^{A_jB_j}} \tag{6.120}$$

$$\kappa^{A_iB_j} = \sqrt{\kappa^{A_iB_i}\kappa^{A_jB_j}}\left[\frac{\sqrt{\sigma_{ii}\sigma_{jj}}}{\sigma_{ij}}\right]^3 \tag{6.121}$$

When there is only one associating component in the mixture such the mixtures studied here where the only associating component is water, then it is irrelevant the cross-associating expression because $\varepsilon^{A_iB_j} = \varepsilon^{A_iB_i}$, and $\kappa^{A_iB_j} = \kappa^{A_iB_i}$, that is, the associating parameters of a pure component are the only ones needed. According to this, five parameters have to be determined for each associating pure component in the mixture: the number of segments in a chain of component m, the segment diameter σ, the segment dispersion energy ε/k, the association energy $\varepsilon^{A_iB_i}/k$, and the association volume $\kappa^{A_iB_i}$. These parameters are determined by a data fit to experimental saturated pressures and liquid densities.

To apply the associating contribution to a specific system, it is necessary to establish the type of association sites, that is, the association scheme. Several association schemes have been proposed, which are dependent on the considered chemical family or pure component. According to the terminology suggested by Huang and Radosz [79], there are three schemes handled in the case of water: 2B, 3B, and 4C schemes, which are shown in Table 6.16. As observed, 2B scheme implies two association sites (A and B sites), whereas 3B scheme implies three association sites (A, B, and C sites), and 4C scheme implies four association sites (A, B, C, and sites). For 2B and 4C schemes, all sites are considered as equivalent; hence, their non-bounded fractions are equal to each other ($X_A = X_B = X_C = X_D$). For 2B scheme, the two pair of electrons are combined into a single site B, while the pair of hydrogen atoms are combined into a single site A. In the case of 3B scheme, only the A and B sites are considered as equivalent; therefore, only their non-bounded fractions are considered equal to each other ($X_A = X_B$), while the remaining non-bounded fraction X_C is equal to $2X_A - 1$. As seen in Table 6.16, there are two possible associating schemes for 3B option: either the two pair of electrons are combined into a single site, or the pair of hydrogen atoms are combined into a single site, which corresponds to the sites labeled as C. In all cases, X_1 is equal to the product of all non-bounded fractions ($X_1 = X_AX_BX_CX_D$). In the symmetric cases (2B and 4C), the non-bounded fractions are assumed equal for all types of sites. This assumption is a statistical consideration, which states that all site types have the same probability of participating in hydrogen bonding. It has been found that the most appropriate scheme for describing the properties of pure water is the 4C scheme [84]. Thus, from the above considerations, 4C scheme satisfies the following relationships:

Table 6.16 Bonding types of water and alcohols molecules [84]

Species	Formula	Type
Alcohol	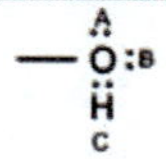	3B
	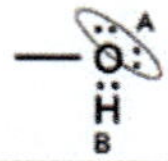	2B
Water	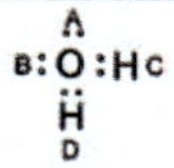	4C
		3B
		3B
		2B

$$\Delta^{AA} = \Delta^{AB} = \Delta^{BB} = \Delta^{CC} = \Delta^{CD} = \Delta^{DD} = 0 \quad (6.122)$$

$$\Delta^{AC} = \Delta^{AD} = \Delta^{BC} = \Delta^{BD} \neq 0 \quad (6.123)$$

$$X_A = X_B = X_C = X_D \quad (6.124)$$

Equation (6.122) expresses that there is no association strength between two equivalent sites; the only one present is between different sites (Eq. (6.123)). In other words, there is only two unknown variables: X_A and Δ^{AC}. Hence, these relationships allows Eq. (6.116) to be solved explicitly for X_A is terms of Δ^{AC} and $\tilde{V}$ as follows:

$$X_A = \frac{\tilde{V}\left(-1 + \sqrt{1 + \frac{8\Delta^{AC}}{\tilde{V}}}\right)}{4} \quad (6.125)$$

Similar explicit expressions for associating schemes different from 4*C* scheme can be found elsewhere [79]. Once the five molecular parameters for the PC-SAFT equation of state (m, σ, ε/k, $\varepsilon^{A_iB_i}/k$, and $\kappa^{A_iB_i}$) are known, Eq. (6.125) allows Z^{ass} and $ln\left(\phi_i^{ass}\right)$ to be only a function of the reduced packing fraction η, because $\tilde{V}$ and η are connected to each other through Eq. (6.112).

One of the biggest uncertainties in the estimation of parameters for a given physical model is the so-called degeneracy; that is, the ability of different sets of parameters to yield the same result. This is the reason why several sets of binary interaction parameters k_{ij} and equation-of-state parameters can be found in the open literature for the same system. From a practical point of view, this means that such parameters might not be suitable for other thermophysical properties or conditions different from those used in the estimation process of such parameters. In this context, the Gibbs free energy minimization for estimating equation-of-state parameters (GFEM) that has been employed before in this manuscript is just an attempt of reducing such a degeneracy; the only considered parameters are those yielding a minimum energy. Concerning the associating parameters of SAFT and related models, Clark et al. [85]. found a large degree of degeneracy in values of the parameters when the parameters model are conventionally determined using saturated pressure and liquid densities data alone. On the other hand, Liang at al. [82] investigated the performance of eight sets of PC-SAFT parameters on the properties of pure water such a speed of sound. They concluded that parameter sets with 4*C* scheme, which are shown Table 6.17, present better performance, but none of such 4C schemes appears superior to the others clearly. Hence, phase equilibriums of binary systems between water and light n-alkanes were studied in order to reach a clearer perspective. However, it was concluded that it is not possible to determine a unique parameter set, even if the prediction of the mutual solubility between water and n-alkanes is used as additional constraint, because mixtures with different hydrocarbons are described better with different parameter sets.

In Figs. 6.47 and 6.48, the saturation pressure and liquid density of pure water is determined using the parameter sets from Table 6.17. As observed, all sets give similar results for the saturation pressure, while the results are rather different for the saturated liquid density. Based on these results, 4*C*2 scheme (AAD of 1.7%) might be the chosen option for estimating purposes. However, as can be seen in Figs. 6.48 and 6.49, the solubility of water in n-octane and n-decane is better described by using 4*C*5 scheme. The reason for such results is that its corresponding parameters are determined by regression to experimental data from not only saturation pressure and liquid density, but also from phase equilibriums between water and light n-alkanes [82]. It is worth underling that the calculations are carried out setting $k_{ij} = 0$, which corresponds to the parameters handled before for non-associating systems. This implies that the results from using 4*C*2 scheme can be improved by adjusting k_{ij} parameters to water-alkane data, as usually done. The capability of prediction of these two sets of parameters is illustrated in Figs. 6.51 and 6.52, where the mutual solubility of water and benzene is compared with the corresponding experimental data. It is

Table 6.17 Parameters of water for the sPC-SAFT equation of state [82]

Scheme	m	σ [Å]	ε/k [K]	ε^{AB}/k [K]	κ^{AB}
4C1	1.5000	2.6273	180.30	1804.22	0.1800
4C2	2.1945	2.2290	141.66	1804.17	0.3894
4C3	3.0000	2.0135	182.92	1259.00	0.8188
4C4	1.5725	2.6270	291.13	1334.20	0.1420
4C5	2.0000	2.3449	171.67	1704.06	0.3048

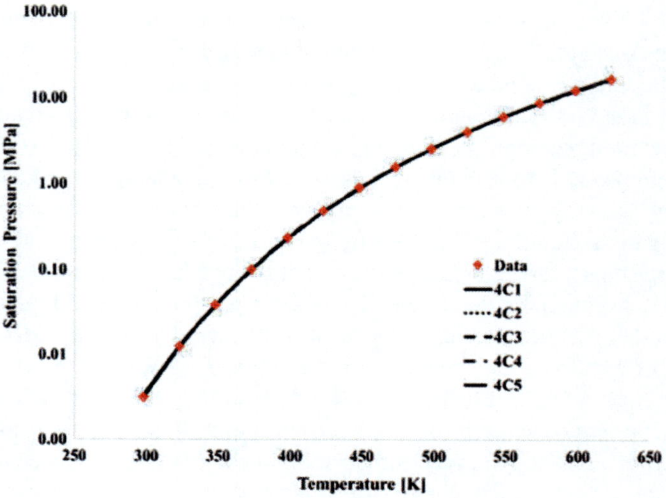

Fig. 6.47 Effect of PC-SAFT parameters with association contribution on the saturation pressure of pure water. Data from Yaw [86]

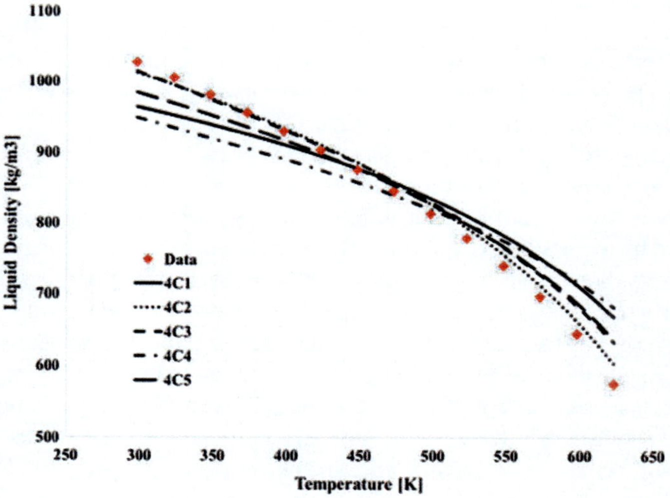

Fig. 6.48 Effect of PC-SAFT parameters with association contribution the liquid density of pure water. Data from Yaw [86]

observed that an additional adjustment is needed in order to improve predictions, irrespective of the selected scheme. Here again, one could include water-aromatic systems in determining the associating parameters of pure water, or on other hand, one could adjust k_{ij} parameters to the experimental data. In conclusion, it is difficult to find predictive models for phase equilibrium calculations and related properties

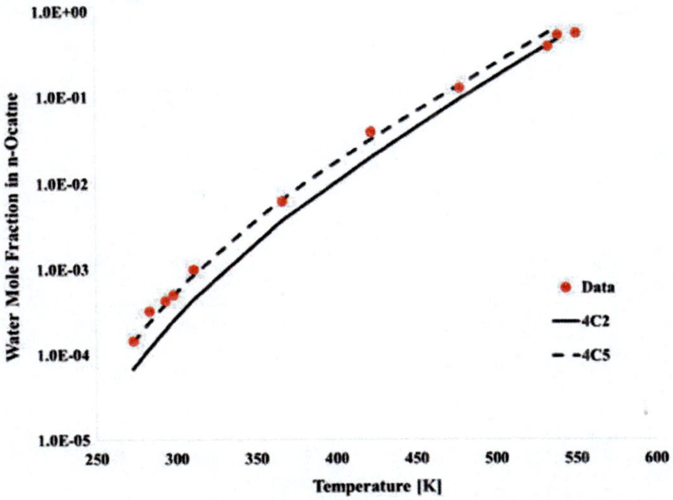

Fig. 6.49 The solubility of water in n-octane using the sPC-SAFT equation of state with the association contribution $\left(k_{ij}=0\right)$. Data from MącZyński et al. [87]

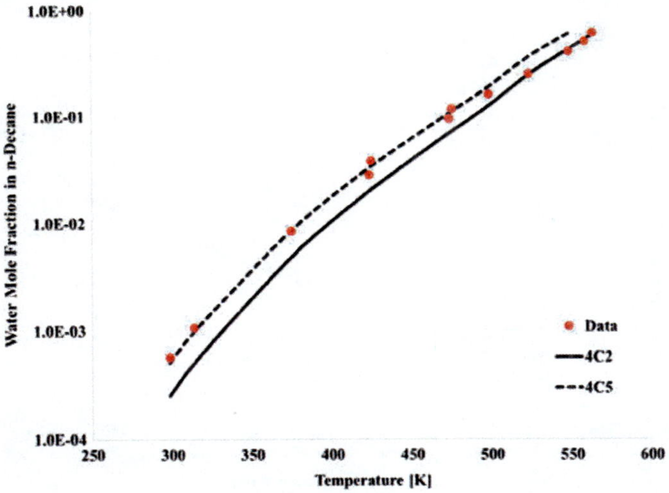

Fig. 6.50 The solubility of water in n-decane using the sPC-SAFT equation of state with the association contribution $\left(k_{ij}=0\right)$. Data from MącZyński et al. [87]

of water-hydrocarbon systems, even though the system is formed by identifiable components (Fig 6.50).

In addition to the PC-SAFT equations of state, associating system can be undertaken using CEoS. The combination of the associating contribution and such equations give rise to the so-called Cubic-plus-Association Equations of state (CPA)

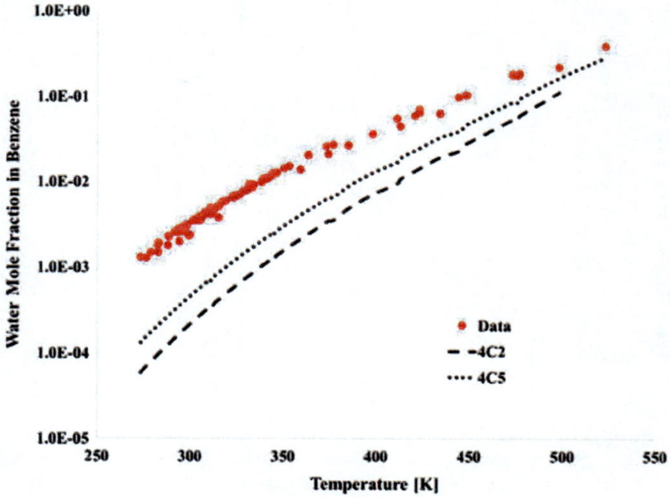

Fig. 6.51 The solubility of water in benzene using the sPC-SAFT equation of state with the association contribution $\left(k_{ij} = 0\right)$. Data from MąсZyński et al. [88]

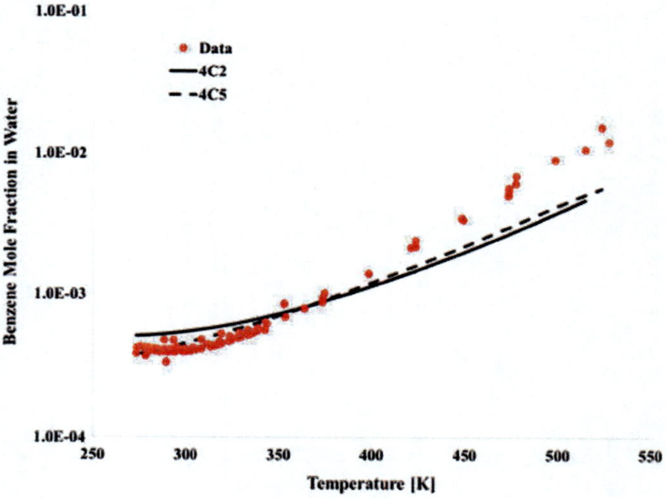

Fig. 6.52 The solubility of benzene in water using the sPC-SAFT equation of state with the association contribution $\left(k_{ij} = 0\right)$. Data from MąсZyński et al. [88]

[89–93]. In this model, the molar residual Helmholtz free energy $\left(\tilde{A}^R/RT\right)$ is assumed to be composed of two contributions:

$$\frac{\tilde{A}^R}{RT} = \frac{\tilde{A}^{phy}}{RT} + \frac{\tilde{A}^{ass}}{RT} \tag{6.126}$$

The physical contribution $\tilde{A}^{phy}/RT$ is determined originally from the SRK equation of state; however, any CEoS can be used in principle. For instance, both the SRK and PR equation of state have been used in predicting the asphaltene phase behavior in live petroleum fluids [75]. Regarding the association contribution, the same expressions exposed before are used, with the exception of the reduced packing fraction η, which is now given by

$$\eta = \frac{b}{4\tilde{V}} \tag{6.127}$$

with b being the co-volume. From this equation and using the linear mixing rule

$$b = \sum_{i=1}^{C} x_i b_i \tag{6.128}$$

Equation (6.114) turns out to be equal to

$$\left[\frac{\partial ln(g)}{\partial n_i}\right]_{T,\tilde{V},n_k} = \frac{b_i}{b}\left[\frac{5\eta - 2\eta^2}{(1-\eta)(2-\eta)}\right] \tag{6.129}$$

To reduce the computation time, it has been applied a simplified hard sphere radial distribution function of the form

$$g = \frac{1}{1 - 1.9\eta} \tag{6.130}$$

On the other hand, the strength of association is now expressed as

$$\Delta^{A_i B_j} = b_{ij}\kappa^{A_i B_j}\left[e^{\frac{\varepsilon^{A_i B_j}}{kT}} - 1\right]g \tag{6.131}$$

with

$$b_{ij} = \frac{b_i + b_j}{2} \tag{6.132}$$

As in the case of the PC-SAFT equation of state, the cross-association energy $\varepsilon^{A_i B_j}$ can be estimated through Eq. (6.119), and the cross-association volume $\kappa^{A_i B_j}$ can be also estimated through either Eqs. (6.120) or (6.121), changing σ_{ij} by b_{ij} as follows:

$$\kappa^{A_i B_j} = \sqrt{\kappa^{A_i B_i}\kappa^{A_j B_j}}\frac{\sqrt{b_{ii} b_{jj}}}{b_{ij}} \tag{6.133}$$

In the original version of the CPA equation of state, the parameter a accounting for attractive forces between molecules is defined using a Soave-type expression:

$$a = a_0\left[1 + c_1\left(1 - \sqrt{\frac{T}{Tc}}\right)\right]^2 \tag{6.134}$$

Therefore, likewise the PC-SAFT equations of state, five adjustable parameters are needed for each component: three for non-associating compounds (a_0, b, and c_1) and two additional parameters for associating compounds ($\varepsilon^{A_iB_i}$, $\kappa^{A_iB_i}$). These five parameters are obtained by adjusting experimental saturation pressure and liquid density as usual. For the specific case of the SRK equation of state with $4C$ scheme, the parameters corresponding to pure water are $b = 1.4515 \times 10^{-5}\, m^3/mol$, $c_1 = 0.67359$, $a_0 = 1.2277 \times 10^{-7}\, m^6\, MPa/mol^2$, $\varepsilon^{AB}/R = 2003.13\, K$, and $\kappa^{AB} = 0.0692$ [94]. Recently, comparisons have been performed between the sPC-SAFT and CPA equations of state for associating systems involving hydrocarbons [67, 82]. It seems that both equations are comparable with each other as long as the equation-of-state parameters are adjusted to experimental information properly. These results may indicate that the association contribution is more important than the non-associating contributions in describing phase equilibrium and related properties for associating systems.

For dealing with water-heavy petroleum, it is customary to assume that there is not association between water and the polar fractions of the petroleum fluid such as aromatic, resins, and asphaltenes. Besides, self-association between such fractions is neglected as well. The only considered association is the self-association between water molecules as in the case of systems formed by identifiable compounds. Thus, the characterization of petroleum fluids can be performed through distribution functions, either correlations or energy minimization for estimating properties of fractions, and regressions to experimental information accordance with the material exposed previously. For instance, calculated and experimental values of the water solubility in MacKay River Bitumen[53] and the corresponding liquid density are illustrated in Figs. 6.53 and 6.54. The experimental simulated distillation of MacKay River Bitumen is shown in the Chap. 5. Assuming that only basic experimental information is available (bulk molecular weight and specific gravity) the GDF is used with the recommended values of $B_M = 1$ and $B_S = 3$. The resulting pseudocomponents (PS) are thermodynamically characterized by GFEM using the PC-SAFT equation of state and the functional groups from Burgess et at. [40]. Therefore, the sPC-SAFT equation of state with $4C5$ scheme from Table 6.17 are used for phase equilibrium calculations. For comparison purposes, GP from Table 6.10 are also included. Finally, k_{ij} parameters are fitted through a function of temperature and molecular weight to match calculations with experiments. In the case of 13 pseudocomponents either specific or generalized, the liquid density data are not included in the regression process because their AAD is about 1%. However, AAD increases as the number of pseudocomponents decreases; hence, the liquid density data are included together with the water solubility data. For 3 and 13 PS as well as 13 GP, AAD are equal

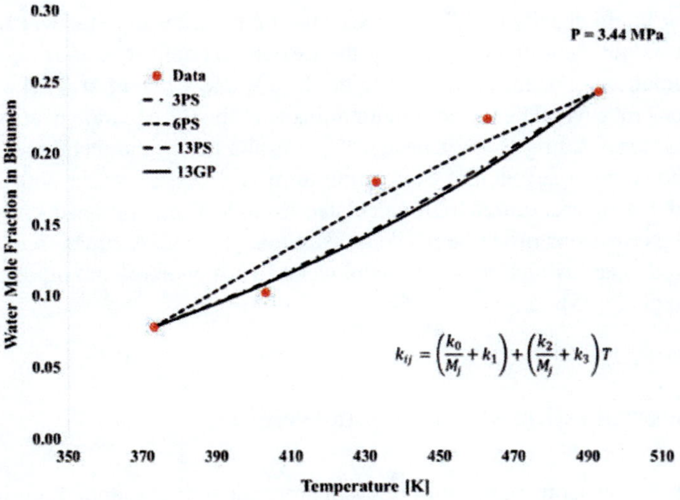

Fig. 6.53 The solubility of water in MacKay River Bitumen [53]. Calculations with the sPC-SAFT equation of state and 4C5 scheme. Characterization with GDF ($B_M = 1$ and $B_S = 3$), GFEM, and GP from Table 6.10

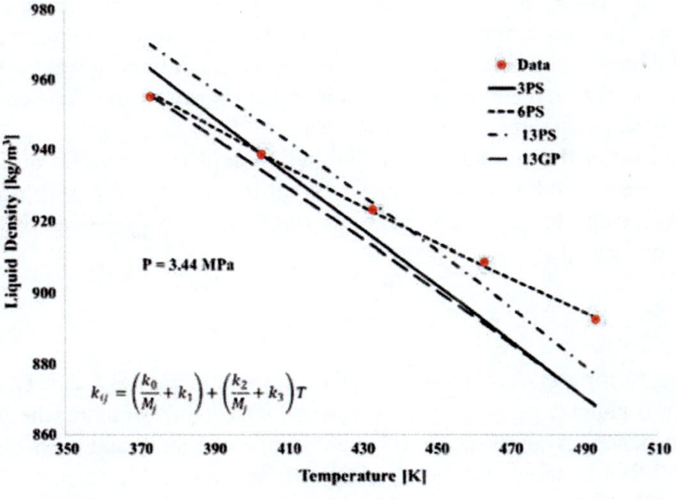

Fig. 6.54 Liquid density of water-bitumen mixtures corresponding to the solubility of water in MacKay River Bitumen from Fig. 6.53. Data from Zirrahi et al. [53]

to 7.8, 8.1 and 8.2 respectively, while it is equal to 6.8% for 6 PS. These errors lies in range of ±10%, which is an acceptable accuracy for the water solubility in bitumen [53]. From this point of view, although all descriptions are good enough, 6 PS option appears as the better option not only because of the solubility, but also because of liquid density (AAD of 0.1%). Despite these results, it is worth pointing out that it is not clear enough whether the association contribution is necessary in these estimations. For instance, Zirrahi et al. [53] and Chen et al. [74] performed calculations for several heavy petroleum fluids with the PR equation of state without association term, fitting k_{ij} parameters; their results lies in the mentioned range of ±10%. However, it seems that association term is necessary when both solubility and liquid density are desired to be calculated from the same framework. Otherwise, empirical expressions might be needed [53]. Finally, the CPA equation of state can also be used in describing the solubility of water in heavy petroleum fluids, as shown by Zirrahi et al. [95].

6.8 Determination of Molecular Weight

As has been shown, one of the most relevant properties of petroleum fluids in general is the molecular weight; many sets of correlations and the energy minimization for estimating equation-of-state parameters are written in terms of the molecular weight. Hence, it is worth analyzing concepts and assumptions behind measurements. One set of experiments is based on the so-called colligative properties: the depression of the freezing point, the elevation of the boiling point, and the osmotic pressure. All of them have one common feature: they do not depend on the nature of solutes present, but they depend only on their composition in solution. In what follows, the component present in the greatest amount is called solvent, while those constituents in relatively small amounts are called solutes.

For the case of the elevation of the boiling point, it is convenient to consider a multicomponent liquid solution, which is in equilibrium with a vapor formed by pure solvent. According to the necessary equilibrium condition expressed by Equation $\mu_{ij} = \mu_{ir}$, we have that

$$\mu_{iL} = \mu_{iV}^{0} \tag{6.135}$$

where the subscripts L and V indicate liquid and vapor phase, respectively, while the superscript 0 indicate pure-component state. Contrary to a mixture, the pure state does not depends on composition; it depends on temperature and pressure only. In the Chap. 2, it was shown that, if an ideal solution was used a reference fluid, the total molar Gibbs free energy $\tilde{G}$ could be determined from

$$\frac{\tilde{G}}{RT} = \sum_{i=1}^{C} x_i \mu_i = \sum_{i=1}^{C} x_i \left[\frac{\mu_i^0}{RT} + ln(a_i) \right] \tag{6.136}$$

From where, we have that

$$\frac{\mu_i}{RT} = \frac{\mu_i^0}{RT} + ln(a_i) \tag{6.137}$$

Here, μ_i is the chemical potential of component i in solution, and a_i is the corresponding activity, whereas μ_i^0 is the chemical potential of component i in pure state. Remembering that $\tilde{G}_i^0 = \mu_i^0$, we can rewrite Eq. (6.135) as follows:

$$ln(a_{iL}) = \frac{\mu_{iV}^0 - \mu_{iL}^0}{RT} = \frac{\Delta\tilde{G}_i^{VL}}{RT} \tag{6.138}$$

To find how temperature composition are related to each other, it is convenient to obtain the total differential, while holding pressure constant, that is,

$$dln(a_{iL}) = \left[\frac{\partial}{\partial T}\left(\frac{\Delta\tilde{G}_i^{VL}}{RT}\right)\right]_P dT \tag{6.139}$$

$\Delta\tilde{G}_i^{VL}$ accounts for the molar Gibbs energy of vaporization of the pure solvent i, that is, the energy associated with a change in state of aggregation from a liquid phase to a vapor phase. It is more convenient to express changes in terms of enthalpy instead of free energy, because the former is the measured variable, and therefore, it is the available one. Specifically, enthalpy is related to the heat exchanged with surroundings where pressure is kept constant during changes in state ($\Delta H = Q$).

On the other hand, in the Chap. 2, it was derived the fundamental relationship

$$\left(\frac{\partial\tilde{G}}{\partial T}\right)_P = -\tilde{S} \tag{6.140}$$

According to Eq. (6.139), it is necessary the partial derivative of the ratio of G to T with respect to temperature

$$\left[\frac{\partial}{\partial T}\left(\frac{\tilde{G}}{RT}\right)\right]_P = \frac{1}{RT}\left(\frac{\partial\tilde{G}}{\partial T}\right)_P - \frac{\tilde{G}}{RT^2} \tag{6.141}$$

The substitution of the Eq. (6.140) into Eq. (6.141), together with the Gibbs free energy definition $\tilde{G} \equiv \tilde{H} - T\tilde{S}$, yields

$$\left[\frac{\partial}{\partial T}\left(\frac{\tilde{G}}{RT}\right)\right]_P = -\frac{\tilde{S}T + \tilde{G}}{RT^2} = -\frac{\tilde{H}}{RT^2} \tag{6.142}$$

Therefore, we arrive at

$$dln(a_{iL}) = -\frac{\Delta\tilde{H}_i^{VL}}{RT^2}dT \tag{6.143}$$

Now, $\Delta\tilde{H}_i^{VL}$ accounts for the molar enthalpy of vaporization of the pure solvent i. The previous expressions are valid for any kind of fluid. Their use requires the knowledge of a_{iL} as a function of temperature, pressure and composition, and of $\Delta\tilde{H}_i^{VL}$ as a function of pressure and temperature. However, the problem can be simplified considerably if dilute solutions are preset, that is, if ideal solutions are assumed. For ideal solutions, the activity of a given component is equal to its composition ($a_{iL} = x_i$); thus, Eq. (6.143) turns out to be

$$dln(x_i) = -\frac{\Delta\tilde{H}_i^{VL}}{RT^2}dT \tag{6.144}$$

On the other hand, because we are interested in very dilute solution, it is advisable to use the total molality of the solutes instead of their mole fractions as measure of the composition. The molality l_j is defined as the number of moles of solute n_j in a specified amount of mass w_i of the solvent:

$$l_j \equiv \frac{n_j}{w_i} = \frac{n_j}{n_i M_i} \tag{6.145}$$

Accordingly, the total molality l_t is

$$l_t = l_1 + l_2 + \ldots = \frac{n_s}{w_i} = \frac{w_s}{M_s w_i} \tag{6.146}$$

The superscript s indicates properties of solutes as a whole. Using these definitions, the mole fraction of solvent in terms of the total molality is given by

$$x_i = \frac{n_i}{n_1 + n_2 + \ldots + n_i} = \frac{n_i}{n_i M_i (l_1 + l_2 + \ldots) + n_i} = \frac{1}{M_i l_t + 1} \tag{6.147}$$

Taking logarithms and differentiating, we obtain that

$$ln(x_i) = -ln(1 + M_i l_t) \tag{6.148}$$

and

$$dln(x_i) = -\frac{M_i}{1 + M_i l_t}dl_t \tag{6.149}$$

Finally, by combining Eq. (6.144) with Eq. (6.149), the partial derivative of temperature with respect to the total molality takes the form of

$$-\left(\frac{\partial T}{\partial l_t}\right)_P = \left(\frac{RT^2}{\Delta\tilde{H}_i^{VL}}\right)\frac{M_i}{1 + M_i l_t} \tag{6.150}$$

Because of the ideal-solution assumption, Eq. (6.150) must be valid just for infinitely dilute solutions. In the limit of dilute solutions $l_t \to 0$ and $T \to T_{bi}$, T_{bi} being the boiling temperature of the pure solvent i. In other words, at the limit of dilute solutions, Eq. (6.150) becomes

$$-\lim_{l_t \to 0} \left(\frac{\partial T}{\partial l_t} \right)_P \equiv K_{bi} = \frac{RM_i T_b^2}{\Delta \tilde{H}_i^{VL}} \tag{6.151}$$

The boiling-point elevation constant K_{bi} depends only on properties of the pure solvent i; it does not depend on either properties or amount of solutes in solution. To analyze further the nature of K_{bi}, it is convenient to write it in terms of entropy instead of enthalpy, for which a relationship between both thermodynamic potentials is needed. Such a relationship is derived starting with the fundamental relationship

$$\left(\frac{\partial H}{\partial S} \right)_P = T \tag{6.152}$$

which can be integrated directly to

$$\Delta H^{VL} = T_b \Delta S^{VL} \tag{6.153}$$

Thus, K_{bi} can be expressed in terms of the molar entropy of vaporization of the pure solvent i $\Delta\tilde{S}_i^{VL}$ as follows:

$$K_{bi} = \frac{RM_i T_{bi}}{\Delta \tilde{S}_i^{VL}} \tag{6.154}$$

As inferred, $K_{bi} > 0$ because $M_i > 0$, $T_{bi} > 0$ and, entropy of gases is always larger than that of liquids $\Delta\tilde{S}_i^{VL} > 0$. Bearing in mind this fact, and using the K_{bi} definition, Eq. (6.151) reduces to

$$\left(\frac{\partial T}{\partial l_t} \right)_P = K_{bi} \tag{6.155}$$

which can be integrated directly to

$$\theta_{bi} \equiv T - T_{bi} = K_{bi} l_t \tag{6.156}$$

Here again, it follows that $\theta_{bi} > 0$ because both $K_{bi} > 0$ and $l_t > 0$. In other words, the presence of solutes with total molality l_t leads to an elevation of the boiling temperature of the solvent, irrespective of the nature of the solutes present in solution. The larger l_t, the larger the elevation of the boiling temperature, leading to more accurate measurements. However, the system departs from the ideality assumption, and consequently, the equations are no longer valid. This means that a solvent with a very large K_{bi} is needed, or the ideal-solution assumption has to be discarded. As will be shown later, both options are used in practice. Finally, the equation for estimating the molecular weight of solutes arises from combining Eq. (6.146) with Eq. (6.156) to give

$$M_s = \frac{w_s K_{bi}}{w\theta_{bi}} = \frac{W_s K_{bi}}{\theta_{bi}} \tag{6.157}$$

where W_s is the weight ratio of solute to solvent (w_s/w). Therefore, the measured values of W_s and θ_{bi} together with the knowledge of K_{bi} for a given solvent allows M_s to be estimated.

Another colligative property that can be used in order to estimate M_s is the depression of the freezing temperature of the solvent. In this case, we consider that the multicomponent liquid solution is in equilibrium with a solid phase formed by pure solvent, that is, $\mu_{iS}^0 = \mu_{iL}$. The derivation of the corresponding expressions is the same as before essentially. Thus, we have that the freezing-point depression constant K_{fi} is now

$$K_{fi} = \frac{RM_i T_{fi}^2}{\Delta\tilde{H}_i^{LS}} = \frac{RM_i T_{fi}}{\Delta\tilde{S}_i^{LS}} > 0 \tag{6.158}$$

$\Delta\tilde{H}_i^{LS}$ and $\Delta\tilde{S}_i^{LS}$ account for the molar enthalpy and entropy of fusion of the pure solvent i, whose freezing temperature is T_{fi} and molecular weight is M_i. θ_{fi} is now defined as

$$\theta_{fi} \equiv T_{fi} - T = K_{fi} l_t > 0 \tag{6.159}$$

Unlike θ_{bi}, $\theta_{fi} > 0$ implies that the freezing temperature of the solvent is decreases $T_{fi} > T$. Finally, M_s can be determined from

$$M_s = \frac{w_s K_{fi}}{w\theta_{fi}} = \frac{W_s K_{fi}}{\theta_{fi}} \tag{6.160}$$

The elevation of the boiling temperature and the depression of the freezing temperature can be illustrated on a phase diagram for pure components. The addition of a small amount of solutes to a pure solvent reduces its saturation pressure during an isothermal process. On the other hand, it increases the boiling temperature during an isobaric process. The combined result is the displacement of the vapor-liquid saturation curve toward the right. Similarly, the isothermal and isobaric processes of the freezing point result in the displacement of the liquid-solid saturation curve toward the left. In general, the effect is bigger on the freezing point than on the boiling point.

In nature, liquid mixtures do not behave as ideal ones, even at the smallest concentrations that can be used in experiments. Therefore, the expressions derived earlier are not useful as such for being applied to the experimental determination of the molecular weight of solutes. Usually, power series in solute concentration with empirical coefficients, which are known as virial equations, are proposed in order to deal with the non-ideal behavior of mixtures. However, such expressions preserve the form of the ideal-solution equations in the limit of dilute solutions so that some thermodynamic consistency is retained. For instance, the expanded form in power series of Eq. (6.156) takes the form of

$$\frac{\theta_{bi}}{W_s} = K_{bi}\left(\frac{1}{M_s} + AW_s + BW_s^2 + \ldots\right) \tag{6.161}$$

which is the fundamental expression in order to measure M_s by using the elevation of the boiling temperature of the solvent. Thus, if all virial coefficients $(A, B, \ldots)$ are set equal to zero, Eq. (6.156) is retrieved and therefore the ideal-solution limit. This means that the virial coefficients account for the non-ideal effects of the solution. Although Eq. (6.161) has been suggested to be used with two terms of the power series ($A \neq 0$ and $B \neq 0$) [61], it seems that it is enough to include only the first term of the power series for most systems involving hydrocarbons [59, 96, 97]. In other words, Eq. (6.161) is reduced to

$$\frac{\theta_{bi}}{W_s} = K_{bi}\left(\frac{1}{M_s} + AW_s\right) \tag{6.162}$$

Therefore, experimental data can be adjusted to a straight line, K_{bi}/M_s and AK_{bi} being the intercept and the slope, respectively. By extrapolating θ_{bi}/W_s to zero W_s, the intercept is calculated and hence M_s is estimated as follows:

$$M_s = \frac{K_{bi}}{(\theta_{bi}/W_s)_{W_s=0}} \tag{6.163}$$

When θ_{bi} is intended to be used, the experimental technique is called vapor phase osmometry, vapor pressure osmometry or more accurately thermoelectric differential vapor pressure depressing (VPO). In this experimental technique, two coupled thermistors are located in a thermostatted chamber that is saturated with vapor of pure solvent. A drop of liquid solvent is placed on one thermistor and another drop of liquid solution of equal size is placed on the other thermistor. As observed from Fig. 6.55, the liquid solution has a lower saturation pressure at a given temperature, that is, it has a lower chemical potential. Therefore, the solvent transfers from the pure solvent thermistor to the solution thermistor until the chemical potential difference disappears between both chambers. Heat of vaporization $\left(\Delta\tilde{H}_i^{VL}\right)$ is released during the process, increasing the temperature of the liquid solution. At equilibrium,

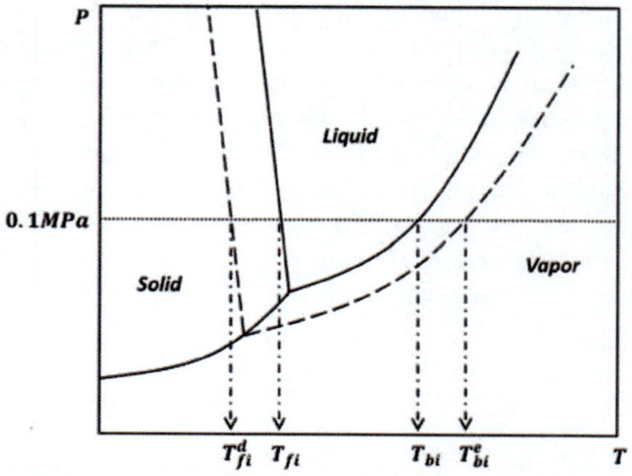

Fig. 6.55 Colligative properties: the elevation of the boiling temperature $\left(T^e_{bi}\right)$ and the depression of the freezing temperature $\left(T^d_{fi}\right)$

the solvent has the same saturation pressure on both sensors, but their temperatures are different. The difference of temperatures gives rise to a voltage difference between thermistors (ΔE); hence, the measured variable is ΔE instead of θ_{bi}.

In practice, the equilibrium state is difficult to attain because of the transfer processes occurring during experiments, such as molecular diffusion and heat conduction within the drops. The equilibrium state is reached when the driving forces of such processes disappear; however, this is not the case in experiments performed at a finite period. To overcome such equilibrium limitations, the system is operated in steady state. Under such condition, the expressions derived previously are retained, but K_{bi} is substituted by an empirical constant K relying on the specific experiment. The equipment is calibrated for a given solvent by using solutes of known molecular weight. In conclusion, Eq. (6.163) is substituted by the following expression in order to determine M_s.

$$M_s = \frac{K}{(\Delta E/W_s)_{W_s=0}} \tag{6.164}$$

Peramanu et al. [59] measured molecular weights of bitumens and their fractions by using the VPO experimental technique at 393.15 K. In their experiments, o-dichlorobenzene component was used as solvent, whose molecular weight is equal to 147.0. To determine the K value of the experiment, the calibration was performed using benzyl as solute, whose molecular weight is equal to 210.2. As observed from

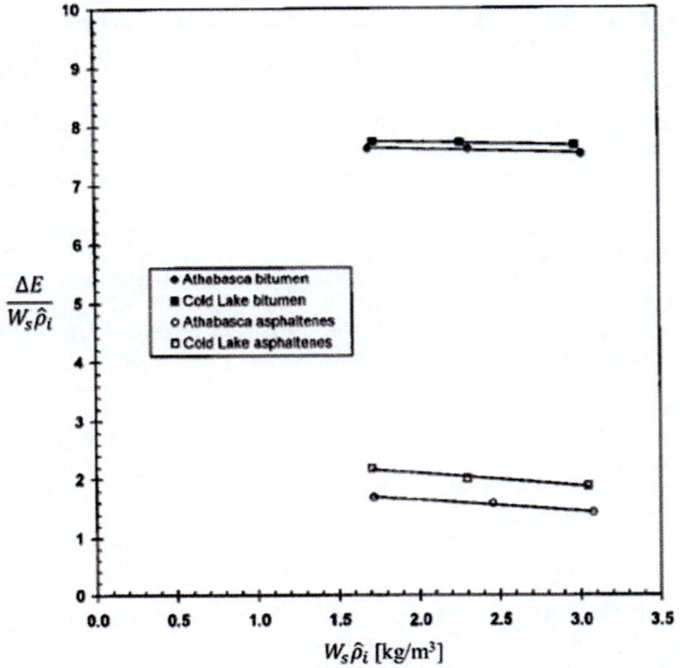

Fig. 6.56 VPO diagram for Athabasca and Cold Lake bitumens and their asphaltene fractions [59]

Fig. 6.56, the experiments response lies in a linear region in accordance with the assumptions made in deriving Eq. (6.164).

References

1. Michelsen ML, Mollerup J (2007) Thermodynamic models: fundamentals & computational aspects, second. Holte, Denmark, TIE-LINE
2. Reneaume JM, Meyer M, Joulia X (1996) A Global MINLP approach for phase equilibrium calculations. Comput Chem Eng 20:S303–S308
3. Meyer CA, Swartz CLE (1998) A regional convexity test for global optimization : application to the phase equilibrium problem. Comput Chem Eng 22(10)
4. Lee YP, Rangaiah GP, Luus R (1999) Phase and chemical equilibrium calculations by direct search optimization. Comput Chem Eng 23:1183–1191
5. Teh YS, Rangaiah GP (2003) Tabu search for global optimization of continuous functions with application to phase equilibrium calculations. Comput Chem Eng 27:1665–1679
6. Jalali F, Seader JD, Khaleghi S (2008) Global solution approaches in equilibrium and stability analysis using homotopy continuation in the complex domain. Chem Eng Sci 32:2333–2345
7. Nerón A, Lantagne G, Marcos B (2012) Computation of complex and constrained equilibria by minimization of the gibbs free energy. Chem Eng Res Des 82:260–271
8. Sun AC, Seider WD (1995) Homotopy-continuation method for stability analysis in the global minimization of the gibbs free energy. Fluid Phase Equilib 103(2):213–249

9. Hua JZ, Brennecke JF, Stadtherr MA (1998) Reliable computation of phase stability using interval analysis: cubic equation of state models. Comput Chem Eng, 1207–1214
10. Bausa J, Marquardt W (2000) Quick and reliable phase stability Test in VLLE flash calculations by homotopy continuation. Comput Chem Eng 24(11):2447–2456
11. Chaikunchuensakun S, Stiel LI, Baker EL (2002) A combined algorithm for stability and phase equilibrium by gibbs free energy minimization. Ind Eng Chem Res 41(16):4132–4140
12. Sofyan Y, Ghajar AJ, Gasem KAM (2003) Multiphase equilibrium calculations using gibbs minimization Sofyan2003.Pdf, 3786–3801
13. Carreón-Calderón B (2012) Theoretical study of vapor-liquid homogeneous nucleation using stability analysis of a macroscopic phase. J Chem Phys, *137*(14)
14. Edgar TF, Himmelblau DM, Lasdon LS (2001) Optimization of chemical processes. Singapore, Second.; McGraw-Hill
15. Jalali-Farahani F, Seader JD (2000) Use of homotopy-continuation method in stability analysis of multiphase reacting systems. Comput Chem Eng 24(8):1997–2008
16. Gritton KS, Seader JD, Lin WJ (2001) Global Homotopy continuation procedures for seeking all roots of a nonlinear equation. Comput Chem Eng 25(7–8):1003–1019
17. Leibovici CF, Jean N (1995) A solution of Rachford-rice equations for Multuplhase systems. Fluid Phase Equilib 112:217–221
18. Leibovici CF, Nichita DV (2008) A new look at multiphase Rachford-rice equations for negative flashes. Fluid Phase Equilib 267(2):127–132
19. Carreón-Calderón B, Soria A, Romero-Martínez A (2010) Modified Rachford-rice equations including interfacial contributions and their application to the nucleation process. AIChE J 56 (7)
20. Iranshahr A, Voskov D, Tchelepi HA (2010) Generalized negative-flash method for multiphase multicomponent systems. Fluid Phase Equilib 299(2):272–284
21. Bazúa ER (1999) Thermodynamic properties with cubic equations of state, 310–314
22. Biegler LT, Grossmann IE, Westerberg AW (1997) Systematic mehods of chemical process design. Prentice Hall, New Jersey, USA
23. Michelsen ML (1994) Calculation of multiphase equilibrium. Comput Chem Eng 18(7):545–550
24. Nichita DV, Gomez S, Luna E (2002) Multiphase equilibria calculation by direct minimization of gibbs free energy with a global optimization method. Comput Chem Eng 26(12):1703–1724
25. Hinojosa-Gómez H, Solares-Ramírez J, Bazúa-Rueda RE (2015) An improved algorithm for the three-fluid-phase VLLE flash calculation. AICHE J 61(9):3081–3093
26. Carreón-Calderón B, Uribe-Vargas V, Ramírez-De-Santiago M, Ramírez-Jaramillo E (2014) Thermodynamic characterization of heavy petroleum fluids using group contribution methods. Ind Eng Chem Res 53(13)
27. Lindeloff N, Michelsen ML (2002) Phase envelope calculations for hydrocarbon-water mixtures. Proc SPE Annu Tech Conf Exhib, 3749–3756
28. Nichita DV (2008) Phase envelope construction for mixtures with many components. Energy Fuels 22(1):488–495
29. Jaubert JN, Avaullee L, Souvay J-F (2002) A Crude oil data bank containing more than 5000 PVT and gas injection data. J Pet Sci Eng 34:65–107
30. Peng DY, Robinson DB (1976) A new two-constant equation of state. Ind Eng Chem Fundam 15(1):59–64
31. Marrero J, Gani R (2001) Group-contribution based estimation of pure component properties. Fluid Phase Equilib 184:183–208
32. Ihmels EC, Gmehling J (2003) Extension and revision of the group contribution method GCVOL for the prediction of pure compound liquid densities. Ind Eng Chem Res 42(2):408–412
33. Danesh A (2003) PVT and phase behaviour of petroleum reservor fluids. Elsevier, The Netherlands
34. Pedersen KS, Christensen PL, Shaikh JA (2015) Phase behavior of petroleum reservoir fluids. Second.; CRC Press, Boca Raton, USA

35. Carreón-Calderón B, Uribe-Vargas V, Ramírez-Jaramillo E, Ramírez-de-Santiago M (2012) Thermodynamic characterization of undefined petroleum fractions using group contribution methods. Ind Eng Chem Res 51:14188–14198
36. Aguilar-Cisneros H, Uribe-Vargas V, Carreón-Calderón B, Domínguez-Esquivel JM, Ramirez-De-Santiago M (2017) Hydrogen solubility in heavy undefined petroleum fractions using group contributions methods. Oil Gas Sci Technol 72 (1)
37. Aguilar-Cisneros H, Carreón-Calderón B, Uribe-Vargas V, Domínguez-Esquivel JM, Ramirez-de-Santiago M (2018) Predictive method of hydrogen solubility in heavy petroleum fractions using EOS/GEand group contributions methods. Fuel 224(January):619–627
38. Aguilar-Cisneros H, Uribe-Vargas V, Carreón-Calderón B (2020) Estimation of Gas solubility in petroleum fractions using PR-UMR and group contributions methods. Fuel 275(April):117911
39. Ma M, Chen S, Abedi J (2015) Binary interaction coefficients of asymmetric CH4, C2H6, and CO2 with high n-Alkanes for the simplified PC-SAFT correlation and prediction. Fluid Phase Equilib 405:114–123
40. Burgess WA, Tapriyal D, Gamwo IK, Wu Y, Mchugh MA, Enick RM (2014) New Group-contribution parameters for the calculation of PC-SAFT parameters for use at Pressures to 276 MPa and Temperatures to 533 K
41. Gross J, Sadowski G (2001) Perturbed-chain SAFT: an equation of state based on a perturbation theory for chain molecules. Ind Eng Chem Res 40(4):1244–1260
42. Von Solms N, Michelsen ML, Kontogeorgis GM (2003) Computational and physical performance of a modified PC-SAFT equation of state for highly asymmetric and associating mixtures. Ind Eng Chem Res 42(5):1098–1105
43. James LA, Rezaei N, Chatzis I (2008) VAPEX, Warm VAPEX and Hybrid VAPEX—the state of enhanced oil recovery for in situ heavy oils in Canada. J Can Pet Technol 47(4):12–18
44. Sheng JJ (2010) modern chemical enhanced oil recovery: theory and practice. Burlington, USA, GPP
45. Lal D, Otto FD, Mather AE (1999) Solubility of hydrogen in athabasca bitumen. Fuel 78(12):1437–1441
46. Haddadnia A, Zirrahi M, Hassanzadeh H, Abedi J (2017) Solubility and thermo-physical properties measurement of CO2- and N2-Athabasca Bitumen systems. J Pet Sci Eng 154(April):277–283
47. Zhao Q, Li Z, Wang S, Lai F, Li H (2019) Phase behavior measurements and modeling for N 2 /CO 2 /extra heavy oil mixtures at elevated temperatures. Ind Eng Chem Res 58(1):428–429
48. Haddadnia A, Sadeghi Yamchi H, Zirrahi M, Hassanzadeh H, Abedi J (2018) New solubility and viscosity measurements for Methane-, Ethane-, Propane-, and Butane-Athabasca bitumen systems at high temperatures up to 260 & #xB0;c. J Chem Eng Data 63(9):3566–3571
49. Saajanlehto M, Uusi-Kyyny P, Alopaeus V (2014) Hydrogen solubility in heavy oil systems: experiments and modeling. Fuel 137:393–404
50. Cai HY, Shaw JM, Chung KH (2001) Hydrogen solubility measurements in heavy oil and bitumen cuts. Fuel 80(8):1055–1063
51. Ma M, Chen S, Abedi J (2016) Modeling the solubility and volumetric properties of H2 and heavy hydrocarbons using the simplified PC-SAFT. Fluid Phase Equilib 425:169–176
52. Saajanlehto M, Alopaeus V (2014) Heavy oil characterization method for PC-SAFT. Fuel 133:216–223
53. Zirrahi M, Hassanzadeh H, Abedi J (2016) Experimental and modeling studies of Mackay river bitumen and water. J Pet Sci Eng 2017(151):305–310
54. Tihic A, Kontogeorgis GM, von Solms N, Michelsen ML (2006) Applications of the simplified perturbed-Chain SAFT equation of state using an extended parameter table. Fluid Phase Equilib 248(1):29–43
55. Kariznovi M, Nourozieh H, Abedi J (2010) Bitumen characterization and Pseudocomponents determination for equation of state modeling. Energy Fuels 24(1):624–633
56. Castellanos Díaz O, Modaresghazani J, Satyro MA, Yarranton HW (2011) Modeling the phase behavior of heavy oil and solvent mixtures. Fluid Phase Equilib 304(1–2):74–85

57. Agrawal P, Schoeggl FF, Satyro MA, Taylor SD, Yarranton HW (2012) Measurement and modeling of the phase behavior of solvent diluted bitumens. Fluid Phase Equilib 334:51–64
58. Horstmann S, Jabłoniec A, Krafczyk J, Fischer K, Gmehling J (2005) PSRK group contribution equation of state: comprehensive revision and extension IV, including critical constants and α-function parameters for 1000 components. Fluid Phase Equilib 227(2):157–164
59. Peramanu S, Pruden BB, Rahimi P (1999) Molecular weight and specific gravity distributions for Athabasca and cold lake bitumens and their saturate, aromatic, resin, and Asphaltene Fractions. Ind Eng Chem Res 38(8):3121–3130
60. Yarranton HW (2005) Asphaltene self-association. J Dispers Sci Technol 26(1):5–8
61. Acevedo S, Gutierrez LB, Negrin G, Pereira JC, Mendez B, Delolme F, Dessalces G, Broseta D (2005) Molecular weight of petroleum Asphaltenes: a comparison between mass spectrometry and vapor pressure osmometry. Energy Fuels 19(4):1548–1560
62. Gonzalez DL, Hirasaki GJ, Creek J, Chapman WG (2007) Modeling of Asphaltene precipitation due to changes in composition using the perturbed chain statistical associating fluid theory equation of state. Energy Fuels 21(3):1231–1242
63. Panuganti SR, Vargas FM, Gonzalez DL, Kurup AS, Chapman WG (2012) PC-SAFT Characterization of crude oils and modeling of Asphaltene phase behavior. Fuel 93:658–669
64. Punnapala S, Vargas FM (2013) Revisiting the PC-SAFT characterization procedure for an improved Asphaltene precipitation prediction. Fuel 108:417–429
65. Li Z, Firoozabadi A (2010) Modeling Asphaltene precipitation by N-Alkanes from heavy oils and bitumens using cubic-plus-association equation of state. Energy Fuels 24(2):1106–1113
66. Li Z, Firoozabadi A (2010) Cubic-plus-association equation of state for Asphaltene precipitation in live oils. Energy Fuels 24(5):2956–2963
67. Alhammadi AA, Vargas FM, Chapman WG (2015) Comparison of cubic-plus-association and perturbed-chain statistical associating fluid theory methods for modeling Asphaltene phase behavior and pressure-volume-temperature properties. Energy Fuels 29(5):2864–2875
68. Johnston KA, Schoeggl FF, Satyro MA, Taylor SD, Yarranton HW (2017) Phase behavior of bitumen and N-Pentane. Fluid Phase Equilib 442:1–19
69. Johnston KA, Satyro MA, Taylor SD, Yarranton HW (2017) Can a cubic equation of state model bitumen-solvent phase behavior? Energy Fuels 31(8):7967–7981
70. Glandt CA, Chapman WG (1995) Effect of water dissolution on oil viscosity. SPE Reserv Eng Soc Pet Eng 10(1):59–64
71. Goral M, Wisniewska-Goclowska B, Skrzecz A, Owczarek I, Blazej K, Haulait-Pirson MC, Hefter GT, Maczynska Z, Szafranski A, Young CL (2005) IUPAC-NIST solubility data series. 81. Hydrocarbons with Water and Seawater—Revised and Updated. Part 3. C6H8-C6H 12 Hydrocarbons with Water and Heavy Water. J Phys Chem Ref Data 34(2):657–708
72. Amani MJ, Gray MR, Shaw JM (2013) Phase behavior of Athabasca bitumen + water mixtures at high temperature and pressure. J Supercrit Fluids 77:142–152
73. Zirrahi M, Hassanzadeh H, Abedi J (2017) Experimental and modeling studies of water, Light n-Alkanes and Mackay river bitumen ternary systems. Fuel 196:1–12
74. Chen Z, Yang D (2018) Quantification of phase behaviour of solvents-heavy oil/bitumen systems in the presence of water at high pressures and elevated temperatures. Fuel 232(May):803–816
75. Shirani B, Nikazar M, Mousavi-Dehghani SA (2012) Prediction of Asphaltene phase behavior in live oil with CPA equation of state. Fuel 97:89–96
76. Poiling BE, Prausnitz JM, O'Conell JP (2001) The properties of gases and liquids, fifth. McGraw-Hill, New York, USA
77. Chapman WG, Gubbins K.E, Jackson G, Radosz M (1989) SAFT: equation-of-state solution model for associating fluids. Fluid Phase Equilib 52(C):31–38
78. Chapman WG, Gubbins KE, Jackson G, Radosz M, Chapman WG, Radosz M, Chapman WG, Jackson G (1990) New reference equation of state for associating liquids. Ind Eng Chem Res 29(8):1709–1721
79. Huang SH, Radosz M (1990) Equation of state for small, large, polydisperse, and associating molecules. Ind Eng Chem Res 29(11):2284–2294

80. Müller EA, Gubbins KE (2001) Molecular-based equations of state for associating fluids: a review of SAFT and related approaches. Ind Eng Chem Res, 2193–2211
81. Michelsen ML, Hendriks EM (2001) Physical properties from association models. Fluid Phase Equilib 180(1–2):165–174
82. Liang X, Tsivintzelis I, Kontogeorgis GM (2014) Modeling water containing systems with the simplified PC-SAFT and CPA equations of state. Ind Eng Chem Res 53(37):14493–14507
83. Gross J, Sadowski G (2002) Application of the Perturbed-Chain SAFT equation of state to associating systems. Ind Eng Chem Res 41(22):5510–5515
84. Von Solms N, Michelsen ML, Passos CP, Derawi SO, Kontogeorgis GM (2006) Investigating models for associating fluids using spectroscopy. Ind Eng Chem Res 45(15):5368–5374
85. Clark GNI, Haslam AJ, Galindo A, Jackson G (2006) Developing optimal wertheim-like models of water for use in Statistical associating fluid theory (SAFT) and related approaches. Molecul Phys 104:3561–3581
86. Yaws CL (1999) Chemical properties handbook. USA, New York
87. Mączyński A, Wiśniewska-Gocłowska B, Góral M (2004) Recommended liquid-liquid equilibrium data. Part 1. Binary alkane-water systems. J Phys Chem Ref Data 33 (2):549–577
88. Góral M, Wiśniewska-Gocłowska B, ; MącZyński A (2004) recommended liquid-liquid equilibrium data. Part 3. Alkylbenzene-water systems. J Phys Chem Ref Data 33(4):1159–1188
89. Kontogeorgis GM, Voutsas EC, Yakoumis IV, Tassios DP (1996) An equation of state for associating fluids. Ind Eng Chem Res 35(11):4310–4318
90. Yakoumis IV, Kontogeorgis GM, Voutsas EC, Hendriks EM, Tassios DP (1998) Prediction of phase equilibria in binary aqueous systems containing Alkanes, cycloalkanes, and alkenes with the cubic-plus-association equation of state. Ind Eng Chem Res 37(10):4175–4182
91. Kontogeorgis GM, Yakoumis IV, Meijer H, Hendriks E, Moorwood T (1999) Multicomponent phase equilibrium calculations for water-methanol-alkane mixtures. Fluid Phase Equilib 158–160:201–209
92. Folas GK, Gabrielsen J, Michelsen ML, Stenby EH, Kontogeorgis GM (2005) Application of the Cubic-Plus-Association (CPA) equation of state to cross-associating systems. Ind Eng Chem Res 44(10):3823–3833
93. Kontogeorgis GM, Michelsen ML, Folas GK, Derawi S, Von Solms N, Stenby EH (2006) Ten Years with the CPA (Cubic-Plus-Association) Equation of State. Part 1. pure compounds and self-associating systems. Ind Eng Chem Res 45(14):4855–4868
94. Kontogeorgis GM, Folas GK (2010) Thermodynamic models for industrial applications: from classical and advanced mixing rules to association theories. Chichister, England, WILEY
95. Zirrahi M, Hassanzadeh H, Abedi J (2015) Prediction of water solubility in petroleum fractions and heavy crudes using cubic-plus-association equation of state (CPA-EoS). Fuel 159:894–899
96. Yarranton HW, Alboudwarej H, Jakher R (2000) Investigation of Asphaltene association with vapor pressure osmometry and interfacial tension measurements. Ind Eng Chem Res 39(8):2916–2924
97. Barrera DM, Ortiz DP, Yarranton HW (2013) Molecular weight and density distributions of Asphaltenes from crude oils. Energy Fuels 27(5):2474–2487

Chapter 7
Chemical Reaction Equilibrium

List of Abbreviations

GFEM	Gibbs free energy minimization
MHV1	Modified Huron-Vidal mixing rule of order one
PR	Peng Robinson equation of state

Nomenclature

a	Atoms
a	Activity
A, B, C, D, E	Constants belong to the heat capacity expression $\tilde{C}_p$
AAD	Average absolute deviations
b_k	Number of moles of element *k*
C	Chemical components
C_p	Capacity at constant pressure
C_V	Capacity at constant volume
e	Chemical elements
f	Fugacity
F_i	Chemical formula
G	Gibbs free energy
H	Enthalpy
I	Integral
K	Equilibrium constant
L(n, λ)	Lagrangian function
n	Moles
P	Pressure
R	Uiversal gas constant
S	Entropy

B. Carreón-Calderón et al., *Thermophysical Properties of Heavy Petroleum Fluids*,
Petroleum Engineering, https://doi.org/10.1007/978-3-030-58831-1_7

T	Temperature
U	Internal energy
z_i	Total mole fraction of the system
ε	Reaction coordinate or degree of progress
λ	Lagrange multipliers
μ	Chemical potential
ν	Stoichiometric coefficient
ν_{ki}	Number of atoms of an element of type *k* forming the component *i*
ϕ	Fugacity coefficient

Superscript

~	Molar property
0	Standard
ig	Ideal gas

Subscript

e	Elements
f	Formation

Chemical equilibrium of petroleum fluids has been studied in less depth than the phase equilibrium, because it is a more difficult condition to be achieved in exploitation, transportation and processing industries. Besides, reactive systems of petroleum fluids, such as the primary conversion processes in refining facilities, involve simultaneous multiple chemical reactions containing a multitude of components, many of which may not be present in the original mixtures. In such processes, the transformation of petroleum fluids into petroleum fractions with higher added value is mainly addressed from a kinetic point of view, where the main purpose is the catalyst development, operating at optimal temperature and pressure. Usually, some generalizations about kinetics and thermodynamics are found concerning existing chemical reactions on the basis of chemical reaction studies made with the pure components [1, 2].

Although chemical equilibrium is not achieved in practice, the kinetic and thermodynamic aspects, whenever possible, should be simultaneously addressed, regardless of the chemical reaction system under consideration, because chemical equilibrium provide limit values of chemical conversion, temperature and pressure in the system. Thus, for example, the catalyst development is limited to the conditions imposed by the chemical equilibrium. In other words, it does not matter temperature, pressure, or the type of catalyst, the maximum conversion that can be achieved in a reactive system is that value provided by the chemical equilibrium condition. As can

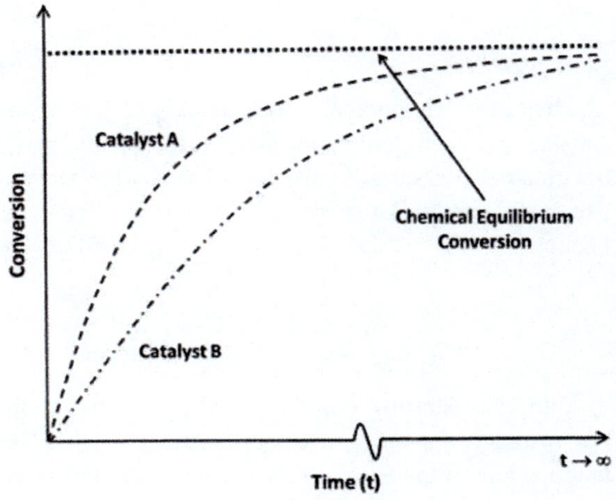

Fig. 7.1 Chemical Equilibrium versus Kinetic Conversion

be observed schematically in Fig. 7.1, the chemical conversion under equilibrium conditions is independent of the time, as explained in Chap. 2; hence, it is reached by chemical reactions in the presence of any catalyst when time goes to infinite.

In what follows, just the fundamentals concepts to determine the effect of temperature, pressure and initial composition on equilibrium conversions are addressed together with some few practical examples that can be verified from the qualitative or quantitative point of view.

7.1 Ideal Gas Heat Capacity

The heat exchanged by a system, in which there are not phase transitions, chemical reactions, and changes in composition, gives rise to changes in the system temperature. When the system is a homogeneous substance of constant composition, the internal energy U of a substance can be expressed as a function of two other state variables as explained in Chap. 2. However, when we are interested in energetic exchanges in the form of work and heat, it is advisable to change entropy by an accessible experimental variable such as the system volume V. Thus, we have that

$$U = U(T, V) \tag{7.1}$$

where its total differential is

$$dU = \left(\frac{\partial U}{\partial T}\right)_V dT + \left(\frac{\partial U}{\partial V}\right)_T dV \tag{7.2}$$

By definition, the heat capacity at constant volume for a closed system is

$$C_V \equiv \left(\frac{\partial U}{\partial T}\right)_V \tag{7.3}$$

Physically, C_V indicates the needed energy for increasing the temperature of a substance per unit temperature as a result of at constant volume. Thus, C_V indicates the greater or lesser difficulty for a given substance to undergo temperature changes because of energy exchanges with the surroundings. Using the C_V definition, Eq. (7.2) becomes

$$dU = C_V dT + \left(\frac{\partial U}{\partial V}\right)_T dV \tag{7.4}$$

Taking as starting point this equation, several special cases can be analyzed. For instance, for any process at constant volume, or when the internal energy is independent of the system volume, the second term vanishes reducing this equation to

$$dU = C_V dT \tag{7.5}$$

which is, for example, satisfied by ideal gases and incomprehensible fluids. Integration of this equation from temperature T_1 to T_2 is

$$\Delta U = \int_{T_1}^{T_2} C_V dT \tag{7.6}$$

It was analyzed in Chap. 2 that the first law of thermodynamics reduces to $dU = dQ$ when $dV = 0$ as long as the expansion-compression work is the only one considered. Therefore, we have that for mechanically reversible processes at constant volume $Q = \Delta U$, with which we obtain

$$Q = \int_{T_1}^{T_2} C_V dT \tag{7.7}$$

Similarly, the enthalpy can be expressed as a function of temperature and pressure:

$$H = H(T, P) \tag{7.8}$$

So that its total differential is

$$dH = \left(\frac{\partial H}{\partial T}\right)_P dT + \left(\frac{\partial H}{\partial P}\right)_T dP \tag{7.9}$$

By definition, the heat capacity at constant pressure for a closed system is now

$$C_P \equiv \left(\frac{\partial H}{\partial T}\right)_P \tag{7.10}$$

Here again, C_P indicates the greater or lesser difficulty for a substance to undergo temperature changes because of energy exchanges with the surroundings, but holding constant pressure. The substitution of the C_P definition into Eq. (7.9) leads to

$$dH = C_P dT + \left(\frac{\partial H}{\partial P}\right)_T dP \tag{7.11}$$

Similar to Eq. (7.4), the second term will be equal to zero in the following circumstances: for any process under constant pressure, and when the enthalpy of the system is independent of pressure, which is fulfilled by ideal gases and approximately by gases at low pressures, solids, and liquids outside the critical region. In this case, we have that Eq. (7.11) simplifies to

$$dH = C_P dT \tag{7.12}$$

whose integral from temperature T_1 to T_2 is

$$\Delta H = \int_{T_1}^{T_2} C_P dT \tag{7.13}$$

From the fundamental equation $dH = TdS + VdP$ studied in Chap. 2, we have that $Q = \Delta H$ for reversible processes at constant pressure ($dP = 0$) because, according to the second law of thermodynamics, $dQ = TdS$. Thus, we arrive at

$$Q = \int_{T_1}^{T_2} C_P dT \tag{7.14}$$

To evaluate this integral, it is necessary to know the functionality of the heat capacity with respect to temperature. The most common relationship is empirical, and has the form of:

$$\frac{C_P}{R} = A + BT + CT^2 + DT^3 + ET^4 \tag{7.15}$$

Here $A, B, C, D, andE$ are specific constants of the particular substance. The term C_P/R is dimensionless, so the units of C_P will depend on the units of the considered R. For instance, if $R = 8.3144725 J/molK$ in the International System of Units, then C_P becomes $\tilde{C}_P$, the molar heat capacity at constant pressure expressed in $J/molK$. Although the above expression can be applied to any state of aggregation, it is generally used for expressing C_P of ideal gases only, because the ideal gas, as illustrated in Chaps. 2, 3, 4 and 6, can be used as reference state to describe the three states of aggregation of a substance either gas, liquid or solid. The heat capacities of ideal gases C_p^{ig} increase smoothly to an upper limit as temperature rises. This limit is reached when all modes of translational, rotational, and vibration molecular motions are fully excited. C_p^{ig} are obtained experimentally from spectroscopic data and from knowledge of molecular structure using methods of statistical mechanics. C_p^{ig} are exactly correct for real gases at zero pressure only, which seldom deviate significantly from the ideal behavior at low pressures; hence, C_p^{ig} usually represents well heat capacities of real gases at low pressures ($P < 1MPa$). In Table 7.1, heat capacities for some relevant elements and components in hydrocarbon mixtures are shown. In this table, the superscript sl indicates solid state, and the superscript ig indicates ideal-gas state. For unidentifiable petroleum fractions, C_p^{ig} can be estimated by making use of the functional-group concept described in Chaps. 3 and 4, and used in Chap. 6. The expressions from the Constantinou-Gani [5] and Joback–Reid [4] methods are shown Table 7.2, and the corresponding group-contribution parameters are shown in Table 7.3. In these tables, υ_{ij} accounts for the occurrence of a functional group i in either a component or pseudocomponent j (petroleum fraction); that is, the number of times that the functional group i appears in the chemical structure or pseudostructure. The parameters C_{PA}^{ij}, C_{PB}^{ij}, C_{PC}^{ij}, and C_{PD}^{ij} correspond to the specific contributions from each functional group to C_p^{ig}.

Table 7.1 Heat Capacities C_p/R [3]

Component	A	Bx10	Cx10^4	Dx10^7	Ex10^{10}
C^{sl}	−0.1000666	0.0419101	−0.0159156	0.0000000	0.0000000
H_2^{ig}	3.0547975	0.0242686	−0.0463638	0.0383428	−0.0105341
$O_2{}^{ig}$	3.5511616	−0.0107041	0.0458033	−0.0392437	0.0106570
S^{sl}	0.2409055	0.1443509	−0.1948653	0.0000000	0.0000000
$CH_4{}^{ig}$	4.2025566	−0.0480572	0.2307305	−0.1840528	0.0472923
$CO_2{}^{ig}$	3.2999126	0.0508932	−0.0235193	0.0048070	−0.0003593
$H_2S{}^{ig}$	4.0745868	−0.0134897	0.0632368	−0.0461810	0.0108583
$C_2H_6{}^{ig}$	3.3851857	0.0522547	0.2278680	−0.2292632	0.0641641
$C_3H_8{}^{ig}$	3.4009414	0.1395159	0.2356977	−0.2798858	0.0825898
i-$C_4H_{10}{}^{ig}$	0.8144844	0.0004107	−0.1235317	−0.0443192	0.0245704
$C_4H_{10}{}^{ig}$	2.4121824	0.3386028	−0.0158074	−0.1137428	0.0410718
i-$C_5H_{12}{}^{ig}$	−0.1059599	0.5712696	−0.2982394	0.0811983	−0.0102644
$C_5H_{12}{}^{ig}$	3.2077840	0.3887684	0.0515006	−0.2001212	0.0673958
$C_6H_{14}{}^{ig}$	3.1179405	0.5042659	−0.0150232	−0.1914255	0.0707009
$C_7H_{16}{}^{ig}$	3.2454292	0.6060163	−0.0538195	−0.2024785	0.0783971
$C_6H_6{}^{ig}$	−3.7727032	0.5708126	−0.3744920	0.1025165	−0.0060766
$C_7H_8{}^{ig}$	−2.8982029	0.6276653	−0.3587364	0.0735105	0.0015125
$C_{10}H_8{}^{ig}$	8.0701548	0.0520046	1.1033786	−1.2050087	0.3715935
$H_2O{}^{ig}$	4.0812018	−0.0101252	0.0359687	−0.0214385	−0.0044421

Table 7.2 Expressions for the estimation of the heat capacity in the ideal-gas standard state from Group Contribution Methods

Group Contribution Method	$\tilde{C}_p^{ig} [J/mol K]$
Joback–Reid [4]	$\sum_i \upsilon_{ij} C_{PA}^{ij} - 37.93 + \left[\sum_i \upsilon_{ij} C_{PB}^{ij} + 0.21\right] T + \left[\sum_i \upsilon_{ij} C_{PC}^{ij} - 3.91x10^{-4}\right] T^2 + \left[\sum_i \upsilon_{ij} C_{PD}^{ij} - 2.06x10^{-7}\right] T^3$
Constantinou-Gani [5]	$\sum_i \upsilon_{ij} C_{PA}^{ij} - 19.7779 + \left[\sum_i \upsilon_{ij} C_{PB}^{ij} + 22.5981\right] (\frac{T-298}{700}) + \left[\sum_i \upsilon_{ij} C_{PC}^{ij} - 10.7983\right] (\frac{T-298}{700})^2$

7.2 Formation Enthalpy

The heat associated with a chemical reaction is the heat exchanged with the environment in the transformation of reactants to products at constant temperature and pressure. The formation heat or formation enthalpy ΔH_{fi} of a given component i is the heat associated with the chemical reaction to form one mole of a pure component from their constituent elements in their stable states of aggregation. In other words, ΔH_{fi} can be defined through

$$H_i = \Delta H_{fi} + \sum_e H_e^i \tag{7.16}$$

$\sum_e H_e^i$ accounts for the total enthalpy of elements forming the component i, and it represents the reference state from which the total enthalpy H_i is established. As reference state, $\sum_e H_e^i$ is independent of any chemical reaction. Thus, if we use Eq. (7.16) to determine H_i for each component in a chemical reaction, then the sum of the enthalpies of the elements forming the reactants must be equal to the sum of the enthalpies of the elements forming the products, cancelling each other. Therefore, $\sum_i H_i^e$ does not appear in the final expression for the chemical reaction; only the appropriate combination of ΔH_{fi} from each component remains, as will be shown later. The enthalpy of formation of a given component at $0.101325 MPa$ $(1atm)$ is the standard enthalpy of formation ΔH_{fi}^0, which are shown in Table 7.4 for some relevant components in hydrocarbon mixtures.

At this point, it is important to state explicitly the standard-state concept in the context of chemical reactions: a standard state is the state of a given component at specific pressure and aggregation state. The standard state remains fixed regardless of temperature, which is always the system temperature; therefore, they are functions of temperature only. For instance, we have that the ideal-gas state can be the standard state for real gases and liquids at any temperature, from which their equilibrium properties can be estimated, as illustrated in Chap. 6. For the specific case of chemical reactions, $\Delta \tilde{H}_{fi}^0$ is arbitrarily set equal to zero for all elements such as O_2 and H_2 at their standard states at any temperature, because numerical values of

Table 7.3 Contributions to $\tilde{C}_p^{ig}$ from functional groups

Functional Group	Joback–Reid [4]				Constantinou-Gani [5]		
	C_{PA}^{ij}	C_{PB}^{ij}	C_{PC}^{ij}	C_{PD}^{ij}	C_{PA}^{ij}	C_{PB}^{ij}	C_{PC}^{ij}
CH_3	19.5	−8.08E-3	1.53E-4	−9.67E-8	35.1152	39.5923	−9.9232
-CH_2- (chain)	−0.909	9.50E-2	−5.44E-5	1.19E-8	22.6346	45.0933	−15.7033
> CH- (chain)	−23.0	2.04E-1	−2.65E-4	1.20E-7	8.9272	59.9786	−29.5143
> C < (chain)	−66.2	4.27E-1	−6.41E-4	3.01E-7	0.3456	74.0368	−45.7848
ACH	−2.14	5.74E-2	−1.64E-6	1.59E-8	16.3794	32.7433	−13.1692
AC	−8.25	1.01E-1	−1.42E-4	6.78E-8	10.4283	25.3634	−12.7283
-CH_2- (cyclic)	−6.03	8.54E-2	−8.00E-6	−1.80E-8	22.6346	45.0933	−15.7033
> CH- (cyclic)	8.67	1.62E-1	−1.60E-4	6.24E-8	8.9272	59.9786	−29.5143
> C < (cyclic)	−90.9	5.57E-1	-9.00E-4	4.69E-7	0.3456	74.0368	−45.7848

Table 7.4 Molar Gibbs free energy and Enthalpy of formation in their standard state at 298.15 K [5]

Component	$\Delta \overset{\sim 0}{H}_{if}$[kJ/mol]	$\Delta \overset{\sim 0}{G}_{if}$[kJ/mol]
CH_4	−74.52	−50.45
CO_2	−393.51	−394.38
H_2S	−20.63	−33.43
C_2H_6	−83.82	−31.86
C_3H_8	−104.68	−24.29
i-C_4H_{10}	−134.99	−21.44
C_4H_{10}	−125.79	−16.57
i-C_5H_{12}	−153.70	−13.86
C_5H_{12}	−146.76	−8.65
C_6H_{14}	−166.92	0.15
C_7H_{16}	−187.80	8.20
C_6H_6	82.88	129.75
C_7H_8	50.17	122.29
$C_{10}H_8$	150.30	223.50
H_2O	−241.81	−228.42

the constituent elements are not of concern, as explained before. $\Delta \overset{\sim 0}{H}_{fi}$ is an important thermodynamic variable because it is the starting point for the estimation of the heat involved in chemical reaction as a function of temperature, as observed from Eq. (7.14).

When $\Delta \overset{\sim 0}{H}_{fi}$ is not available, as in the case of unidentifiable petroleum fluids, there are group contribution methods that can be used for estimation purposes as long as the corresponding chemical structure is known. For instance, in Fig. 7.2, the estimation of $\Delta \overset{\sim 0}{H}_{fi}$ for 25 identifiable components is illustrated. Here, three methods of group contribution are evaluated: the Joback–Reid [4], Marrero-Gani [6] and Albahri-Aljasmi [7] methods. The expressions for each method in terms of occurrences υ_{ij} of the functional group i in the component j are shown Table 7.5, where h_f^{ij} are contributions to $\Delta \overset{\sim 0}{H}_{fi}$ because of the presence of the functional group i. The h_f^{ij} values are shown in Table 7.6 for the used functional groups in the thermodynamic characterization of petroleum fractions. In Fig. 7.2, it can be seen that the AAD is below 20% in almost all cases except cyclohexane and phenanthrene. The average absolute deviations (AAD) for each case are 7.4, 4.8 and 4.4% for Albahri-Aljasmi, Marrero-Gani and Joback–Reid, respectively.

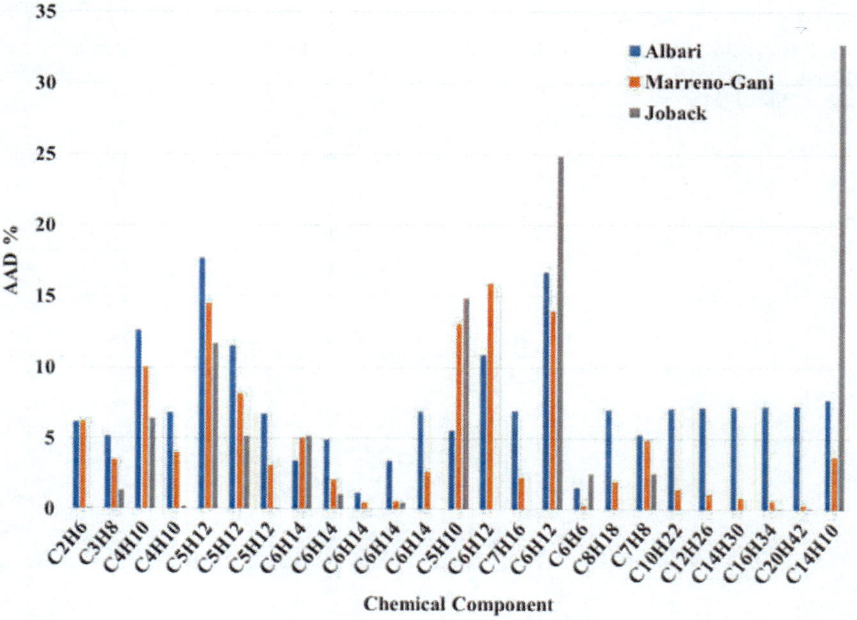

Fig. 7.2 Average absolute deviation percent (AAD %) of the estimation of enthalpy of formation for pure components

Table 7.5 Expressions for the estimation of $\Delta \overset{\sim 0}{H}_f$ and $\Delta \overset{\sim 0}{G}_f$ at 298.15 K from Group Contribution Methods

Group Contribution Method	$\Delta \overset{\sim 0}{H}_f [kJ/mol]$	$\Delta \overset{\sim 0}{G}_f [kJ/mol]$
Joback–Reid [4]	$68.29 + \sum_i \upsilon_{ij} h_f^{ij}$	$53.88 + \sum_i \upsilon_{ij} g_f^{ij}$
Marrero-Gani [6]	$5.549 + \sum_i \upsilon_{ij} h_f^{ij}$	$-34.967 + \sum_i \upsilon_{ij} g_f^{ij}$
Albahri-Aljasmi [7]	$23.084 + \sum_i \upsilon_{ij} h_f^{ij}$	n/a

7.2.1 *Heat Combustion*

As an application of the formation enthalpy to chemical reactions involving petroleum fluids, we have the combustion reactions. These are chemical reactions between hydrocarbons and oxygen at gaseous state mainly, where great amount of heat is released to the surroundings. The reaction rates are so fast that chemical equilibrium is reached in practice. The measurements of heat of combustion are used to determine the heats of formation of all organic compounds that contain only carbon, hydrogen and oxygen.

Table 7.6 Contributions to $\Delta \overset{\sim 0}{G}_f$ and $\Delta \overset{\sim 0}{H}_f$ from functional groups

Functional Group	Albahri-Aljasmi [7]	Marrero-Gani [6]		Joback–Reid [4]	
	h_f^{ij}	g_f^{ij}	h_f^{ij}	h_f^{ij}	g_f^{ij}
CH_3	−51.2687	2.878	−42.479	−76.45	−43.96
$-CH_2-$ (chain)	−19.048	8.064	−20.829	−20.64	8.42
$>CH-$ (chain)	3.142	8.254	−7.122	29.89	58.36
$>C<$ (chain)	21.8887	16.413	8.928	82.23	116.02
ACH	10.1903	26.732	12.861	2.09	11.3
AC	24.5625	20.379	20.187	46.43	54.05
$-CH_2-$ (cyclic)	−20.9319	13.287	−18.575	−26.8	−3.68
$>CH-$ (cyclic)	16.8999	6.107	−12.464	8.67	40.99
$>C<$ (cyclic)	56.8728	−0.193	−2.098	79.72	87.88

By definition the heat of combustion arises from the combustion reaction where one mole of the substance is burned on the reactive side, with as much oxygen as necessary to burn the substance completely. Organic components that only have carbon, hydrogen and oxygen are burned to the formation of gas carbon dioxide and liquid water. For instance, any alkane component reacts with oxygen according to the following general combustion reaction:

$$C_nH_{2n+2} + \left(\frac{3n+1}{2}\right)O_2 \rightarrow nCO_2 + (n+1)H_2O \tag{7.17}$$

Figure 7.3 illustrates errors in the form of the Absolute Average Deviation percent (AAD %) of the prediction of the heat of combustion for the petroleum fractions from the Chevron data bank [8], which contains experimental heats of combustion for 420 petroleum fractions with API gravities ranging from 0.5 to 88 degrees. For the calculations illustrated in Fig. 7.3, the chemical pseudostructures of the petroleum fractions are obtained by performing the Gibbs free energy minimization (GFEM) with the Peng-Robinson equation of state [9] and both group contribution methods Marreno-Gani [6] and Joback –Reid [4], as explained in Chap. 3, and used widely in Chap. 6. Using these calculated pseudostructures, the formation enthalpy are obtained from the Marrero-Gani and Joback–Reid expressions in Tables 7.5 and 7.6.

Once chemical pseudostructures are determined, the atomic balance was performed, that is, the total number of carbon and hydrogen atoms are obtained for each petroleum fraction. Additionally, the stoichiometric balance is performed considering oxygen as reactant, while water and carbon dioxide are considered as products of the combustion reaction according to the chemical reaction given by Eq. (7.17). In the case of petroleum fractions, it is important to remember that chemical

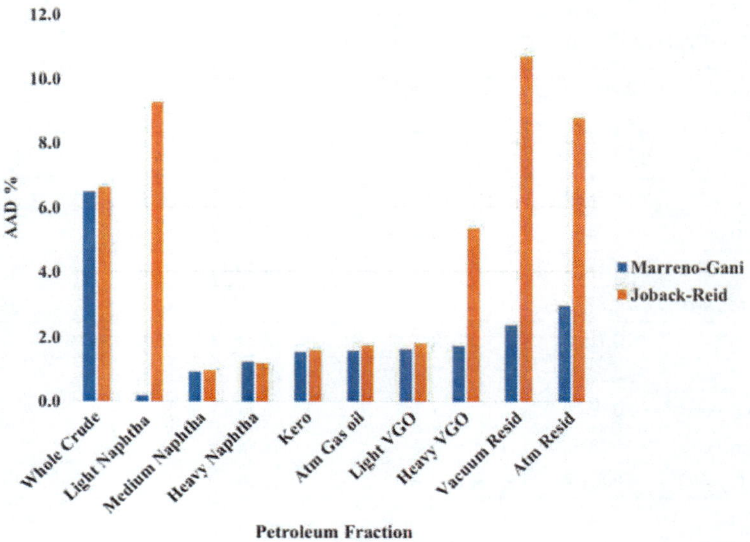

Fig. 7.3 Estimation of the heat combustion of petroleum fractions, using chemical pseudostructures from the Gibbs free energy minimization procedure

pseudostructures are made up of functional groups, which in turn are formed by non-integer numbers of atoms. Hence, the stoichiometric balance may not be performed straightforwardly. Such balance can be performed through the concept of Atomic Matrix.

An Atomic Matrix A consists of a mathematical matrix of size either (e, C) or (C, e), where C is the number the chemical components, and e is the number the chemical elements in the reacting system. The array inputs are given by a_{ki}, which is the number of atoms of the element k present in each molecule of the component i. For the specific case of the combustion reaction of petroleum fractions described by a chemical pseudostructure of the form C_xH_y, the Atomic Matrix is

$$A = \begin{pmatrix} & C_xH_y & O_2 & H_2O & CO_2 \\ C & x & 0 & 0 & 1 \\ H & y & 0 & 2 & 0 \\ O & 0 & 2 & 1 & 2 \end{pmatrix} \tag{7.18}$$

As seen, the number of components (C_xH_y, O_2, H_2O, and CO_2) is equal to 4, while the number of chemical elements (C,H, and O) is equal to 3. In the chemical pseudostructure C_xH_y, x and y stand for the non-integer number of carbon and hydrogen atoms calculated from the GFEM procedure.

To perform the stoichiometric balance from an Atomic Matrix, it is necessary to determine the rank r of the matrix, which is denoted by $rank(A) = r$. This $rank(A)$ accounts for the number of linearly independent either columns or rows in the matrix, and it is related to the number of independent chemical reactions ($\#ir$) as follows:

$$\#ir = C - r \tag{7.19}$$

To determine their ranks every matrix must be transformed into a form such that in the $k-th$ row ($k = 2, 3, \ldots, e$), at least the first $k-1$ inputs are equal to zero. The number of rows different from zero in the transformed matrix is equal to the rank r of the matrix. For instance,

$$A = \begin{pmatrix} & C_xH_y & O_2 & H_2O & CO_2 \\ C & 26.4330838131707 & 0 & 0 & 1 \\ H & 43.2548126231137 & 0 & 2 & 0 \\ O & 0 & 2 & 1 & 2 \end{pmatrix} \tag{7.20}$$

whose reduced staggered form is

$$A = \begin{pmatrix} & C_xH_y & O_2 & H_2O & CO_2 \\ C & 1 & 0 & 0 & 0.03783137 \\ H & 0 & 1 & 0 & 1.40909729 \\ O & 0 & 0 & 1 & -0.81819459 \end{pmatrix} \tag{7.21}$$

Here, we have three rows different from zero; therefore, $rank(A) = 3$, with which $\#ir = 4-3 = 1$. In other words, there is only one independent combustion reaction that can be built with these components. It is worth underlying that 13 digits are necessary in the inputs for C_xH_y in order to avoid an ill-conditioned matrix. The matrix is called ill-conditioned when a small change in the constant inputs results in a large change in the solution.

The stoichiometric coefficients of the combustion reaction are obtained from the off-diagonal inputs from the reduced staggered form of the matrix. In other words, the combustion reaction of the petroleum fraction with chemical pseudostructure C_xH_y is given by

$$0.03783137C_xH_y + 1.40909729O_2 \rightarrow CO_2 + 0.81819459H_2O \tag{7.22}$$

The stoichiometric balance calculated from the Atomic Matrix is independent of the form of the matrix. For instance, alternatively, the matrix could have been written as

$$A = \begin{pmatrix} & C_xH_y & CO_2 & O_2 & H_2O \\ C & 26.4330838131707 & 1 & 0 & 0 \\ H & 43.2548126231137 & 0 & 0 & 2 \\ O & 0 & 2 & 2 & 1 \end{pmatrix} \tag{7.23}$$

where the last three columns were exchanged. In this case, the reduced staggered form of the matrix is now

$$A = \begin{pmatrix} & C_xH_y & CO_2 & O_2 & H_2O \\ C & 1 & 0 & 0 & 0.04623762 \\ H & 0 & 1 & 0 & -1.22220313 \\ O & 0 & 0 & 1 & 1.72220313 \end{pmatrix} \tag{7.24}$$

From where, the balanced combustion reaction turns out be equal to

$$0.04623762 C_xH_y + 1.72220313 O_2 \rightarrow 1.22220313 CO_2 + H_2O \tag{7.25}$$

The final balanced reaction arises from dividing all coefficients by that corresponding to the petroleum fraction, that is,

$$C_xH_y + 37.25 O_2 \rightarrow 26.43 CO_2 + 21.63 H_2O \tag{7.26}$$

This result is achieved irrespective of the used balanced reaction either that given by Eq. (7.22) or that given by Eq. (7.25). Finally, we apply the reaction heat equation:

$$\Delta \tilde{H}_{combustion} = \sum_{i=1}^{C} \nu_i \Delta \tilde{H}_{fi} \tag{7.27}$$

For the specific example given by the balance reaction Eq. (7.26), the stoichiometric coefficients are $\nu_{C_xH_y} = -1$, $\nu_{O_2} = -37.25$, $\nu_{CO_2} = 26.43$, and $\nu_{H_2O} = 21.63$, whereas the standard enthalpy of formation expressed in kJ/mol are $\Delta H^0_{fC_xH_y} = -331.88$ using the Marreno-Gani method, $\Delta H^0_{fO_2} = 0.0$, $\Delta H^0_{fCO_2} = -393.51$, and $\Delta H^0_{fH_2O} = -241.81$, resulting in $\Delta \tilde{H}_{combustion} = -15298.9 kJ/mol$.

As observed from Fig. 7.3, calculations from both methods agree well with experiments for intermediate petroleum fractions. In contrast, better outcomes are obtained by using the Marreno-Gani method for the heaviest petroleum fractions. This result may be generalized to any kind of equilibrium problem where cubic equations of state are intended to be used: the heavier the petroleum fraction, better outcomes are obtained by using the Marreno-Gani method instead of the Joback–Reid method.

The previous results confirm what has already been established previously, having a correct thermodynamic characterization of petroleum fluids, by means of the contribution of groups, makes it possible to obtain properties derived from their chemical structure, as is the case of the enthalpy of combustion.

7.3 Degree of Advancement

To express chemical equilibrium criteria in terms of measurable variable, it is advisable to start with systems formed by a single phase in the presence of a single chemical reaction of the form

$$\nu_1 F_1 + \nu_2 F_2 + \cdots \rightarrow \nu_3 F_3 + \nu_4 F_4 + \tag{7.28}$$

where ν_i is the stoichiometric coefficient of a component i in the chemical reaction, whose chemical formula is F_i. By convection, ν_i is positive for a product (+), it is negative (−) for a reactant, and it is equal to zero for any inert species. The stoichiometric coefficients indicate that the number of the atoms in the reactants side must be equal to the number of atoms in the products side. Therefore, when changes in the number of moles n_i of a given component i occur as a result of the chemical reaction, they are proportional to the stoichiometric coefficientsν_i. Besides, this change is not independent of changes in the number of moles from the other components. All these changes are connected with each other through the so-called reaction coordinate or degree of progressε, which represent the extent or progress to which the reaction has proceeded. In differential form, we have that

$$d\varepsilon \equiv \frac{dn_1}{\nu_1} = \frac{dn_2}{\nu_2} = \frac{dn_3}{\nu_3} = \frac{dn_4}{\nu_4} = \cdots = \frac{dn_i}{\nu_i} \tag{7.29}$$

From which, we have that

$$dn_i = \nu_i d\varepsilon \tag{7.30}$$

Because ν_i is a dimensionless number, ε has units of moles. For the initial state of the system before the reaction takes places, $\varepsilon = 0$, and for the final equilibrium state $\varepsilon = \varepsilon_e$, the degree of advance at equilibrium conditions. Between these two extremes values, irreversible chemical reactions, as those illustrated schematically in Fig. 7.1, take place. Equation (7.30) may be conceived as a mass constraint that has to be fulfilled by each chemical reaction.

The integration of Eq. (7.30) from an initial state to any arbitrary state ε is

$$\int_{n_{i0}}^{n_i} dn_i = \nu_i \int_0^{\varepsilon} d\varepsilon \tag{7.31}$$

or

$$n_i = n_{i0} + \nu_i \varepsilon \tag{7.32}$$

where n_{i0} are the initial number of moles of component i before the reaction takes place. From this relationship, we have that the summation over all components leads to

$$n = n_0 + \nu \varepsilon \tag{7.33}$$

where $n = \sum_{i=1}^{C} n_i$, $n_0 = \sum_{i=1}^{C} n_{i0}$, and $\nu = \sum_{i=1}^{C} \nu_i$. Thus, the mole fraction x_i of component i is related to ε by

$$x_i \equiv \frac{n_i}{n} = \frac{n_{i0} + \nu_i \varepsilon}{n_0 + \nu \varepsilon} \tag{7.34}$$

7.4 Formation Free Energy and Chemical Equilibrium Constant

In Chap. 6, the phase equilibrium criteria for systems in the absence of chemical reactions are widely discussed. Here, we will focus on establishing the applicability of the equilibrium criteria for systems formed by a single phase in the presence of chemical reactions. The simultaneous phase and chemical equilibrium will be analyzed later.

As analyzed widely in Chap. 2, the total Gibbs free energy G of a closed system at constant temperature and pressure decreases during an irreversible process, until the conditions for the equilibrium state are reached, which corresponds to the minimum value of G. From a mathematical point of view, this corresponds to a stationary point, that is,

$$(dG)_{T,P} = 0 \tag{7.35}$$

In other words, in a mixture of chemical components, when an equilibrium state is not present, and chemical reactions take place at constant temperature and pressure, the total Gibbs free energy of the system will decrease. Once the equilibrium state is reached, subsequent changes do not occur, and the system remains in this state as long as temperature and pressure remain fixed. It is worth remembering that the equilibrium criteria expressed by Eq. (7.35) remains valid irrespective of the path followed by the system towards the equilibrium state, as illustrated in Fig. 7.1. In other words, the chemical equilibrium state is independent of the specific chemical reactions occurring in the system.

On the other hand, the fundamental relation of thermodynamics in terms of G for single phase systems was derived in Chap. 2:

$$dG = -SdT + VdP + \sum_{i=1}^{C} \mu_i dn_i \tag{7.36}$$

Thus, dn_i can be replaced by product of $\nu_i d\varepsilon$ by using Eq. (7.30), that is,

$$dG = -SdT + VdP + \sum_{i=1}^{C} \mu_i \nu_i d\varepsilon \tag{7.37}$$

Because G is a state function, the right side of this equation is an exact differential equation, and it follows that:

$$\left(\frac{\partial G}{\partial \varepsilon}\right)_{T,P} = \sum_{i=1}^{C} \nu_i \mu_i \tag{7.38}$$

The quantity $\sum_i \upsilon_i \mu_i$ represents the rate of change of total Gibbs energy of the system with respect to the reaction coordinate at constant temperature and pressure. Because G must be minimum at the chemical equilibrium state, where all partial derivatives of G must be equal to zero according the mathematical definition of a stationary point, it follows that

$$\sum_i \nu_i \mu_i = 0 \tag{7.39}$$

To express chemical potentials μ_i in terms of measurable variables, the fugacity f_i was introduced in Chap. 2. There, it is shown that μ_i is given by

$$\mu_i(T, P, \boldsymbol{x}) = \mu_i^0 + RTln\left(\frac{f_i}{f_i^0}\right) \tag{7.40}$$

Here, μ_i^0 is the chemical potential of pure component i at the reference fugacity of the pure component f_i^0, whose value are arbitrarily assigned. In terms of the activity a_i, which is defined as $a_i \equiv f_i/f_i^0$ in the same Chap. 2, one can write Eq. (7.40) as

$$\mu_i(T, P, \boldsymbol{x}) = \mu_i^0 + RTln(a_i) \tag{7.41}$$

By substituting this equation into the equilibrium criteria expressed by Eq. (7.39), we arrive at

$$\sum_{i=1}^{C} \nu_i\left[\mu_i^0 + RTln(a_i)\right] = 0 \tag{7.42}$$

which can be rewritten as

$$\sum_{i=1}^{C} \nu_i \mu_i^0 + \sum_{i=1}^{C} \nu_i RTln(a_i) = 0 \tag{7.43}$$

Or equivalently

$$ln\left(\prod_{i=1}^{C} a_i^{\nu_i}\right) = \frac{-\sum_{i=1}^{C} \nu_i \mu_i^0}{RT} \tag{7.44}$$

where Π means the product of all components i. In exponential form, this equation turns out be equal to:

$$\prod_{i=1}^{C} a_i^{\nu_i} = e^{-\sum_{i=1}^{C} \nu_i \mu_i^0/RT} \tag{7.45}$$

On the right hand side from this equation, one can define the so-called equilibrium constant K for the chemical reaction as follows:

$$-RTlnK = \sum_{i=1}^{C} \nu_i \mu_i^0 \equiv \Delta \tilde{G}^{o} \tag{7.46}$$

Considering that μ_i^0 is a pure-component property relying on temperature only at fixed pressure, K depends only on temperature as well. The term $\Delta \tilde{G}^{o}$ is the conventional way of representing the quantity $\sum_{i=1}^{C} \nu_i \mu_i^0$, which is called the standard Gibbs free energy change of the reaction. As seen, $\Delta \tilde{G}^{0}$ in is the difference between the Gibbs free energies of the products and the reactants, when each one is in its standard state at the system temperature and at fixed pressure. Therefore, the $\Delta \tilde{G}^{0}$ value is fixed for a given reaction, once the temperature has been established, and it is independent of pressure and the equilibrium composition.

Activities a_i provide the connection between the equilibrium states and the standard states of the individual components. Standard states are arbitrary but must always be at equilibrium temperature. The selected standard states are no necessary to be the same for all components taking part in a reaction. However, for a given component, the standard state represented by μ_i^0 must be the same state as for the fugacity f_i^0, on which the activity a_i is based. Hence, it is convenient to set μ_i^0 equal to $\Delta \tilde{G}^{0}_{fi}$, the molar Gibbs free energy of formation at the ideal-gas standard state, with which $f_i^0 = P^0$, and $a_i = Px_i\phi_i/P_0$, ϕ_i being the fugacity coefficient as used to determine phase equilibriums in Chap. 6. Similar to $\Delta \tilde{H}^{0}_{fi}$, $\Delta \tilde{G}^{0}_{fi}$ is the molar Gibbs free energy associated with the chemical reaction to form one mole of component from their constituent elements at their stable states of aggregation. In Table 7.4, $\Delta \tilde{G}^{0}_{fi}$ values for some relevant components in hydrocarbon mixtures are shown, whereas, for unidentifiable petroleum fractions, the $\Delta \tilde{G}^{0}_{fi}$ values can be estimated from the information given in Tables 7.5 and 7.6. Here again, $\Delta \tilde{G}^{0}_{fi}$ is arbitrarily set equal to zero for all elements at their standard states at any temperature.

7.4.1 Effect of Temperature on the Equilibrium Constant

When the determination of molecular weight was addressed in Chap. 6, it was found the following relationship between the molar Gibbs free energy and enthalpy:

$$\left[\frac{\partial}{\partial T}\left(\frac{\tilde{G}}{RT}\right)\right]_P = -\frac{\tilde{H}}{RT^2} \tag{7.47}$$

Which can be written in terms of pure-component properties at its standard state as follows:

$$\tilde{H}_i^o = -RT^2\frac{d(\mu_i^0/RT)}{dT} \tag{7.48}$$

The use of total derivatives is appropriate because the standard state is only a function of temperature. If we multiply Eq. (7.48) on both sides of equality by ν_i, and we perform the summation over all components, we obtain

$$\sum_{i=1}^{C} \nu_i \tilde{H}_i^0 = -RT^2\frac{d\left(\sum_{i=1}^{C} \nu_i \mu_i^0/RT\right)}{dT} \tag{7.49}$$

If $\Delta\tilde{H}^0 \equiv \sum_{i=1}^{C} \nu_i \tilde{H}_i^0$, and $\Delta\tilde{G}^0 \equiv \sum_{i=1}^{C} \nu_i \mu_i^0$, then

$$\Delta\tilde{H}^0 = -RT^2\frac{d\left(\Delta\tilde{G}^0/RT\right)}{dT} \tag{7.50}$$

Because temperature at the standard state is that of the equilibrium mixture, standard properties of reaction such as $\Delta\tilde{G}^0$ and $\Delta\tilde{H}^0$ vary with the equilibrium temperature as well. Therefore, the dependency of $\Delta\tilde{G}^0$ on temperature is given by Eq. (7.50), which can be rewritten as:

$$\frac{d(\Delta\tilde{G}^0/RT)}{dT} = \frac{-\Delta\tilde{H}^0}{RT^2} \tag{7.51}$$

According to the definition of equilibrium constant given by Eq. (7.46), we have that

$$\frac{\Delta\tilde{G}^0}{RT} = -lnK \tag{7.52}$$

with which, Eq. 7.51 can be written as

$$\frac{d\ln K}{dT} = \frac{\Delta \overset{\sim 0}{H}}{RT^2} \tag{7.53}$$

This differential equation gives the effect of temperature on the equilibrium constant, and hence on the equilibrium conversion. If $\Delta \overset{\sim 0}{H}$ is negative, which means that the reaction is exothermic (heat is released to the surroundings during the chemical reaction), then the equilibrium constant decreases with increasing temperature. Conversely, K increases with temperature for an endothermic reaction (heat is absorbed from the surroundings during the chemical reaction).

If ΔH^0, the change in the standard enthalpy (heat) of reaction, is assumed to be independent of temperature, the integration of Eq. (7.53), from a temperature T_1 to an arbitrary temperature T, is

$$ln\frac{K}{K_1} = -\frac{\Delta \overset{\sim 0}{H}}{R}\left(\frac{1}{T} - \frac{1}{T_1}\right) \tag{7.54}$$

To establish a more rigorous development of the effect of temperature on the equilibrium constant, we have from Chap. 2 that G was defined in terms of other thermodynamic potentials as follows:

$$G \equiv U + PV - TS = H - TS \tag{7.55}$$

which can be written for a pure component at its standard state as follows:

$$\overset{\sim 0}{G_i} = \overset{\sim 0}{H_i} - T\overset{\sim 0}{S_i} \tag{7.56}$$

When this relation is multiplying by ν_i, the summation over all components results in

$$\Delta \overset{\sim 0}{G} = \Delta \overset{\sim 0}{H} - T\Delta \overset{\sim 0}{S} \tag{7.57}$$

where $\Delta \overset{\sim 0}{S} \equiv \sum_{i=1}^{C} \nu_i \overset{\sim 0}{S_i}$, the standard-entropy change of reaction. Therefore, this expression expressed the relationship between standard properties changes of reaction. The standard-enthalpy change of reaction ΔH^0 is related to temperature through Eq. (7.14) written in terms of the standard- heat-capacity change of reaction $\left(\Delta \overset{\sim 0}{C}_P \equiv \sum_{i=1}^{C} \nu_i \overset{\sim 0}{C}_{pi}\right)$:

$$\Delta \overset{\sim 0}{H} = \Delta \overset{\sim 0}{H}_0 + R \int_{T_0}^{T} \frac{\Delta \overset{\sim 0}{C}_P}{R} dT \tag{7.58}$$

On the other hand, the dependence of $\Delta \overset{\sim 0}{S}$ with respect to temperature can be obtained similarly. In Chap. 2, a fundamental thermodynamic expression relating entropy to enthalpy was derived; the expression is

$$dH = TdS + VdP \tag{7.59}$$

From which, it can be obtained directly, the following expression for a pure component at its standard state at constant pressure:

$$d\overset{\sim 0}{S}_i = \overset{\sim 0}{C}_{Pi} \frac{dT}{T} \tag{7.60}$$

where the definition of $\tilde{C}_P$ given by Eq. (7.10) is used. Here again, when this relation is multiplying by ν_i, the summation over all components results in

$$d\Delta \overset{\sim 0}{S}_i = \Delta \overset{\sim 0}{C}_P \frac{dT}{T} \tag{7.61}$$

whose integral is given by

$$\Delta \overset{\sim 0}{S} = \Delta \overset{\sim 0}{S}_0 + R \int_{T_0}^{T} \frac{\Delta \overset{\sim 0}{C}_P}{R} \frac{dT}{T} \tag{7.62}$$

$\Delta \overset{\sim 0}{S}$ and $\Delta \overset{\sim 0}{S}_0$ are the standard-entropy changes of reaction at temperature T and at reference temperature T_0, respectively. The combination of Eqs. (7.57), (7.58), and (7.62) results in

$$\Delta \overset{\sim 0}{G} = \Delta \overset{\sim 0}{H}_0 + R \int_{T_0}^{T} \frac{\Delta \overset{\sim 0}{C}_P}{R} dT - T \Delta \overset{\sim 0}{S}_0 - RT \int_{T_0}^{T} \frac{\Delta \overset{\sim 0}{C}_P}{R} \frac{dT}{T} \tag{7.63}$$

$\Delta \overset{\sim 0}{S}_0$ can be eliminated from this equation by using

$$\Delta \overset{\sim 0}{S}_0 = \frac{\Delta \overset{\sim 0}{H}_0 - \Delta \overset{\sim 0}{G}_0}{T_0} \tag{7.64}$$

which is derived from Eq. (7.57). Thus, we arrive at

$$\Delta \overset{\sim 0}{G} = \Delta \overset{\sim 0}{H}_0 - \frac{T}{T_0}\left(\Delta \overset{\sim 0}{H}_0 - \Delta \overset{\sim 0}{G}_0\right) + R\int_{T_0}^{T} \frac{\Delta \overset{\sim 0}{C}_P}{R} dT - RT\int_{T_0}^{T} \frac{\Delta \overset{\sim 0}{C}_P}{R}\frac{dT}{T} \tag{7.65}$$

Finally, if we divide by RT, we obtain

$$\frac{\Delta \overset{\sim \circ}{G}}{RT} = \frac{\Delta \overset{\sim 0}{G}_0 - \Delta \overset{\sim 0}{H}_0}{RT_0} + \frac{\Delta \overset{\sim 0}{H}_0}{RT_0} + \frac{1}{T}\int_{T_0}^{T} \frac{\Delta \overset{\sim 0}{C}_P}{R} dT - \int_{T_0}^{T} \frac{\Delta \overset{\sim 0}{C}_P}{R}\frac{dT}{T} \tag{7.66}$$

For illustrative purposes, the equilibrium constant of the hydrogenation of three aromatic compounds (benzene, toluene, and naphthalene) is determined at three different temperatures. Calculations are compared with experimental results reported in the open literature. The hydrogenation chemical reactions, for each case are shown by chemical reactions given by Eqs. (7.67), (7.68), and (7.69) for benzene, toluene, and naphthalene, respectively.

$$C_6H_{6(g)} + 3H_{2(g)} \rightarrow C_6H_{12(g)} \tag{7.67}$$

$$C_7H_{8(g)} + 3H_{2(g)} \rightarrow C_7H_{14(g)} \tag{7.68}$$

$$C_{10}H_{8(g)} + 2H_{2(g)} \rightarrow C_{10}H_{12(g)} \tag{7.69}$$

To illustrate the procedure, the equilibrium constants were calculated using two approaches: first, pure-component data from Tables 7.1 and 7.4 are used; second, the Joback–Reid group contribution method is used with information from Tables 7.2 and 7.5. As observed in Fig. 7.4, results from both methods are comparable to each other.

Once the equilibrium constant K is available, the equilibrium composition can be calculated by means of Eq. (7.45), which can be written as a function of fugacity coefficients and pressure as follows:

$$\prod_{i=1}^{C}\left(\frac{Px_i\phi_i}{P_0}\right)^{\nu_i} = K \tag{7.70}$$

Or equivalently as

$$\prod_{i=1}^{C}(x_i\phi_i)^{\nu_i} = \left(\frac{P}{P_0}\right)^{-\nu} K \tag{7.71}$$

where $\nu \equiv \sum_{i=1}^{C} \nu_i$. If the mixture is formed by ideal gases ($\phi_i = 1$) and $P = P_0$, then this relationship simplifies to

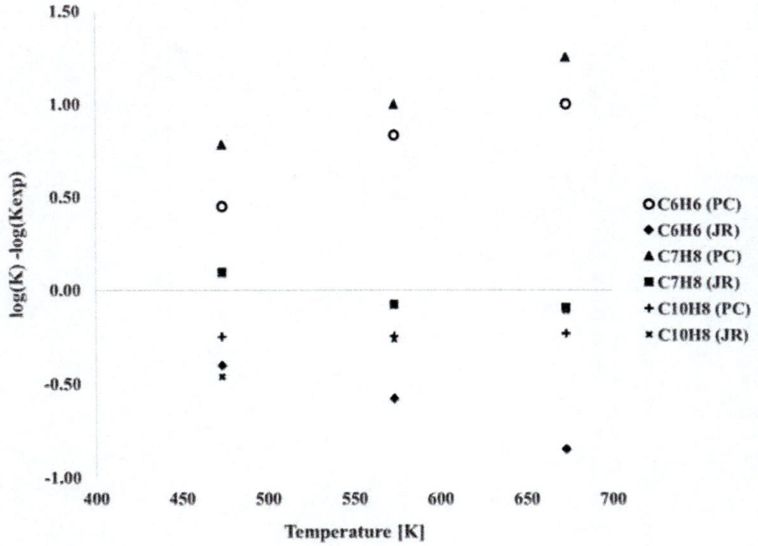

Fig. 7.4 Difference between experimental data [1] $\log(K_{exp})$ and calculated values of log(K) of the equilibrium constant K. PC: pure-component data. JR: the Joback–Reid group contribution method

$$\prod_{i=1}^{C} x_i^{\nu_i} = K \tag{7.72}$$

Although, strictly speaking, this equation is only valid for mixtures of ideal gases and liquids, it has been usually used as starting point for experimental measurements, where all effects different from the composition are included in the equilibrium constant. For instance, if the mole fraction of the components involved in the hydrogenation of benzene as expressed by the chemical reaction Eq. (7.67) are intended to be calculated, then from Eq. (7.34) we can write

$$x_{C6H6} = \frac{1-\varepsilon}{4-3\varepsilon} \tag{7.73}$$

$$x_{H2} = \frac{3-3\varepsilon}{4-3\varepsilon} \tag{7.74}$$

$$x_{C6H12} = \frac{\varepsilon}{4-3\varepsilon} \tag{7.75}$$

Here, $\nu_{H_2} = -3$, $\nu_{C_6H_6} = -1$, and $\nu_{C_6H_{12}} = 1$, resulting in $\nu = -3-1+1 = -3$. Besides, the initial number of moles is assumed to be at stoichiometric proportions; that is, n_{i0} is set equal a 1 and 3 for benzene and hydrogen respectively, and 0 for cyclohexane, resulting in $n_0 = 4$ for the whole mixture at the initial state. The substitution of the mole fractions expressed by Eqs. (7.73), (7.74), and (7.75) into

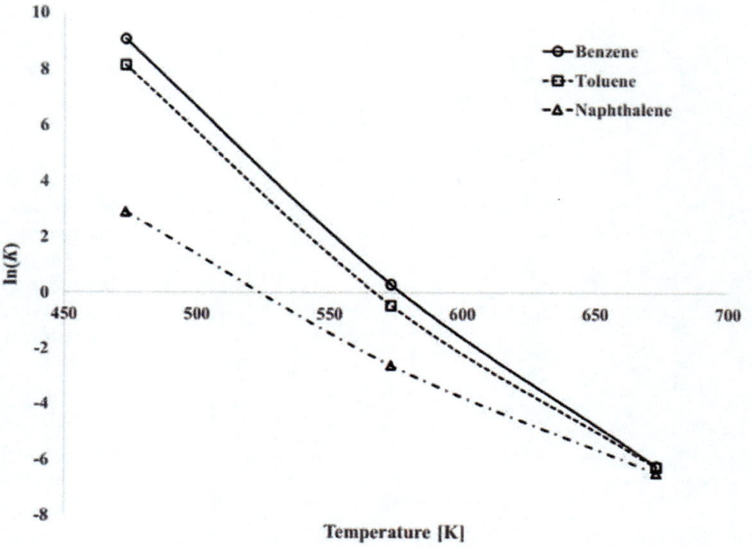

Fig. 7.5 Experimental equilibrium constants for the hydrogenation of aromatic components as a function of temperature. Data from Ali [1]

Eq. (7.72) results in

$$K = \frac{x_{C_6H_{12}}}{x_{C_6H_6}x_{H_2}^3} = \frac{\frac{\varepsilon}{4-3\varepsilon}}{\left(\frac{1-\varepsilon}{4-3\varepsilon}\right)\left(\frac{3-3\varepsilon}{4-3\varepsilon}\right)^3} = \frac{\varepsilon(4-3\varepsilon)^3}{27(1-\varepsilon)^4} \tag{7.76}$$

Experimental equilibrium constants for the hydrogenation of benzene, toluene, and naphthalene are depicted as a function of temperature in Fig. 7.5. Using these values, Eq. (7.76) can be solved in order to find the reaction coordinate ε_e as a function of temperature. The outcomes are shown in Fig. 7.6, where analogous calculations for toluene and naphthalene are also depicted. As observed, the higher temperature, the smaller ε_e; that is, aromatic components are less hydrogenated as temperature increases, reaching a minimum value about 670 K. Above this temperature, hydrogenation of these aromatic components does not take place.

According to what have been exposed before, if a chemical-reaction scheme between petroleum fractions is proposed, and the corresponding chemical pseudostructures are available or can be estimated as in Chaps. 3 and 4, then equilibrium constants, reaction coordinate ε_e, and the associated compositions can be estimated.

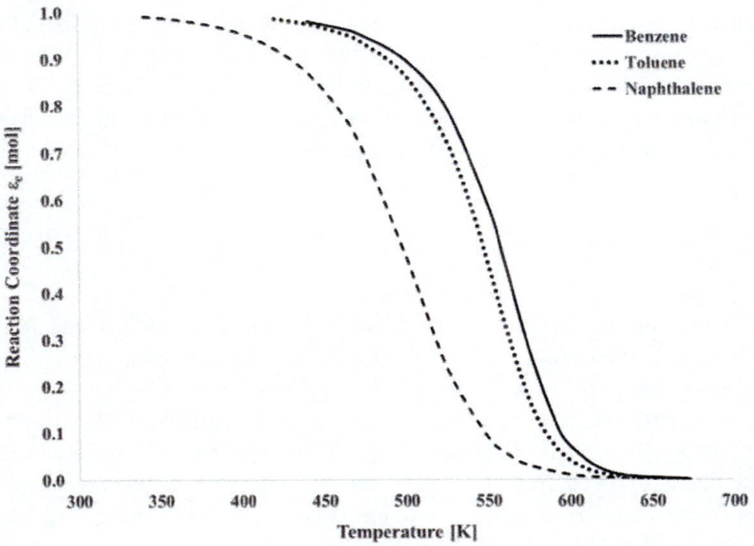

Fig. 7.6 Reaction Coordinate as a function of temperature for the hydrogenation of aromatic components

7.5 The Non-stoichiometric Method

In the previous sections, the chemical equilibrium has been determined in terms of a specific number of chemical reactions, and their corresponding equilibrium constants. These chemical reactions have to be balanced for the mass conservation to be satisfied. As illustrated, a balanced chemical reaction or chemical equation implies that the stoichiometric coefficient of each involved component is known. The stoichiometric coefficients indicate that the number of the atoms in the reactants side must be equal to the number of atoms in the products side. Thus, the methods for determining the chemical equilibrium, where stoichiometric coefficients are needed, are in general called stoichiometric methods.

Unlike stoichiometric methods, the chemical equilibrium can be determined without the need for specific chemical reactions and their corresponding stoichiometric coefficients through the so-called non-stoichiometric methods. These methods consist of minimizing the free energy of the system directly subject to the conservation of chemical elements; in contrast to the phase equilibrium calculations, which are subject to the conservation of the total number of moles for each component n_{it}. The expression of the total Gibbs free energy of the system to be minimized was derived in Chap. 2 and used in Chap. 6, which is

$$\frac{G}{RT} = \sum_{j=1}^{\sigma} \sum_{i=1}^{C} n_{ij} \frac{\mu_{ij}}{RT} \tag{7.77}$$

As usual, μ_{ij} and n_{ij} account for the chemical potential and the number of moles of the component i in the phase j. According to this expression, we search for a multiphase equilibrium, but the amount of chemical elements must be held constant now. In other words, for simultaneous phase and chemical equilibrium, where the total number of components n_t is unknown, the following constraint must be satisfied:

$$\sum_{i=1}^{C}\sum_{j=1}^{\sigma} a_{ki} n_{ij} - b_k = 0 \tag{7.78}$$

a_{ki} is the number of atoms of the element k present in each molecule of the component i. For instance, in the case of methane, we have that $a_{C,CH_4} = 1$ and $a_{H,CH_4} = 4$, while in the case of carbon dioxide we have that $a_{C,CO_2} = 1$ and $a_{O,CO_2} = 2$. Notice that a_{ki} is an integer number for these identifiable components, whereas it is a non-integer number for pseudocomponents in accordance with the characterization procedure of the petroleum fraction. When a given element is not present in the component, then $a_{ij} = 0$. For instance, we have that $a_{O,CH_4} = 0$ and $a_{H,CO_2} = 0$. In Eq. (7.78), b_k is the total number of moles of element k in the whole system, and therefore, it is the value to be held constant during calculations. For a given initial number of moles n_{it}^0 in the whole system, b_k can be calculated through

$$b_k = \sum_{i=1}^{C} a_{ki} n_{it}^0 \tag{7.79}$$

with which, we have that the conservation of chemical elements turns into

$$\sum_{i=1}^{C} a_{ki}\left(\sum_{j=1}^{\sigma} n_{ij} - n_{it}^0\right) = 0 \tag{7.80}$$

To find the minimum value of this constrained optimization problem, we introduce again a Lagrangian function $L(\boldsymbol{n}, \boldsymbol{\lambda})$ as a function of the number of moles and the Lagrange multipliers $\boldsymbol{\lambda}$ as follows:

$$L = \sum_{j=1}^{\sigma}\sum_{i=1}^{C} n_{ij}\frac{\mu_{ij}}{RT} - \sum_{k=1}^{e}\lambda_k\left(\sum_{i=1}^{C}\sum_{j=1}^{\sigma} a_{ki}n_{ij} - b_k\right) \tag{7.81}$$

The superscript e corresponds to the total number of elements in the system. The total differential of the $L(\boldsymbol{n}, \boldsymbol{\lambda})$ is

$$dL = \sum_{j=1}^{\sigma}\sum_{i=1}^{C}\left(\frac{\mu_{ij}}{RT} - \sum_{k=1}^{e}\lambda_k a_{ki}\right)dn_{ij} - \sum_{k=1}^{e}\left(\sum_{i=1}^{C}\sum_{j=1}^{\sigma} a_{ki}n_{ij} - b_k\right)d\lambda_k \tag{7.82}$$

From which, the partial derivatives with respect to the independent variables, which must vanish on an stationary point, are

$$\left(\frac{\partial L}{\partial n_{ij}}\right)_{\lambda, n_{lm;l\neq i,m\neq j}} = \frac{\mu_{ij}}{RT} - \sum_{k=1}^{e} a_{ki}\lambda_k = 0 \tag{7.83}$$

$$\left(\frac{\partial L}{\partial \lambda_k}\right)_{n,\lambda_{l;l\neq k}} = \sum_{i=1}^{C}\sum_{j=1}^{\sigma} a_{ki}n_{ij} - b_k = 0 \tag{7.84}$$

These equations form a set of $\sigma C + e$ non-linear equations in order to determine $\sigma C + e$ unknown variables: σC number of moles n_{ij} plus e Lagrange multipliers. The former set of partial derivatives give rise to the equality of chemical potential of components between phases, while the latter just expresses that the conservation of chemical elements must be satisfied. Both sets constitute the necessary conditions for a system to be at physical and chemical equilibrium. When the Gibbs free energy is evaluated at the stationary point according to the above partial derivatives, the following result is reached

$$\frac{G}{RT} = \sum_{j=1}^{\sigma}\sum_{i=1}^{C} n_{ij}\sum_{k=1}^{e} a_{ki}\lambda_k = \sum_{k=1}^{e}\lambda_k\sum_{i=1}^{C}\sum_{k=1}^{\sigma} a_{ki}n_{ij} = \sum_{k=1}^{e}\lambda_k b_k \tag{7.85}$$

From which, a physical meaning may be given to λ_k: the Additivity Rules shows that the Gibbs free energy of the system can expressed as the summation over all components of the product between number of moles and chemical potentials, for example Eq. (7.77). Then, if b_k is considered as the composition of the element k in the system, then λ_k corresponds to its chemical potential.

As has been done in previous sections and chapters, to express the necessary equilibrium conditions in more practical expressions, we make use of the chemical potential in terms of variables measurable (temperature, pressure and composition) and the fugacity coefficient; that is,

$$\frac{\mu_{ij}}{RT} = \frac{\mu_i^0}{RT} + ln\left(\frac{P}{P^0}\right) + ln\left(x_{ij}\phi_{ij}\right) \tag{7.86}$$

It is worth remembering that μ_i^0 is the chemical potential of pure component i as ideal gas at reference pressure P^0; hence, μ_i^0 is a function of temperature only. This reference state is used when the phase behavior of all phases is described by a single equation of state such a Cubic Equation of State (CEoS). However, when the phase behavior of a given phase is not described for an equation of state properly, a reference state can be established for such a phase. In this case, f_i^0, which is the reference fugacity of the pure component i at the system pressure and temperature, takes the place of P^0. Therefore, μ_i^0 accounts now for the chemical potential of the pure component i in phase j, being a function of both temperature and pressure as well. For instance, we can chose this reference state for liquid phases that are described by an activity coefficient models properly. Bearing this in mind, we have that Eq. (7.86) transforms Eq. (7.83) into

$$\frac{\mu_i^0}{RT} + ln\left(\frac{P}{P^0}\right) + ln\left(x_{ij}\phi_{ij}\right) - \sum_{k=1}^{e} a_{ki}\lambda_k = 0 \tag{7.87}$$

Because of the mole fraction definition $x_{ij} = n_{ij}/n_j$, and the mass balance $n_j = \sum_{l=1}^{C} n_{lj}$, this equation can be written as

$$\frac{\mu_i^0}{RT} + ln\left(\frac{P}{P^0}\right) + ln(n_{ij}) - ln\left(\sum_{l=1}^{C} n_{lj}\right) + ln(\phi_{ij}) - \sum_{k=1}^{e} a_{ki}\lambda_k = 0 \tag{7.88}$$

When we are interested in the chemical equilibrium only, this means that a single phase is present $\sigma = 1$, with which $n_{ij} = n_{it}$, and $\phi_{ij} = \phi_{it}$. Thus, we have that

$$\frac{\mu_i^0}{RT} + ln\left(\frac{P}{P^0}\right) + ln(n_{it}) - ln\left(\sum_{l=1}^{C} n_{lt}\right) + ln(\phi_{it}) - \sum_{k=1}^{e} a_{ki}\lambda_k = 0 \tag{7.89}$$

In this case, the conservation of chemical elements takes the form of

$$\sum_{i=1}^{C} a_{ki} n_{it} - b_k = 0 \tag{7.90}$$

On the other hand, chemical equilibrium calculations should be harmonized with phase equilibrium calculations. This means that the same results should be reached from methods described in Chap. 6, and from the solution of the above equations. To find out whether such a harmonization is present n_t is set equal to one, which is the assumption in phase equilibrium calculations. Thus, if $n_t = 1$ then $\sum_{l=1}^{C} n_{lt} = 1$ and $z_i = n_{it}$, z_i being the total mole fraction of the system. Using this assumption, we arrive at

$$\frac{\mu_i^0}{RT} + ln\left(\frac{P}{P^0}\right) + ln(z_i) + ln(\phi_{it}) - \sum_{k=1}^{e} a_{ki}\lambda_k = 0 \tag{7.91}$$

Together with

$$\sum_{i=1}^{C} a_{ki} z_i - b_k = 0 \tag{7.92}$$

For instance, the solution of these equations should correspond to that solution for the incipient point calculations, where $n_t = 1$ for the considered single phase either liquid or vapor.

Unlike phase equilibrium calculations without chemical reactions, the reference chemical potential μ_i^0 is required in the equations derived above. The fact that its value is established arbitrarily illustrates that absolute values of thermodynamic potentials are a matter of convenience, because energy and mass exchanges with the surroundings are the only magnitudes that can be measured during changes of state. From a practical point of view, this implies that properties values must be all consistent with each other regardless of criteria for establishing their absolute values. In other words, properties values from the different treatments must lead to the same chemical and phase equilibrium results. However, when estimation methods for different properties are employed in calculations, inconsistencies can occur; hence, it is best to check important values by using different routes [5].

As performed in previous sections, in the case of chemical equilibrium, μ_i^0 can be set equal to $\Delta\tilde{G}_{if}$, the molar Gibbs free energy of formation for the component i. Therefore, μ_i^0 is equal to zero for all elements such as O_2 and H_2 at their standard states, because $\Delta\tilde{G}_{if} = 0$ for all elements at any temperature.

On the other hand, by rewriting Eq. (7.66) in terms of just formation properties and performing the corresponding integrals, the estimation of $\Delta\tilde{G}_{if}$ for components at any temperature can be calculated from

$$\Delta\tilde{G}_{if} = \Delta\tilde{H}_{if}^{0} + \frac{T}{T_0}\left(\Delta\tilde{G}_{if}^{0} - \Delta\tilde{H}_{if}^{0}\right) + I_1 - TI_2 \tag{7.93}$$

where the molar Gibbs free energy $\Delta\tilde{G}_{if}^{0}$ and Enthalpy $\Delta\tilde{H}_{if}^{0}$ of formation at standard state ($T_0 = 298.15K$ and $P_0 = 0.101325MPa$) can be estimated as explained in the previous sections. The symbols I_1 and I_2 represents the following integrals

$$I_1 = A(T - T_0) + \frac{B}{2}(T^2 - T_0^2) + \frac{C}{3}(T^3 - T_0^3) + \frac{D}{4}(T^4 - T_0^4) + \frac{E}{5}(T^5 - T_0^5) \tag{7.94}$$

$$I_2 = Aln\left(\frac{T}{T_0}\right) + B(T - T_0) + \frac{C}{2}(T^2 - T_0^2) + \frac{D}{3}(T^3 - T_0^3) + \frac{E}{4}(T^4 - T_0^4) \tag{7.95}$$

$$\Lambda_i = \Lambda_i - \sum_{k=1}^{e} \nu_{ki}\Lambda_k;\ \Lambda = A, B, C, D, E \tag{7.96}$$

ν_{ki} is the number of atoms of an element of type k forming the component i. For instance, $\nu_{C,CH_4} = 1$ and $\nu_{H_2,CH_4} = 2$ for methane, and $\nu_{C,CO_2} = 1$ and $\nu_{O_2,CO_2} = 1$ for carbon dioxide. The A, B,C,D, and E constants belong to the heat capacity expression $\tilde{C}_p$ as a function of temperature as shown in Table 7.1 for some components and elements. Thus, for example, $\Delta\tilde{G}_{CH_4,f}$ is calculated as a function of temperature through Eq. (7.93) taking into account $\tilde{C}_p$ of methane and its corresponding elements C and H_2. The former in solid state, and the latter in ideal-gas state.

As in the stoichiometric method described in the previous section, the above procedure can also be applied to petroleum fractions or pseudocomponents as long as the corresponding properties can be estimated, and the chemical structure is provided such as the pseudostructure from the characterization procedure of GFEM. For instance, Fig. 7.7 illustrates calculated values of $\Delta\tilde{G}_{if}$ for pseudocomponents from the 10.3°API reservoir fluid characterized in Chap. 6. In this figure two approaches are compared with each other: the former corresponds to the use of Eq. (7.93), where $\Delta\tilde{H}^0_{if}$, $\Delta\tilde{G}^0_{if}$, and $\tilde{C}_{pi}$ are estimated from GFEM are described for the stoichiometric method in previous sections, while the latter corresponds to the solution of Eqs. (7.91), and (7.92), where $\mu_i^0 = \Delta\tilde{G}_{if}$. In this last case, calculations were performed on the bubble point of the Reservoir Fluid by using the Peng Robinson equation of state (PR) with the Modified Huron-Vidal mixing rule of order one ($MHV1$). On this bubble point, temperature ($386.15K$), pressure ($5.29MPa$), and total composition of the fluid z_i are known; therefore, $\Delta\tilde{G}_{if}$ and λ_k are the only

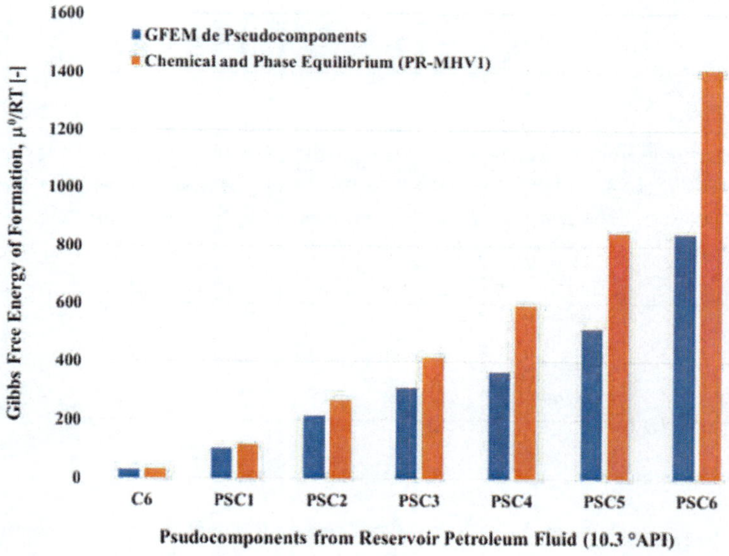

Fig. 7.7 Comparison between $\Delta\tilde{G}_{if}$ for pseudocomponents by using chemical pseudostructures according to GFEM and by solving the necessary chemical equilibrium conditions on the bubble point at 386.15 K of a 10.3°API Reservoir Fluid

variables to be calculated from this system of equations. The $\Delta\widetilde{G}_{if}$ values of the identifiable components of the reservoir fluid were calculated by using Eq. (7.93) with the parameters from Tables 7.1 and 7.4. The aim of the comparison from Fig. 7.7 is to determine whether phase equilibrium calculations are consistent with chemical equilibrium calculations. It they are harmonized, then $\Delta\widetilde{G}_{if}$ from both approaches should be equal to each other. As observed from this figure, calculations from both approaches compare well to each other for the lightest pseudocomponents, but the differences increase as the pseudocomponent becomes heavier. This tendency is confirmed in Fig. 7.8, where the differences are shown as a function of more bubble point temperatures for the same reservoir fluid. From these figures, we may infer that the thermodynamic characterization performed for phase equilibrium calculations should not be used for chemical equilibrium calculations directly. Therefore, if chemical equilibrium is intended to be determined, then the thermodynamic characterization procedures described in Chaps. 3 and 4 should be modified somehow so that both chemical and phase equilibrium situations can be determined properly.

From Figs. 7.9, 7.10 and 7.11, simultaneous chemical and valor-liquid equilibrium are illustrated for the same reservoir fluid from Fig. 7.7 and Fig. 7.8. The used $\Delta\widetilde{G}_{if}$ values are those from Fig. 7.8 corresponding to the solution of Eqs. (7.91) and (7.92). For interpolation and extrapolation purposes, these values were fitted to a potential function of the form $\Delta\widetilde{G}_{if} = \alpha_i T^{\beta_i}$ for each pseudocomponent.

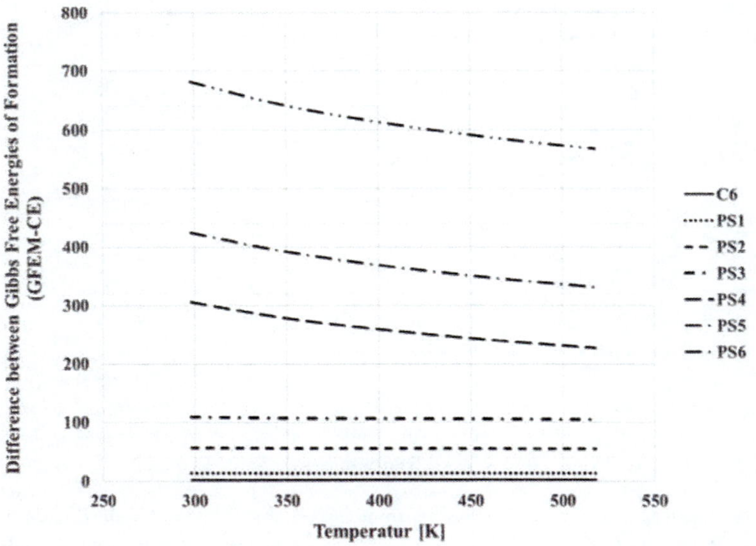

Fig. 7.8 Differences between $\Delta\widetilde{G}_{if}$ for pseudocomponents by using chemical pseudostructures according to GFEM and by solving the necessary chemical equilibrium conditions on the bubble-point curve of a 10.3°API Reservoir Fluid

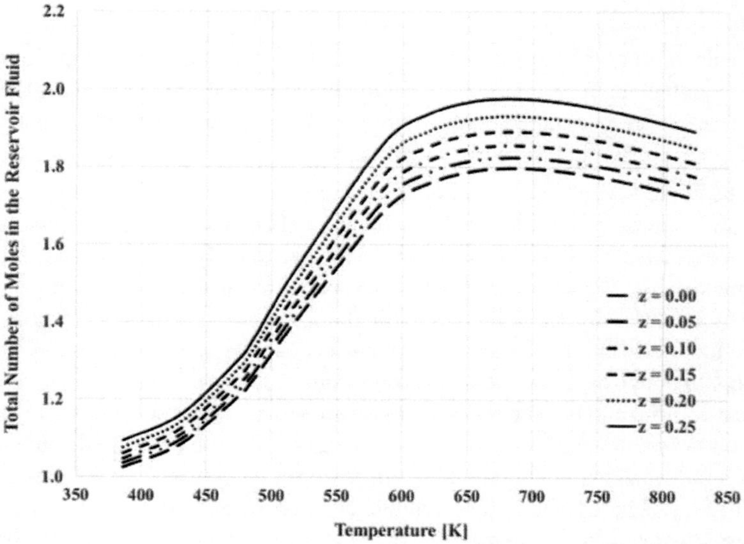

Fig. 7.9 Total number of moles from Chemical and Vapor–Liquid Equilibrium for 10.3°API Reservoir Fluid at 5.29 MPa as a function of the total mole fraction of H_2 (z)

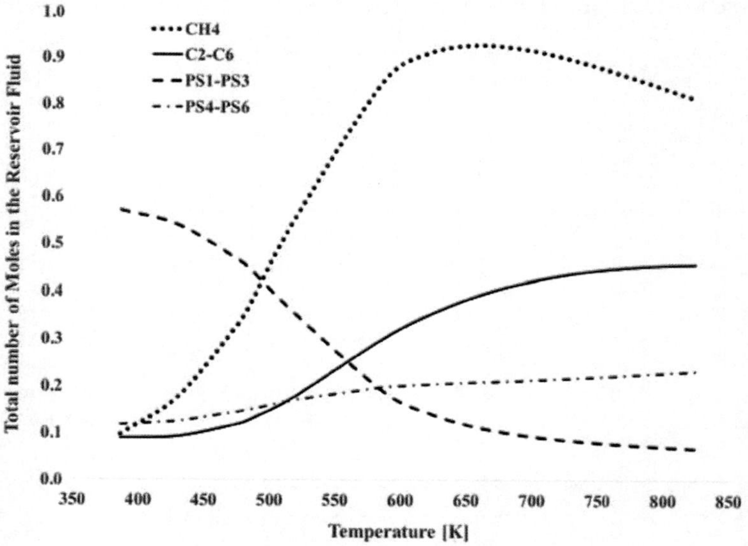

Fig. 7.10 Number of moles of components from Chemical and Vapor–Liquid Equilibrium for 10.3°API Reservoir Fluid at 5.29 MPa and at total mole fraction of H_2 equal to zero ($z = 0$)

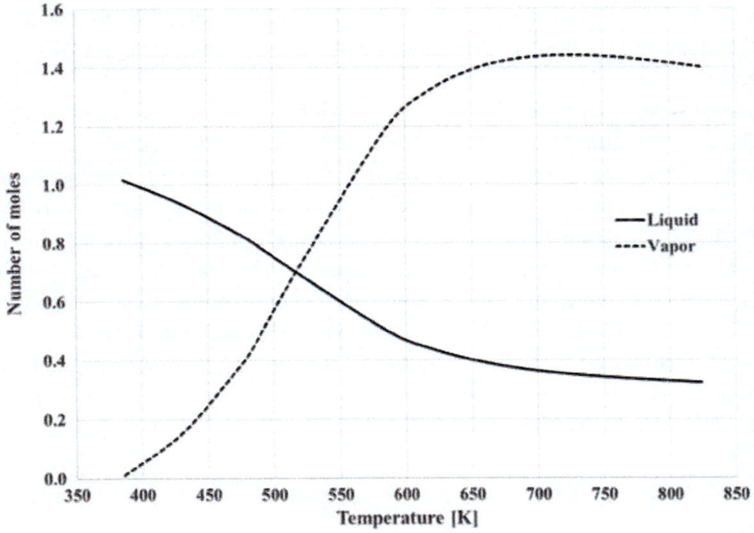

Fig. 7.11 Number of moles of liquid and vapor phases from Chemical and Vapor–Liquid Equilibrium for 10.3°API Reservoir Fluid at 5.29 MPa and at total mole fraction of H_2 equal to zero (z = 0)

The values corresponding to the identifiable components were calculated as before through Eq. (7.93). Equilibrium calculations were performed from solving the system of equations formed by the conservation of chemical elements (Eq. (7.78)) and the necessary equilibrium conditions (Eq. (7.88)). The fugacity coefficients were calculated by using the PR equation of state and the MHV1 mixing rule. In Fig. 7.9, the total number of moles n_t is shown as a function of temperature and mole fraction of hydrogen z_{H_2} in the mixture; $z_{H_2} = 0$ corresponds to the original reservoir fluid composition. As observed, n_t increases rapidly as temperature increases, reaching a maximum value around $654.15K$ (381° C). Above this maximum temperature, n_t decreases as temperature increases. It is worth pointing out that n_t would remain constant with respecto to temperature when vapor–liquid equilibrium calculations are considered only, that is, $n_t = n_t^0$ regardless of temperature. On the other hand, n_t increases with increasing hydrogen concentration, not only because of the addition of hydrogen to the system, but also as result of the thermodynamic chemical distribution of elements in components and pseudocomponents, as can be seen in Fig. 7.10. In this figure, the hydrogen mole fraction is not shown, because it is zero below the maximum value of $654.15K$, that is, a complete hydrogenation of aromatic components takes place below this temperature. On the other hand, above this temperature, hydrogen starts to appear, showing very small mole fractions; therefore, hydrogen consumption begins to be incomplete above such a temperature. Besides, the formation of methane is preferential below such a temperature, whereas it begins to decline above it. In the same Fig. 7.10, we see that the formation of gases and heaviest pseudocomponents is at the expense of a decrease in the formation of the lightest pseudocomponents.

Finally, from Fig. 7.11, it is observed that an increase in temperature leads to higher production of vapor phase at the expense of a decrease in formation of the liquid phase.

From an experimental point of view, it is not possible to reach an actual equilibrium state, because very long times and perfect mixtures would be required, which are not easy to accomplish from a practical point of view. For instance, upgrading of Hamaca Heavy Oil [10] in the presence of gases such as methane, nitrogen, hydrogen and argon have been carried out in the absence of catalyst (thermal conditions) [11–13]. In these experiments, temperature ranges from 653.15 to 693.15 K (380 to 420 °C) at 11 MPa of pressure, while the reaction residence time ranges from one to four hours. It is worth pointing out that an intense aromatization, similar to that observed in calculations from Fig. 7.10, occurs for all the upgrading reactions. On contrary, temperatures below such an interval lead to hydrogenation of aromatic components (non-aromatization), as illustrated in Fig. 7.6.

References

1. Ali SA (2007) Thermodynamic aspects of aromatic hydrogenation. Pet Sci Technol 25(10):1293–1304.
2. Nazarova G, Ivanchina E, Ivashkina E, Kiseleva S, Stebeneva V (2015) Thermodynamic analysis of catalytic cracking reactions as the first stage in the development of mathematical description. Procedia Chem. 15:342–349
3. Yaws CL (1999) Chemical properties handbook. New York, USA
4. Joback KG, Reid RC (1987) Estimation of pure-component properties from group-contributions. Chem Eng Commun 57:233–243
5. Poiling BE, Prausnitz JM, O'Conell JP (2001) The properties of gases and liquids, vol 5. McGraw-Hill, New York, USA
6. Marrero J, Gani R (2001) Group-contribution based estimation of pure component properties. Fluid Phase Equilib 184:183–208
7. Albahri TA, Aljasmi AF (2013) Thermochimica acta SGC method for predicting the standard enthalpy of formation of pure compounds from their molecular structures. Thermochim Acta 568:46–60
8. Chevron. https://www.chevrontexaco.com/crudemarketing.
9. Peng DY, Robinson DB (1976) A new two-constant equation of state. Ind Eng Chem Fundam 15(1):59–64
10. Venturini GJ, Mamora DD (2004) simulation studies of steam-propane injection for the Hamaca heavy oil field. J Can Pet Technol 43(9):40–46
11. Ovalles C, Hamana A, Rojas I, Bolívar RA (1995) Upgrading of extra-heavy crude oil by direct use of methane in the presence of water. Deuterium-labelled experiments and mechanistic considerations. Fuel 74(8):1162–1168
12. Ovalles C, Filgueiras E, Morales A, Rojas I, Carlos de Jesus J, Berrios I (1998) Use of a dispersed molybdenum catalyst and mechanistic studies for upgrading extra-heavy crude oil using methane as source of hydrogen. Energy Fuels 12(2):379–385.
13. Ovalles C, Filgueiras E, Morales A, Scott CE, Gonzalez-Gimenez F, Pierre Embaid B (2003) Use of a dispersed iron catalyst for upgrading extra-heavy crude oil using methane as source of hydrogen. Fuel 82(8):887–892

Chapter 8
Rheology

List of Abbreviations

AAD	Absolute Average Deviation
API	American Petroleum Institute
EF	Expanded Fluid
GFEM	Gibbs free energy minimization
GNF	Generalized Newtonian Fluid
LVE	Linear viscoelastic
MR	Marrero-Gani Group Contribution Method
PC-SAFT	Perturbed Chain Statistical Association Fluid Theory
SAOS	Small-Amplitude Oscillatory Shear
WLF	Williams, Landel and Ferry equation

Nomenclature

$\underline{\underline{\dot{\gamma}}}$	Rate of deformation tensor
$\underline{\underline{I}}$	Identity tensor
$I_{\dot{\gamma}}$	First invariant of the rate of deformation tensor
$II_{\dot{\gamma}}$	Second invariant of the rate of deformation tensor
$III_{\dot{\gamma}}$	Third invariant of the rate of deformation tensor
$\underline{\underline{\Pi}}$	Total stress tensor
$\underline{\underline{\tau}}$	Stress tensor
ΔX	Deviation function for parameters of EF Viscosity Model
$\nabla \underline{v}$	Velocity gradient
°API	API degrees
A	Area or Model parameter for an Arrhenius type of equation for viscosity
a	Model parameters

B. Carreón-Calderón et al., *Thermophysical Properties of Heavy Petroleum Fluids*, Petroleum Engineering, https://doi.org/10.1007/978-3-030-58831-1_8

a_n	$n = 0 \ldots 6$; Model parameters for Eqs. (8.26)-(8.28)
a_T	Shift factor for the time-temperature superposition principle
b	Model parameters
b_n	$n = 0 \ldots 4$; Model parameters for the Expanded Fluid-Based Viscosity Correlation
c	Model parameters
c_2 *and* c_3	Parameters for the EF Viscosity Model of pure components
c_2^i, c_2^{ii} *and* c_2^{iii}	Parameters for the c_2 parameter as a function of temperature
d	Model parameters
d_n	$n = 1, 2$; Model parameters for Eq. {8.58)
F_c	Empirical correction factor, Eq. (8.33)
f_{min}	Objective function of data fitting
F_s	Tangential force
G'	Elastic or storage modulus
G''	Viscous or loss modulus
G^*	Complex modulus
g_1	Model parameters, Eq. (8.59)
h_2	Model parameters, Eq. (8.59)
h_n	$n = 1, 2$; Model parameters, Eq. (8.59)
K	Power law type consistence parameter
k_{ij}	Binary interaction parameter
K_W	Watson factor
M	Molecular Weight
n	Power law type index parameter
P	Thermodynamic pressure
R	Universal gas constant
S	Specific gravity
t	Time
T_*	Dimensionless temperature for Eq. (8.31)
T_0	Reference temperature
T_b	Mean average boiling temperature
T_f	Freeze point temperature
U	Velocity
$\widetilde{V}_c$	Critical molar volume
v_x	x component of velocity
w	Weight fraction for mixing rules
x_i	Mole fraction of component i
Z_{RA}	Parameter for Rackett equation
γ	Deforamtion or strain
$\dot{\gamma}_{xy}$	xy component of shear-rate, rate of strain, rate of deformation
γ_0	Strain wave amplitude
δ	Out of phase angle
δ_r	Dimensionless dipole
ΔE	Energy term of an Arrhenius type of equation for viscosity
η	Non-Newtonian viscosity

η'	Dynamic complex viscosity
η''	Elastic contribution to the complex viscosity
η^*	Complex viscosity
η_0	First Newtonian region (low shear rates)
η_∞	Viscosity plateau at high shear rates
κ	Correction factor for highly associating components, Eq. (8.33)
λ	Characteristic time—fitting parameter for Eqs. (8.54), (8.55), or relaxation time for the LVE part
λ_{tx}	Structure parameter for thixotropy
μ	Newtonian viscosity
ν	Kinematic viscosity
ν_{310}	Kinematic viscosity at 310 K
ν_{372}	Kinematic viscosity at 372 K
$\hat{\rho}$	Fluid density
ρ_s^0	Mass density in the compressed state at vacuum conditions
σ	Shear stress
τ_0	Stress wave amplitude in the LVE tests
τ_0'	Stress coefficient of the elastic part of the stress in SAOS
τ_0''	Stress coefficient of the viscous part of the stress in SAOS
τ_y	Yield stress
τ_{yx}	Shear stress, component xy
ω	Acentric factor or Frequency of oscillations in the LVE tests
Ω_μ	Viscosity integral collision

Subscripts

atm	Evaluated at atmospheric pressure
calc	Calculated value
exp	Experimental value
H_2 O	Properties of water
ij	Components of a tensor
r	Parameters EF Viscosity Model of n-alkane components

Superscripts

atm	Evaluated at atmospheric pressure
c	Evaluated at critical conditions
ig	Ideal gas properties
sat	Evaluated at the saturation point
T	Transpose of a tensor

Symbol

∇ Gradient operator

8.1 Newtonian Model

Viscosity is a transport property considered of great importance for exploitation, transportation and processing of petroleum fluids. Extraction and delivery are greatly affected by viscosity. Flow through porous media or reduced area sections are highly dependent on this physical property. However, as has been exposed in previous chapters, petroleum fluids are very complex mixtures with an extremely large number of unidentifiable components. Hence, unlike phase equilibrium problems where there are relatively well-established methods, the prediction of viscosity still requires additional research, because there is not general accepted method to estimate it, particularly in the case of heavy petroleum fluids.

There have been numerous empirical and semi-empirical methods that have been proposed for estimating the viscosity of petroleum fluids. In these methods, the accuracy of viscosity models is routinely improved with the aid of at least one viscosity experimental datum, allowing the model parameters to be adjusted for obtaining better approximations of the remaining experimental values. In other words, the models turn out to be specific for the considered petroleum fluid, under conditions of the available experimental values; therefore, it is difficult to talk about predictive methods for the entire range of petroleum fluids and conditions of interest. This is particularly true when heavy petroleum fluids are involved, because viscosity may lose its conventional definition of Newtonian fluid under certain operational conditions, turning them into non-Newtonian fluids. For instance, when light components are released from the liquid phase or temperature decreases during exploitation, transportation and processing. In what follows, basic concepts for defining the Newtonian viscosity concept are introduced, and then empirical methods for estimating the Newtonian liquid viscosity of heavy petroleum fluids are described. These methods have been chosen because of their relative simplicity and because they have shown to be predictive to some extent.

8.1.1 Newtonian Constitutive Equation

The so-called viscosity μ is a scalar physical property that accounts for the internal fluid friction undergone by its molecules as consequence of fluid motion. As a measure of friction, the inverse of the fluid viscosity $(1/\mu)$ accounts for the fluidity or mobility, that is, the larger the fluid viscosity, the lower the fluid mobility. On the contrary, the lower the fluid viscosity, the larger the fluid mobility.

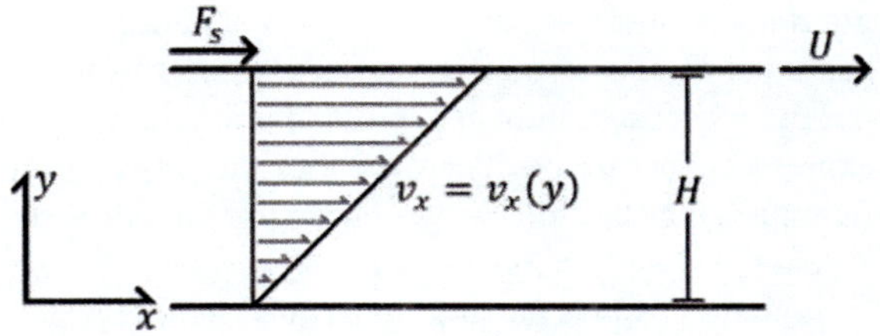

Fig. 8.1 Laminar velocity profile $v_x(y)$ for a fluid contained between two infinite plates as a result of applying a shear force to the upper plate

The fluid viscosity connects the force applied to the fluid for it to move with the resulting deformation. In Fig. 8.1, for example, a force is applied to the upper plate, yielding a laminar velocity profile (unless U is relatively high) for the fluid contained between the two plates. In this hypothetical experiment, the shear force F_s divided by the plate area A, defines the shear stress $\left(\tau_{yx} \equiv F_s/A\right)$, which is related to the fluid velocity v_x as follows:

$$\tau_{yx} = \mu \frac{dv_x}{dy} \tag{8.1}$$

This equation is often called Newton's law of viscosity, which states that the shear stress τ_{yx} is proportional to the velocity gradient or the rate of strain $\left(\dot{\gamma}_{xy} \equiv dv_x/dy\right)$, where the proportionality constant is the fluid viscosity μ (sometimes called dynamic viscosity). Note that Eq. (8.1) is a scalar expression and it should not be used to define what a Newtonian fluid is, because there are some non-Newtonian (specifically Boger) fluids[1] that can be well represented with this scalar version. In the International System of Units, τ_{yx} has units of pressure, that is, Pascal (Pa) and $\dot{\gamma}_{xy}$ has units of seconds raised to minus one $\left(s^{-1}\right)$. For Newtonian fluids, the value of the fluid viscosity depends on other fluid properties such as temperature, pressure, and composition, but it does not depend on the rate of strain $\left(\mu \neq \mu\left(\dot{\gamma}_{xy}\right)\right)$. When the fluid viscosity is a function of the rate of strain, then it is said that the fluid is non-Newtonian.

An important way to express the fluid viscosity is through the so-called kinematic viscosity ν, which is defined as a ratio of the fluid viscosity to the fluid mass density $\hat{\rho}$, that is,

$$\nu \equiv \frac{\mu}{\hat{\rho}} \tag{8.2}$$

In the International System of Units, μ has units of Pascal-second (Pa·s). Because the viscosity expressed in these units frequently turns out to be a very small number, it is more common to use mPa·s, which is numerically equal to a commonly used unit for viscosity, the centipoises (1cP = mPa·s. The poises (P) are defined as one

[1]Boger fluids present in general, high elasticity, while keeping constant (or almost constant) shear viscosity; then, it would not be possible to distinguish a Newtonian from a Boger fluid by just analyzing a common flow curve.

gram per centimeter per second g/cm s; hence, 1cP = 100P. On the other hand, ν has units of square meters per second m^2/s in the International System of Units, but the most used units to expressed it are the centistokes (cSt), which arises from expressed viscosity in cP and the mass density in grams per cubic centimeter g/cm^3. Summarizing, the following conversion factors apply to dynamic viscosity: 1P ≡ 1g/cm·s = 100cP = Pa·s.

Equation (8.1) is an empirical equation; that is, it was originally based on experimental observations instead of being derived from fundamentals principles. In a 3-dimensional space, there are nine possible stress τ_{ij} components connected with nine shear-rate $\dot{\gamma}_{ij}$ components, for which a Newtonian constitutive equation can be deduced. To collect all these equations into just one single expression, the Newtonian constitutive equation can be written using tensorial notation in the following way:

$$\underline{\underline{\tau}} = \mu\left[\nabla\underline{v} + \left(\nabla\underline{v}\right)^T\right] = \mu\underline{\underline{\dot{\gamma}}} \tag{8.3}$$

For Cartesian coordinates, $\underline{\underline{\tau}}$ and $\underline{\underline{\dot{\gamma}}}$ has the tensorial components

$$\underline{\underline{\tau}} = \begin{pmatrix} \tau_{xx} & \tau_{xy} & \tau_{xz} \\ \tau_{yx} & \tau_{yy} & \tau_{yz} \\ \tau_{zx} & \tau_{zy} & \tau_{zz} \end{pmatrix} \tag{8.4}$$

and

$$\underline{\underline{\dot{\gamma}}} = \begin{pmatrix} \dot{\gamma}_{xx} & \dot{\gamma}_{xy} & \dot{\gamma}_{xz} \\ \dot{\gamma}_{yx} & \dot{\gamma}_{yy} & \dot{\gamma}_{yz} \\ \dot{\gamma}_{zx} & \dot{\gamma}_{zy} & \dot{\gamma}_{zz} \end{pmatrix} \tag{8.5}$$

where $\tau_{ij} = \tau_{ji}$ and $\dot{\gamma}_{ij} = \dot{\gamma}_{ji}$. In other words, $\underline{\underline{\tau}}$ and $\underline{\underline{\dot{\gamma}}}$ are symmetric tensors. In terms of the velocity gradients, $\dot{\gamma}_{ij}$ is given by

$$\dot{\gamma}_{ij} = \left(\frac{\partial v_i}{\partial x_j} + \frac{\partial v_j}{\partial x_i}\right) \tag{8.6}$$

Now, using the tensor expression of Newton's law of viscosity, it can be said that a Newtonian fluid is such that its flow response can be described by Eq. (8.3). Although such a definition is correct, it may be difficult to find a physical sense. A common description for a Newtonian fluid is that, given fixed temperature and pressure, its viscosity remains constant independently of flow/deformation conditions. Generally, Newtonian fluids are constituted by small molecules and typical examples are almost all gases (for the very few exceptions see [1, 2]), water, toluene, benzene, to mention a few liquids. In Eqs. (8.3)–(8.6), τ is the stress, μ the viscosity, $\dot{\gamma}$ is the rate-of-deformation, $\nabla\underline{v}$, the velocity gradient, the sub-indexes *ij* refer to the specific tensor component and the super-index T, to the transpose. Also, one underline and two underlines refer to vector and tensor quantities, respectively.[2]

Finally, it is worth mentioning that $\underline{\underline{\tau}}$ together with the thermodynamic pressure P, as it was handled in the previous chapters, give rise to the total stress tensor $\underline{\underline{\Pi}}$

$$\underline{\underline{\Pi}} = \underline{\underline{\tau}} + P\underline{\underline{I}} \tag{8.7}$$

Here, $\underline{\underline{I}}$ is the identity tensor. In this context, the thermodynamic pressure, which acts perpendicularly (normal) to a surface, is an isotropic physical magnitude; that is, it is a force that acts equally in every direction.

8.1.2 API Procedure

In contrast to other methods, where pure-like properties are needed, the required input parameters for the API procedure [3] are volumetric properties of the mixture as a whole; that is, average values of the boiling point T_b and specific gravity S are needed. Here, liquid kinematic viscosity ν of unidentifiable petroleum fractions are estimated at 310.93 K and 372.04 K. In the former case, the proposed correlations are

$$\nu_{310} = \nu_{ref} + \nu_{cor} \tag{8.8}$$

$$log\left(\nu_{ref}\right) = -1.35579 + 1.46890x10^{-3}T_b + 2.71676x10^{-6}T_b^2 \tag{8.9}$$

$$\begin{aligned} log(\nu_{cor}) = {} & 34.931 - 0.15919T_b + 2.18218x10^{-4}T_b^2 \\ & - 5.91330x10^{-8}T_b^3 + K_W\big(-2.92649 + 1.25713x10^{-3}T_b \\ & - 1.65223x10^{-6}T_b^2 + 4.37037x10^{-9}T_b^3\big) \end{aligned} \tag{8.10}$$

where K_W accounts for the Watson factor, which is defined as

$$K_W \equiv \frac{\sqrt[3]{1.8T_b}}{S} \tag{8.11}$$

In the case of the kinematic viscosity at 372.04 K, the corresponding correlation is:

$$log(\nu_{372}) = -1.79301 + 4.33928x10^{-4}T_b + 0.5113log(T_b\nu_{310}) \tag{8.12}$$

[2]An alternate way to express Eq. (8.3) is: $\underline{\underline{\tau}} = 2\mu\underline{\underline{d}}$, where $\underline{\underline{d}} = \frac{1}{2}\left(\nabla\underline{v} + \left[\nabla\underline{v}\right]^T\right)$.

In these expressions, T_b, the mean average boiling temperature is in degrees Kelvin. Once the kinematic viscosity is determined at the above two temperatures, it can be found at other temperatures from

$$\nu = Z - 0.7 - e^{-0.7487-3.295(Z-0.7)+0.6119(Z-0.7)^2-0.3193(Z-0.7)^3} \tag{8.13}$$

Z is in turn calculated from

$$Z = 10^{10^{log[log(Z_{310})]+B[log(T)-log(310.93)]}} \tag{8.14}$$

where

$$B = \frac{log[log(Z_{310})] - log[log(Z_{372})]}{log[log(310.93)] - log[log(372.04)]} \tag{8.15}$$

together with

$$Z_{310} = \nu_{310} + 0.7 + e^{-1.47-1.84\nu_{310}-0.5\nu_{310}^2} \tag{8.16}$$

$$Z_{372} = \nu_{372} + 0.7 + e^{-1.47-1.84\nu_{372}-0.5\nu_{372}^2} \tag{8.17}$$

In the above expressions, ν is expressed in cSt. Figure 8.2 illustrates the estimation of kinematic viscosity as a function of temperature for heavy petroleum cuts

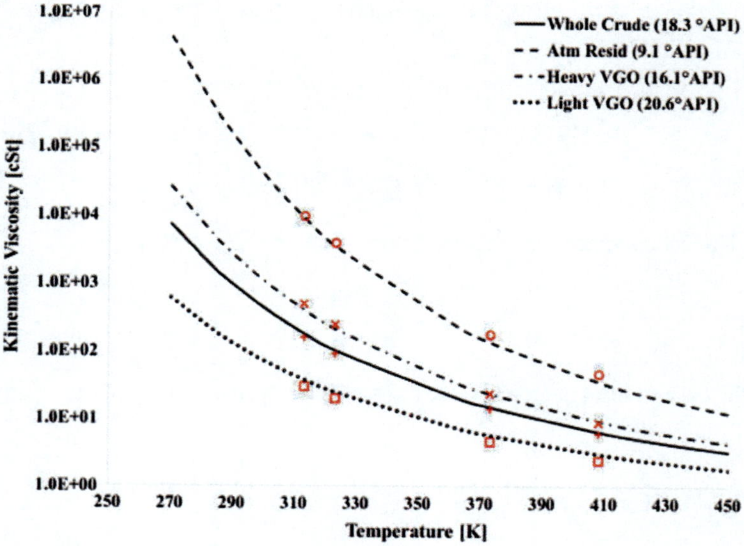

Fig. 8.2 Estimation of the kinematic viscosity for Eocene crude [4] using the API procedure at low pressures

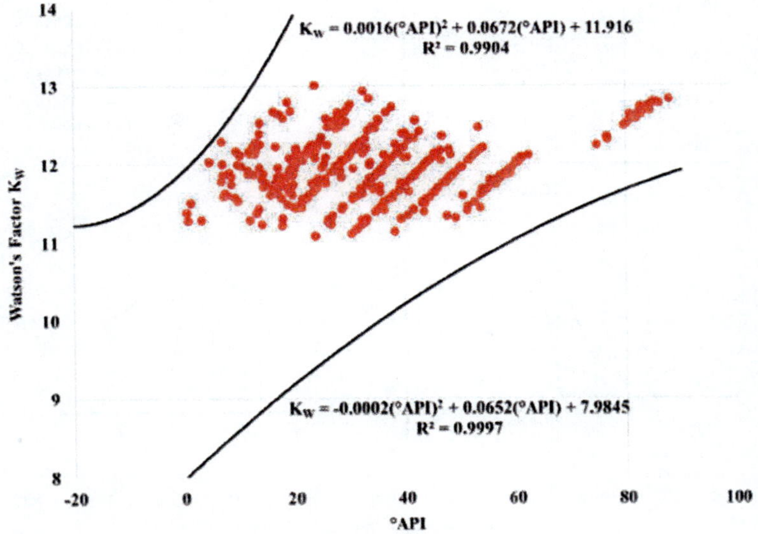

Fig. 8.3 Applicability of the API procedure at low pressures as a function of the Watson's factor and API gravity. Data from Chevron [4]

from Eocene oil [4]. The specific gravity and normal boiling temperature used in calculations are the experimental values.

Figure 8.3 shows the range of applicability of the API procedure in terms of the Watson factor and API gravity. In this figure, petroleum cuts from the Chevron's data bank [4], which contains experimental information of 420 petroleum cuts, are placed. As observed, in principle, the kinematic viscosity of all of them could be estimated by using the API procedure. It is claimed that the procedure provides an overall average error of 14% [3]. However, when it is applied, errors indicate that the API procedure cannot be used always, as can be observed from Fig. 8.4 and Fig. 8.5. In these figures, errors are expressed as the Absolute Average Deviation percent (AAD %). Specifically, large errors are obtained when temperature approaches to the freeze temperature T_f for the heaviest petroleum cuts, as illustrated in Fig. 8.4. In terms of the dimensionless temperature $(T - T_f)/T_f$, large errors are obtained when it is less than 0.2 approximately. It is worth noticing that such errors decrease as temperature increases. One may anticipate that, for such low temperatures, the heaviest petroleum cuts behave as non-Newtonian fluids. Therefore, the API procedure should not be used for these petroleum cuts because it is limited to Newtonian fluids. The description of non-Newtonian petroleum fluids will be discussed later.

From Figs. 8.2, 8.3 and 8.4, the effect of pressure is not taken into account because the procedure described above is intended to be applicable near atmospheric pressure, where petroleum fractions are liquids. For higher pressures, the following expression can be used in order to correct the atmospheric value [3]:

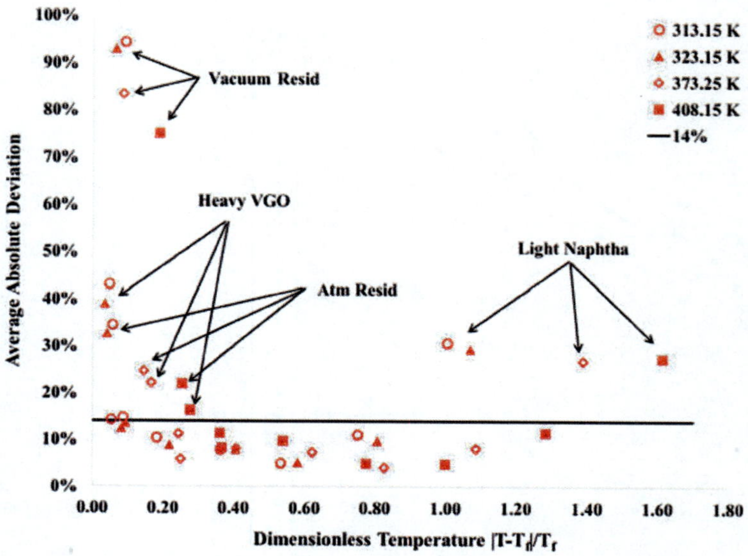

Fig. 8.4 Applicability of the API procedure at low pressures as a function of the dimensionless temperature $|T - T_f|/T_f$, T_f being the freeze point

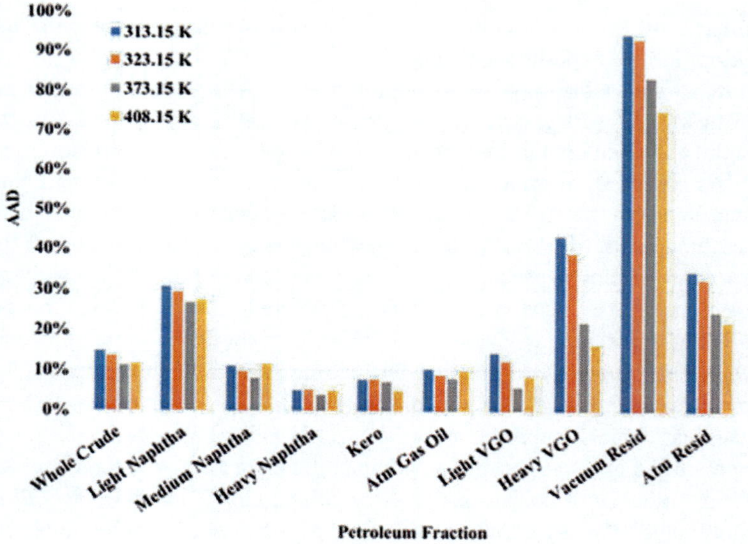

Fig. 8.5 Errors (AAD %) in the estimation of kinematic viscosity of petroleum cuts using the API procedure. Data from Chevron [4]

$$log\left(\frac{\mu}{\mu_{atm}}\right) = \frac{P}{145038}\left(-0.0102 + 0.04042\mu_{atm}^{0.181}\right) \tag{8.18}$$

In this correlation, pressure P is expressed in MPa, while dynamic viscosity μ is expressed in mPa·s. The subscript *atm* indicates atmospheric pressure.

Although the API procedure was intended to be applicable for petroleum refining purposes, that is, for petroleum fluids which are depleted of light hydrocarbon such as methane and ethane, it can be applied to reservoir fluids as well, where such light hydrocarbons are present in significant amounts. In this case, the definition of specific gravity S is modified slightly as follows:

$$S \equiv \frac{\hat{\rho}\left(P^{sat}\right)}{\hat{\rho}_{H_2O}} \tag{8.19}$$

Here, $\hat{\rho}$ and $\hat{\rho}_{H_2O}$ are the mass densities at 288.71 K of the petroleum fluid and the water, respectively. P^{sat} is the saturation pressure of the petroleum fluid at 288.71 K; $\hat{\rho}_{H_2O}$, is equal to 999.02 kg/m^3 at atmospheric pressure. If $P^{sat} < P^{atm}$, then vapor and liquid phases are present at 288.71 K; therefore $\rho\left(P^{sat}\right)$ must be changed by $\rho\left(P^{atm} = 0.101325\text{MPa}\right)$ so that just a liquid phase is present, with which the original definition of specific gravity is retrieved. According to the above definition, predictions of the fluid viscosity require a suitable estimation of the mass density.

Regarding the normal boiling temperature T_b, it is worth remembering that it is defined as the temperature at which a pure component suffers a phase transition from vapor to liquid and vice versa under atmospheric pressure. As illustrated in Chap. 6, the situation is more complex for mixtures because the same phase transition occurs on a temperature interval from bubble temperature to dew temperature. In other words, it is not a well-defined property for mixtures. Hence, the T_b values used in the API procedure are average values from a range of saturation temperatures. These average values are usually determined from experimental distillation curves as those illustrated in Chap. 5. As in the case of the molecular weight, T_b values are the expected values of the distillation curve, where mass or volume fractions are the probability density functions. In other words, T_b values are estimated as the weight or volume average from the distillation curve. For petroleum reservoir fluids, experimental distillation curves are not available usually; hence, they have to be calculated for the API procedure to be applied. To calculate such distillation curves, the theory of the multicomponent simple distillation process can be used [5, 6]. This method consists of solving vapor-liquid equilibriums differentially, which implies that, as in the case of the mass density, the corresponding vapor-liquid phase equilibriums of the reservoir fluid under study have to be correctly calculated before the API procedure can be applied. Once the distillation curve is calculated in terms of the cumulative weight percent, it is suggested to set T_b equal to 55th percentile; which can be modified in the interval from 50 to 59 for purposes of adjustment to viscosity experimental values. Figure 8.6 shows the predictions of viscosity from isothermal

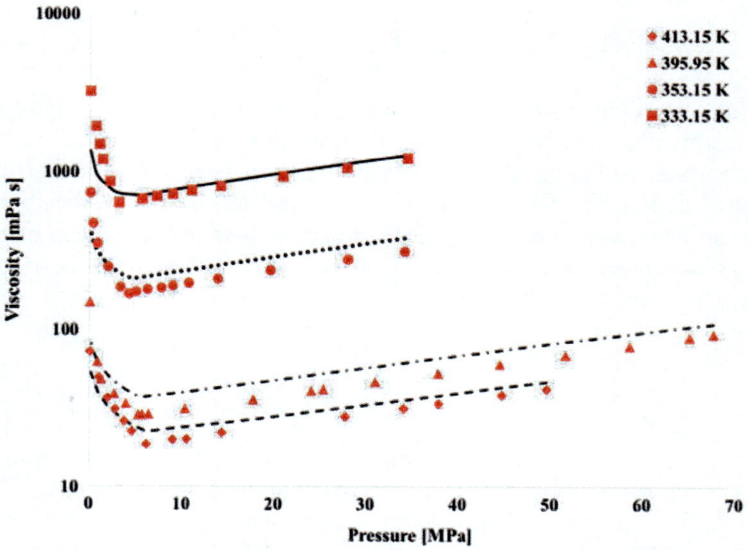

Fig. 8.6 Viscosity of a reservoir fluid (9.9 °API) from isothermal differential vaporizations. Calculations using the API procedure

differential vaporizations of Reservoir Fluid of 9.9 °API studied in Chap. 6, and Fig. 8.7 shows the corresponding algorithm of the API procedure.

8.1.3 Expanded Fluid-Based Viscosity Correlation

The Expanded Fluid-Based Viscosity correlation [7–12] is an encouraging model for estimating the fluid viscosity that has already be applied to heavy petroleum fluids [12, 13]. In principle, this correlation traces continuously the full range of single-phase viscosity in gas, liquid, critical or supercritical regions, it can be used to predict the fluid viscosity of either pure components or mixtures, and it is compatible with the thermodynamic fluid characterization used for phase equilibrium calculations of petroleum fluids. The correlation has a very simple form as can be seen:

$$\mu - \mu^{ig} = 0.165\left(e^{c_2\beta} - 1\right) \tag{8.20}$$

where β is given by

$$\beta = \left[e^{\left(\hat{\rho}_s^*/\hat{\rho}\right)^{0.65} - 1} - 1\right]^{-1} \tag{8.21}$$

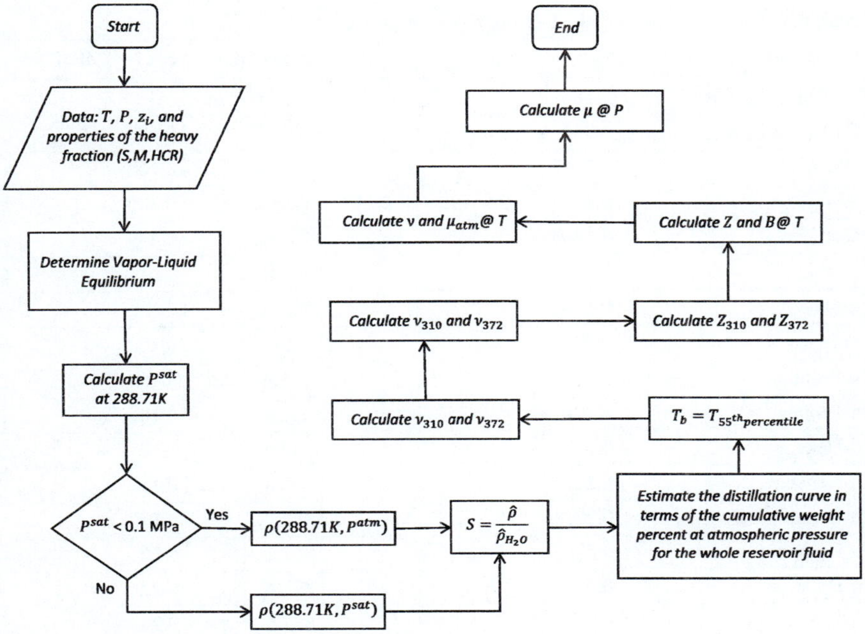

Fig. 8.7 Flowchart of the API procedure applied to reservoir fluids

Here, μ^{ig} is the ideal-gas viscosity, $\hat{\rho}$ is the mass density of the fluid, $\hat{\rho}_s^*$ is the mass density of the fluid in the compressed state, and c_2 is a specific parameter of the fluid. The correlation form was chosen so that if $\hat{\rho} \to \hat{\rho}_s^*$, then $\mu \to \infty$, and if $\hat{\rho} \to 0$, then $\mu \to \mu^{ig}$. Besides, the form of the correlation was conceived so that the effect of temperature on the fluid viscosity was taking into account through the mass density and the ideal-gas viscosity only. Concerning pressure, its effect on the fluid viscosity is introduced through the mass density and through

$$\hat{\rho}_s^* = \hat{\rho}_s^0 e^{c_3 P} \tag{8.22}$$

where, ρ_s^0 is the mass density of the fluid in the compressed state at vacuum conditions, c_3 is another specific parameter of the fluid, and P is the pressure.

Thus, the viscosity model has three temperature-independent parameters for each fluid either a pure component or mixture: c_2, c_3, and ρ_s^0. For some associating components such as water and methanol, however, the parameter c_2 is proposed to be a function of temperature as follows [10]:

$$c_2 = c_2^i + c_2^{ii} e^{c_2^{iii} T} \tag{8.23}$$

For relevant pure components in hydrocarbons mixtures, the specific parameters for the viscosity correlation are shown in Table 8.1. On the other hand, in the case

Table 8.1 Parameters for the EF viscosity model of pure components [7, 10]

Component	c_2^i	c_2^{ii}	c_2^{iii}	ρ_s^0[kg/m^3]	$c_3 x 10^3$[MPa]$^{-1}$
Hydrogen Sulfide	0.188			1194.6	
Carbon Dioxide	0.236			1617.7	0.187
Nitrogen	0.1147			1012.4	0.100
Water	0.1463	99.519	−23.1	1197.0	0.300
Methanol	0.1463	0.6301	−4.22	1045.0	0.449
Methane	0.100			540.0	
Ethane	0.156			724.0	
Propane	0.174			778.0	0.100
n-Butane	0.190			813.0	0.15
n-Pentane	0.198			837.0	0.18
n-Hexane	0.205			849.1	0.18
n-Heptane	0.213			857.8	0.17
n-Octane	0.221			862.7	0.17
n-Decane	0.236			868.1	0.20
n-Dodecane	0.249			871.4	0.22
n-Tetradecane	0.265			875.5	0.24
n-Hexadecane	0.278			878.6	0.28
n-Eicosane	0.306			885.5	
3-Methylpentane	0.206			858.5	
2,2-Dimethylpentane	0.214			850.1	
2-Methylhexane	0.210			861.2	
2,2,4-Trimethylpentane	0.218			855.4	0.15
Cyclopentane	0.215			930.0	0.18
Cyclohexane	0.237			922.1	0.165
Methycyclohexane	0.229			926.0	
Cyclooctane	0.269			957.0	
Benzene	0.226			1066.4	0.135
Toluene	0.223			1056.0	0.14
Ethylbenzene	0.227			1052.0	0.14
Styrene	0.191			1058.0	
Naphthalene	0.211			1141.0	
1-Methylnaphthalene	0.225			1138.0	0.14
2-Methylnaphthalene	0.222			1130.0	
Phenanthrene	0.256			1228.0	
Tetralin	0.217			1087.0	0.16

Table 8.2 Constants for the fluid-specific parameters departure function ΔX [11]

Parameters	b_0	b_1	b_2	b_3	b_4
$\Delta\rho_s^0$	0	14640	739	0	0.67
Δc_2	0.4925	−191900	−0.371	83930	2.67

of the unidentifiable petroleum fractions, the parameter c_3 can be estimated by using the following correlation in terms of the molecular weight [11]:

$$c_3 = \frac{2.8 \times 10^{-7}}{1 + 3.23e^{-1.54x10^{-2}M}} \tag{8.24}$$

Regarding the parameters c_2, and ρ_s^0, they can be estimated from a departure functions ΔX, where viscosity and density for the n-alkanes family from methane (CH_4) to n-tetratetracontane ($n - C_{44}H_{90}$) are used to establish reference parameters as follows [11]:

$$\Delta X = \left(b_0 + \frac{b_1}{M^{b_4}}\right)(S - S_r)^2 + \left(b_2 + \frac{b_3}{M^{b_4}}\right)(S - S_r) \tag{8.25}$$

In this expression, ΔX accounts for either c_2 or ρ_s^0, that is, $\Delta c_2 = c_2 - c_{2r}$ or $\Delta\rho_s^0 = \rho_s^0 - \rho_{sr}^0$. The model constants b_0, b_1, b_2, b_3, and b_4 are shown in Table 8.2 for both parameters.

The reference parameters S_r, c_{2r}, and ρ_{sr}^0 corresponding to the n-alkanes family are, in turn, calculated from

$$S_r = a_0 + \frac{a_1}{\sqrt{M}} + \frac{a_2}{M} + \frac{a_3}{M^2} + \frac{a_4}{M^3} \tag{8.26}$$

together with

$$c_{2r} = (a_0 + a_1 M)e^{a_2/M + a_3 M} + a_4 ln(M) \tag{8.27}$$

and

$$\rho_{sr}^0 = \left(\frac{a_0}{M} + a_1 M^{a_2}\right)e^{a_3 M} + \frac{a_4}{1 + a_5 e^{a_6 M}} \tag{8.28}$$

The correlation constants are shown in Table 8.3 for the three reference properties of the n-alkanes family.

In addition to the specific parameters of the fluid, the input data to the correlation are the mass density, pressure and the ideal-gas viscosity. For identifiable components, it has been suggested the empirical correlation provided by Yaws [14] in order to estimate the ideal-gas viscosity, which is given by

Table 8.3 Constants for the fluid-specific parameters of the n-alkanes family [11]

Property	a_0	a_1	a_2	a_3	a_4	a_5	a_6
S_r	0.845393	0.1419	−16.6	−41.27	2535	n/d	n/d
ρ^0_{sr}	−4775	3.984	0.4	−0.001298	938.3	0.08419	−0.00106
c_{2r}	0.09353	0.000442	−333.4	−0.000166	0.0477	n/d	n/d

$$\mu^{ig} = A + BT + CT^2 \tag{8.29}$$

Here, the constants A, B, and C whose values are shown in Table 8.4 for relevant pure components in hydrocarbons mixtures, are specific for each identifiable component either at pure state or in a mixture.

Regarding unidentifiable petroleum fractions, it has been suggested the Chung method for estimating the ideal-gas viscosity [15].

$$\mu^{ig} = 4.0785x10^{-3} \frac{F_c\sqrt{MT}}{\widetilde{V}_c \Omega_\mu} \tag{8.30}$$

μ^{ig} is expressed in mPa·s, M is the molecular weight, $\widetilde{V}_c$ is the critical molar volume expressed in cm^3/mole, and Ω_μ corresponds to the viscosity integral collision, which takes the following form

$$\Omega_\mu = \frac{1.16145}{T_*^{0.14874}} + 0.52487e^{-0.7732T_*} + 2.16178e^{-2.43787T_*} \tag{8.31}$$

The dimensionless temperature T_* is determined by using the critical temperature T_c as follows:

$$T_* \equiv 1.25593 \frac{T}{T_c} \tag{8.32}$$

Finally, the empirical correction factor F_c is a function of the acentric factor ω, the dimensionless dipole δ_r, and a correction factor κ for highly associating components. The expression is

$$F_c = 1 - 0.2756\omega + 0.059035\delta_r^4 + \kappa \tag{8.33}$$

In the case of hydrocarbons, it is assumed that $\delta_r = 0$ and $\kappa = 0$.

Once the specific parameters of each either identifiable component or unidentifiable petroleum fraction has been determined, the viscosity model can be applied to mixtures by using mixing rules for each parameter. The following mixing rules are based on the one-fluid approximation, which assumes that the form of the viscosity model remains unchanged for a mixture; the mixture effects such as composition are taking into account only through the mixing parameters. The recommended mixing rules are [9]:

Table 8.4 Constants for the ideal-gas viscosity correlation [14]

Component	A[mPa·s]	B[mPa·s/K]	C[mPa·s/K^2]
Hydrogen Sulfide	−1.4839	5.1000E-2	−1.2600E-5
Carbon Dioxide	1.1811	4.9838E-2	−1.0851E-5
Nitrogen	4.2606	4.7500E-2	−9.8800E-6
Water	−3.6826	4.2900E-2	−1.6200E-6
Methanol	−1.4236	3.8935E-2	−6.2762E-6
Methane	0.3844	4.0112E-2	−1.4303E-5
Ethane	0.0514	3.3449E-2	−7.1071E-6
Propane	−0.5462	3.2722E-2	−1.0672E-5
n-Butane	−0.4946	2.9001E-2	−6.9665E-6
n-Pentane	−0.3202	2.6746E-2	−6.6178E-6
n-Hexane	−0.8222	2.6229E-2	−5.7366E-6
n-Heptane	−1.0378	2.4401E-2	−5.4003E-6
n-Octane	0.394	1.6640E-2	1.4470E-5
n-Decane	−0.7297	1.8506E-2	−4.8008E-7
n-Dodecane	−1.2217	1.8099E-2	−8.9949E-7
n-Tetradecane	−1.0397	1.5700E-2	1.0229E-7
n-Hexadecane	−1.3585	1.6007E-2	−5.5846E-7
n-Eicosane	−0.7990	1.1866E-2	2.7167E-6
3-Methylpentane	−0.7066	2.6489E-2	−6.6209E-6
2,2-Dimethylpentane	−0.0362	2.3381E-2	−5.0753E-6
2-Methylhexane	−0.0928	2.2971E-2	−5.5133E-6
2,2,4-Trimethylpentane	0.4636	2.0622E-2	−3.2545E-5
Cyclopentane	−0.2321	3.0131E-2	−1.2230E-5
Cyclohexane	0.1190	2.4542E-2	−3.8334E-6
Methycyclohexane	−0.3574	2.4991E-2	5.3652E-6
Cyclooctane	−0.7878	2.3495E-2	−3.1777E-6
Benzene	−0.0151	2.5706E-2	−8.9797E-7
Toluene	0.1787	2.3566E-2	−9.3508E-7
Ethylbenzene	−0.4267	2.4735E-2	−5.4264E-6
Styrene	−1.0035	2.5191E-2	−3.7932E-6
Naphthalene	−1.6789	2.5406E-2	−3.5495E-6
1-Methylnaphthalene	−0.7548	2.1310E-2	−4.0458E-6
2-Methylnaphthalene	−0.6045	2.1481E-2	−4.2980E-6
Phenanthrene	0.2085	1.7139E-2	−3.9740E-6
Tetralin	−1.5394	2.5754E-2	−3.2295E-6

$$\rho_s^0 = \left[\sum_{i=1}^{C} \sum_{j=1}^{C} \frac{w_i w_j}{2} \left(\frac{1}{\rho_{si}^0} + \frac{1}{\rho_{sj}^0} \right) (1 - k_{ij}) \right]^{-1} \tag{8.34}$$

$$\frac{c_2}{\rho_s^0} = \sum_{i=1}^{C} \sum_{j=1}^{C} \frac{w_i w_j}{2} \left(\frac{c_{2i}}{\rho_{si}^0} + \frac{c_{2j}}{\rho_{sj}^0} \right) (1 - k_{ij}) \tag{8.35}$$

$$c_3 = \left(\sum_{i=1}^{C} \frac{w_i}{c_{3i}} \right)^{-1} \tag{8.36}$$

Similar to equations of state in predicting phase equilibrium situations, k_{ij} is a binary interaction parameter between the components i and j. This interaction parameter can be used to adjust calculations to experimental viscosity values. In the case of mixtures formed by hydrocarbons only, it has been illustrated that $k_{ij} = 0$ leads to predictions within the experimental uncertainties [9]; hence, this is the assumed default value for such type of mixtures.

Concerning the ideal-gas viscosity for a mixture, the Wilke method [16] is suggested, where the mixing rules is expressed as:

$$\mu^{ig} = \sum_{i=1}^{C} \frac{x_i \mu_i^{ig}}{\sum_{j=1}^{C} x_j \varphi_{ij}} \tag{8.37}$$

together with

$$\varphi_{ij} = \frac{\left[1 + \left(\frac{M_j}{M_i} \right)^{0.25} \sqrt{\frac{\mu_i^{ig}}{\mu_j^{ig}}} \right]^2}{\sqrt{8 \left(1 + \frac{M_i}{M_j} \right)}} \tag{8.38}$$

where x_i is the mole fraction of component i, and μ_i^{ig} is the corresponding ideal-gas viscosity.

As may be inferred from Eq. (8.21), predictions from the Expanded Fluid-Based Viscosity correlation are strongly dependent on the estimation of the mass density $\hat{\rho}$ of the petroleum fluids. For instance, deviations of 3% in the estimation of $\hat{\rho}$ may result in errors three or four times larger in the viscosity prediction. Either Cubic or PC-SAFT equations of state can be used for density estimation purposes as long as errors are lesser than 1.0%. However, for quick estimations, the modified Rackett equation [17] may be used to calculate the liquid density of petroleum fractions at their saturations points or at low pressure not far from the atmospheric pressure. The modified Rackett equation (MR) is

$$\frac{1}{\hat{\rho}} = \left(\frac{RT_c}{1000MP_c} \right) Z_{RA}^{[1+(1-T/T_c)^{2/7}]} \tag{8.39}$$

$\hat{\rho}$ is the mass density expressed in kg/m^3, R is the gas constant (8.3144725 J/mol K), M is the molecular weight expressed in g/mol, T_c and P_c are the critical temperature and pressure expressed in K and MPa, respectively. The parameter Z_{RA} is determined from the knowledge of the specify gravity S of the petroleum fraction at 288.71 K as follows:

$$Z_{RA} = \left(\frac{S \hat{\rho}_{H_2O} R T_c}{1000 P_c M} \right)^{[1+(1-288.71/T_c)^{2/7}]} \quad (8.40)$$

Z_{RA} lies in the interval from 0.22 to 0.33. Figure 8.8 illustrates the prediction of the Expanded Fluid-Based Viscosity correlation for the petroleum cuts from the Chevron data bank [4]. In this figure, $\hat{\rho}$ was estimated from the modified Rackett equation, where the critical properties were estimated from a Gibbs free energy minimization (GFEM) using Peng-Robinson equation of state [18], and Molecular Weight, Specific Gravity and Hydrogen-Carbon Ratio as constraints, as explained in Chaps. 3 and 5. These results show that the Expanded Fluid-Based Viscosity correlation is a useful tool for viscosity prediction about 30% of the experimental viscosity measurements [11, 12].

At this point, viscosity estimations are routinely improved by performing adjustments of model parameters according to the available experimental values. However, before proceeding in such a way, it is recommended to analyze whether the characterization of the petroleum fraction can be changed so that viscosity predictions

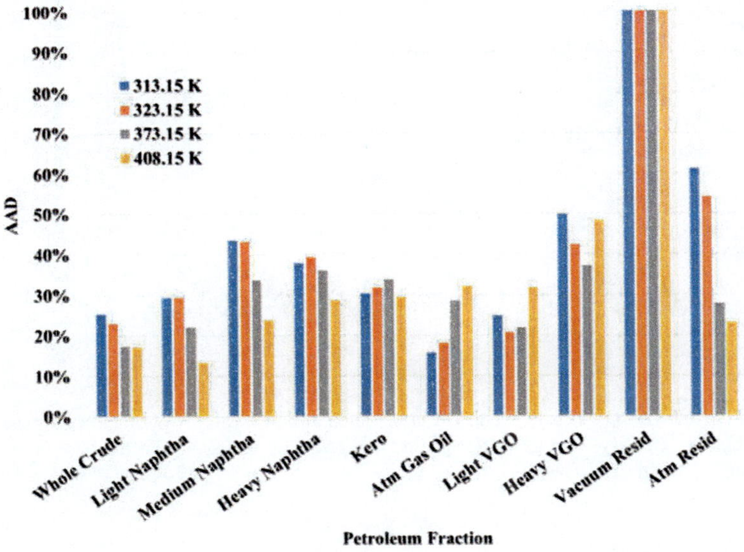

Fig. 8.8 Errors (AAD %) in the estimation of kinematic viscosity of petroleum cuts using the EF correlation. Density from the MR equation and critical properties from the Gibbs free energy minimization with three constraints. Data from Chevron [4]

improve. For instance, the values of the molecular weight can be adjusted by up to $\pm 10\%$, holding the weight composition constant as explained in Chap. 5. As another option, the normal boiling temperature can be introduced as four constraint in the GFEM problem. Although convergence turns out to be more difficult to achieve, and more computational time is spent in calculations, results can be improved as observed in Fig. 8.9, where the Soreide correlation [19] is used to estimated T_b. In Fig. 8.9, the total AAD, where vacuum residue is not taken into account, is reduced from 32.3% in Fig. 8.8 to 24.7% in Fig. 8.9.

Finally, if modifications of the characterization do not produce better results, then estimations can be improved to match the experimental viscosities by adjusting the correlation parameters c_2, c_3, and ρ_s^0 using single multipliers to each parameter. The following recommendations are suggested [11]: the parameter c_2 is recommended to be adjusted first, because of higher uncertainties in its estimation. If enough experimental values are available to justify two-parameter adjustment of the correlation, adjustments are possible through the simultaneous application of single multipliers to the parameters c_2, and ρ_s^0. Finally, the adjustment of the parameter c_3 is essential for effects of high pressures.

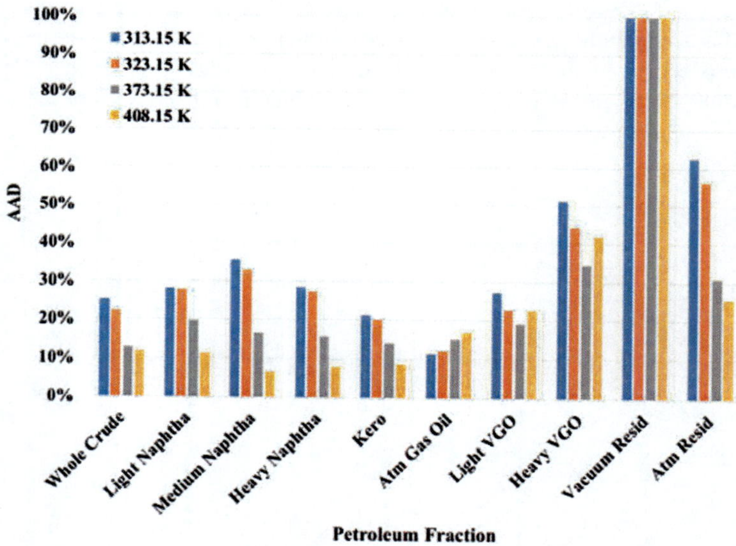

Fig. 8.9 Errors (AAD %) in the estimation of kinematic viscosity of petroleum cuts using the EF correlation. Density from the MR equation and critical properties from the GFEM with four constraints. Data from Chevron [4]

8.2 Non-Newtonian Fluids

Now, petroleum could behave as a Newtonian fluid if it is light enough (normally above 14 °API); even heavy oil may be adequately represented by Newton's law of viscosity at high temperatures. However, oil transport in pipes typically occurs at ambient (293 K) or lower temperatures and a non-Newtonian response may be expected in such cases.

This section is about *Rheology*, which is the science that studies the flow and deformation of matter. Now, from a rheological viewpoint, all materials can be classified Newtonian, non-Newtonian and ideal elastic solids. In turn, non-Newtonian fluids may be considered inelastic or viscoelastic. Figure 8.10 presents such classification.

In general, is hard to find a purely thixotropic fluid; also, a material with yield stress commonly displays thixotropy, and some elastic effects when behaves as solid but also as a liquid. Then, this classification is not definitive, just to give a glimpse of the complex rheological phenomena presented by many real materials.

8.2.1 *Inelastic Models*

As it was mentioned earlier, heavy oils are frequently non-Newtonian fluids, this is, it is expected that their viscosity varies not only with temperature and pressure but also with flow (kinematic) conditions. Now, the first approach to model this departure from Newton's law of viscosity is through the Generalized Newtonian Fluid (GNF) constitutive relations [20].

The GNF class of models retain the form of Eq. (8.3) but allowing the viscosity to vary according to the local rate-of-deformation. One standard change in notation is that instead of μ, now the non-Newtonian viscosity is denoted by η and it is assumed that depends on the flow kinematics [20, 21]. However, note that the flow kinematics are given by the six different quantities of the rate-of-deformation tensor and the functionality of the non-Newtonian viscosity parameter must consider all of them

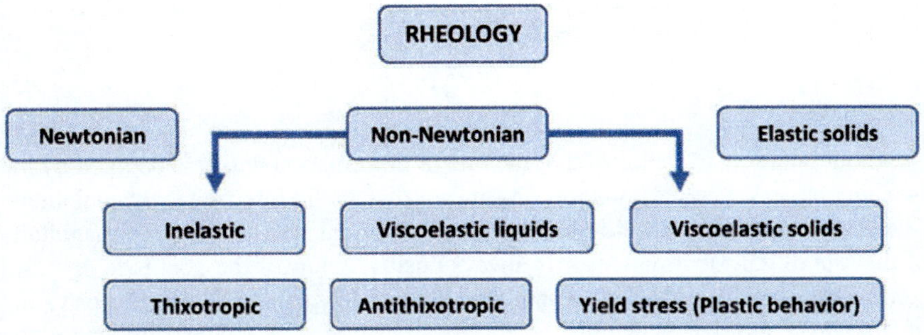

Fig. 8.10 Classification of materials rheological response

in order to represent adequately any flow condition. To this end, we must recall that for a second order tensor, there are three associated scalar quantities known as the invariants. Given any second order tensor, the value of each invariant is independent to the frame of reference used to express the tensor. In particular, as stated by Middleman [21], the second invariant of the rate-of-deformation tensor $\left(II_{\dot{\gamma}}\right)$ is used due to the fact that the first invariant $\left(I_{\dot{\gamma}}\right)$ is always zero for incompressible fluids while the third invariant $\left(III_{\dot{\gamma}}\right)$ also vanishes for 2D flows such as the Simple Shear Flow.

The second invariant for the rate-of-deformation tensor is:

$$II_{\dot{\gamma}} = \underline{\underline{\dot{\gamma}}} : \underline{\underline{\dot{\gamma}}}^T = \sum_{i=1}^{3}\sum_{j=1}^{3} \dot{\gamma}_{ij}\dot{\gamma}_{ij} \tag{8.41}$$

Now, as mentioned above, the invariants are scalar quantities and the second $\left(II_{\dot{\gamma}}\right)$ is related to the magnitude of the rate-of-deformation tensor:

$$\dot{\gamma} = \left|\underline{\underline{\dot{\gamma}}}\right| = \sqrt{\frac{1}{2} II_{\dot{\gamma}}} \tag{8.42}$$

To expand the sums on Eq. (8.41) for the Cartesian frame, the sub-index 1 refers to x, and the sub-indexes 2 and 3 to y and z respectively.

$$\dot{\gamma} = \sqrt{\frac{1}{2}\left(\dot{\gamma}_{11}^2 + \dot{\gamma}_{22}^2 + \dot{\gamma}_{33}^2 + 2\dot{\gamma}_{12}^2 + 2\dot{\gamma}_{13}^2 + 2\dot{\gamma}_{23}^2\right)} \tag{8.43}$$

where the symmetry of $\underline{\underline{\dot{\gamma}}}$ has been used; then, applying Eqs. (8.42) and (8.43) for the Simple Shear Flow, given by:

$$\begin{pmatrix} 0 & \tau_{xy} & 0 \\ \tau_{yx} & 0 & 0 \\ 0 & 0 & 0 \end{pmatrix} = \mu \begin{pmatrix} 0 & \frac{dv_x}{dy} & 0 \\ \frac{dv_x}{dy} & 0 & 0 \\ 0 & 0 & 0 \end{pmatrix} \tag{8.44}$$

We get:

$$\dot{\gamma} = \sqrt{\frac{1}{2}\left(2\dot{\gamma}_{xy}^2\right)} = \left|\frac{dv_x}{dy}\right| \tag{8.45}$$

About notation, if we know that the rate of deformation tensor is denoted by the two underscores, to write it without any, means that we are referring to the magnitude (a scalar) of the vector. In addition, and more important is to note that the magnitude of the rate of deformation tensor is always positive. Now, if the goal is to program some code which requires the computation of non-Newtonian inelastic viscosity, the magnitude of the shear-rate must be computed using Eq. (8.43), at all points in the domain. If the problem to solve is in cylindrical or spherical frames, Eq. (8.43) must be applied also, but the sub-indexes 1, 2 and 3 will refer to (r, θ, z) or (r, θ, ϕ), respectively.

In this section, we recap some of the most common and useful inelastic non-Newtonian models which all have an expression similar to Eq. (8.3), but with the change in the viscosity coefficient:

$$\underline{\underline{\tau}} = \eta(\dot{\gamma})\underline{\underline{\dot{\gamma}}} \tag{8.46}$$

One characteristic of the following models is that there are inelastic, which means that, mathematically, they ignore any reaction time that could be presented by the fluid when flow conditions change. For such models, a change in flow kinematics will produce an instantaneous change in viscosity.

8.2.1.1 Ostwald-de Waele (Power Law) Model

Perhaps the most used non-Newtonian model since it is the simplest deviation from Newton's law of viscosity. Here, the viscosity instead of being constant, now depends on a power of $\dot{\gamma}$. The viscosity function for the power law equation is:

$$\eta(\dot{\gamma}) = K\dot{\gamma}^{n-1} \tag{8.47}$$

It is quite common to express this model in terms of a scalar stress:

$$\tau_{12} = K\dot{\gamma}^{n-1}\dot{\gamma}_{12} \tag{8.48}$$

However, as mentioned previously, this scalar expression given in Eq. (8.15) is only valid for very few specific cases such as Simple Shear Flow, and for such specific flows $\dot{\gamma} = |\dot{\gamma}_{12}|$. For Eqs. (8.47) and (8.48), K, and n are the consistency and the index parameters, respectively; note also that for these equations, the term $\dot{\gamma}^{n-1}$ is always positive but $\dot{\gamma}_{12}$ in Eq. (8.48) can be positive or negative, depending on the local value of the derivative of velocity. Three cases are to be mentioned:

if $n > 1$, the fluid is said to be dilatant (shear-thickening), which means that for higher values of the shear-rate, the viscosity increases;

if $n = 1$, the fluid is Newtonian and the consistency parameter K becomes μ and,

for $n < 1$, the fluid is pseudo-plastic (shear-thinning) and a reduction of viscosity is expected for increasing values of the deformation rate. Both values of K and n must be positive for all cases.

8.2.1.2 Sisko's Model

Presented in 1958 [22] to fit behavior of lubricating greases. It allows to model fluids with a power law zone with a constant (Newtonian) region at high shear-rates.

$$\eta(\dot{\gamma}) = K\dot{\gamma}^{n-1} + \eta_\infty \tag{8.49}$$

where K, η_∞ and n are parameters to adjust with n being a power law type index η_∞, the Newtonian limit at high shear-rates.

8.2.1.3 Bingham's Model

Bingham's model was the first to include the so-called yield stress (τ_y), which for a fluid obeying this model, is a stress that must be exceeded to produce flow [23]. The yield stress is a parameter that occurs frequently in oil industry [24]. In terms of viscosity, this model is expressed by [20]:

$$\eta(\dot{\gamma}) = \begin{cases} \infty & \tau \leq \tau_y, \\ \tau_y \dot{\gamma}^{-1} + \eta_\infty & \tau > \tau_y, \end{cases} \tag{8.50}$$

again η_∞ is the Newtonian limit at high shear-rates. Note that Eq. (8.50) says that if the stress does not reach the yield value, the fluid presents a solid-like behavior or gel; this solid-like material may be a very soft solid, but still the material will not level out as a liquid would do. Then, if the material is below the yield and it is inelastic, the infinite viscosity means that the magnitude of the deformation-rate is zero. It may be convenient also to express the Bingham model in terms of the stress; assuming an inelastic response when the yield point has been negotiated, the model can be expressed by [25]:

$$\underline{\underline{\tau}} = \left[\frac{\tau_y}{\dot{\gamma}} + \eta_\infty\right]\underline{\underline{\dot{\gamma}}} \quad \frac{1}{2}II_\tau > \tau_y^2 \tag{8.51}$$

and below the yield point $\left(\frac{1}{2}II_\tau < \tau_y^2\right)$, the model is just $\underline{\underline{\dot{\gamma}}} = \underline{\underline{0}}$. In addition, we can think of Bingham's fluid as a special case of Sisko's model, with $n = 0$, in Eq. (8.49), see Barnes et al. [26].

The Herschel-Bulkley model [25] is based on the Bingham model together with a power law response, making it more flexible for data fitting, by adding a third adjustment parameter, i.e. the Power law exponent. The model is expressed as:

$$\eta(\dot{\gamma}) = \frac{\tau_y}{\dot{\gamma}} + K\dot{\gamma}^{n-1} \quad \tau > \tau_0 \tag{8.52}$$

Another common rheological equation with the yield stress phenomenon is the Casson model [25]:

$$\tau^{1/2} = \tau_y^{1/2} + (K\dot{\gamma})^{1/2} \tag{8.53}$$

Casson, Herschel-Bulkley and Binghams models are commonly use in oil applications were the yield stress phenomena may be present, the most common oils for which the yield stress is expected are waxy oils. The paraffins may form a gel-like structure when the oil is at rest or close to it. However, it also quite common that heavy and extra heavy oils do not present such a plastic response. For this fluids, models as Cross and Carreau-Yasuda are frequently employed.

8.2.1.4 Cross Model

This model incorporates a first Newtonian region, followed by a Power law zone and, for very high shear rates, a second Newtonian plateau. Such a behavior is expected to be followed by many fluids, including extra-heavy oils, if viscosity measurements could be carried out for largely enough intervals of shear-rate. Viscosity of many heavy and extra heavy oils may be adequately fitted by models considering a first Newtonian region instead of a yield point. In some cases, the measurements also show or at least hint of the presence of the second Newtonian region, which make models like Cross, Carreau and Carreau-Yasuda even more suitable for such fluids.

The equation for Cross model is given by [26],

$$\frac{\eta(\dot{\gamma}) - \eta_\infty}{\eta_0 - \eta_\infty} = \frac{1}{\left(1 + \left[\lambda\dot{\gamma}\right]^n\right)} \tag{8.54}$$

where η_0 and η_∞ represent the first and second Newtonian regions, respectively; λ has time units related to the inverse of the shear-rate where the Power law zone begins and n is a non-dimensional parameter related to the slope of shear-thinning.

8.2.1.5 Carreau-Yasuda Model

This model is a modification of the proposed by Carreau, which includes an extra parameter for adjustment [27]. This model can be written as [28]:

$$\frac{\eta(\dot{\gamma}) - \eta_\infty}{\eta_0 - \eta_\infty} = \left(1 + \left[\lambda\dot{\gamma}\right]^a\right)^{\frac{n-1}{a}} \tag{8.55}$$

As mentioned before, the Carreau-Yasuda equation is of the class of models that can reproduce the two Newtonian regions and the Power law zone; here, the parameter a controls the sharpness of the transition from the first Newtonian limit to the power law zone, while the inverse of λ corresponds to the beginning of the power law region.

It is quite common to express some of the models presented here in terms of the stress; however, the η versus $\dot{\gamma}$ curve is chosen here since it makes it much easier to identify which model should be tried to fit the data, by comparing the experimental values to the trends presented in Fig. 8.11. Note that this is a log-log plot.

Frequently, texts in rheology present a number of rheological models, some of the presented here are of common use. Obviously, there are many more equations to fit non-Newtonian viscosity, some adequate for fluids with yield point, others for

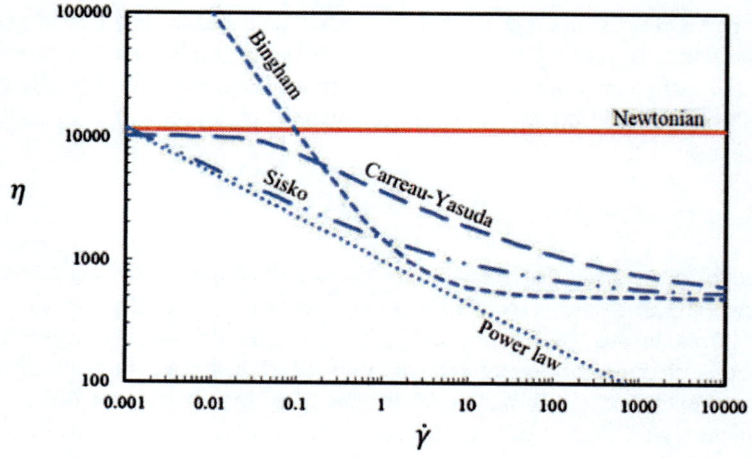

Fig. 8.11 Typical viscosity curves of common inelastic models

data showing the first, second or both Newtonian regions. From the many sources available, Refs. [25, 29] may be quite useful.

8.2.1.6 Viscosity Curves

Selecting the right model and its parameter fitting to reproduce experimental viscosity data is essential for simulation and analysis. In general, to adjust viscosity versus shear-rate at a fixed temperature can be done reasonably easy with a spreadsheet, once the right model has been chosen. The objective function to minimize is the relative error,

$$f_{\min} = \sum_{i}^{nt} \left| \frac{\eta_{\exp_i} - \eta_{\text{calc}_i}}{\eta_{\exp_i}} \right| \tag{8.56}$$

However, if viscosities at different temperatures are also available, it may be desired to have a model which allows computation of viscosity as function of both variables, $\dot{\gamma}$ and T.

In general, extra heavy oils can be fitted with equations with the characteristics of Cross and Carreau models. It is possible to propose an $\eta(\gamma, T)$ expression by taking into account that the temperature dependence on viscosity is $\mu = A\exp(\Delta E/RT)$ for organic substances; then, choosing the right model, an analysis must be done to observe the dependence of each parameters with temperature, which some of these parameters might show the exponential dependence also. A multivariable regression analysis would improve the quality of the fit.

However, a stablished procedure to obtain a $\eta(\gamma, T)$ expression was presented by Ramírez-González [30], which make use of the *Time-Temperature superposition principle* to represent viscosity curves at several temperatures. The Time-Temperature superposition was first used in polymer science to study temperature dependence of linear viscoelastic modules. It is observed that plots at different temperatures of the relaxation modulus versus time, present very similar shapes, but shifted to the left or to the right. Same behavior is observed for the elastic modulus vs. frequency and for all viscoelastic properties of polymers [31]. The principle can be used even for polymer viscosities which are measured far from the linear viscoelastic (LVE) region.

In Ramírez-González work, the shift factor $a_T(T)$ is approximated by Eq. (8.57), and its correlation with temperature, according to the WLF (Williams, Landel and Ferry) equation is in Eq. (8.58). a_T is the shift required to "overlap" the viscosity curves at different temperatures.

$$a_T(T) = \frac{\eta_0(T)}{\eta_0(T_0)} \tag{8.57}$$

$$log(a_T) = -\frac{d_1(T - T_0)}{d_2 + (T - T_0)} \tag{8.58}$$

The viscosity master curves consist on (η/η_0) versus $(\dot{\gamma}\eta_0)$ plots. These curves are then adjusted to an empirical viscosity equation, which in Ramírez-González [30] research is:

$$\frac{\eta(T)}{\eta_0(T)} = \frac{h_1}{1 + h_2\dot{\gamma}\eta_0} + g_1 \tag{8.59}$$

In Eqs. (8.57)–(8.59), η_0 is a reference viscosity, corresponding to its zero shear-rate value, T_0 is a reference temperature and d_1, d_2, h_1, h_2 and g_1 are adjustment parameters. The detailed procedure can be found in reference [30].

Published viscosity and other rheological measurements of dead oil are almost entirely performed on "clean" samples; this is, procedures to remove solids and water are applied to the test fluids. Here, the procedure to obtain the master curves was applied to untreated oils, allowing to obtain the rheological properties closer to the real flow conditions in pipes; however, water content and solids in suspension may affect the quality of measurements. To overcome in some extent this problem, the equilibration time of each sample was determined at the lowest temperature used in the experimental determinations. The equilibration time is the time needed for a complex material to "erase" the effect of sample handling or previous deformations. Even though the extra heavy oils measured here did not show the yield stress phenomenon, which would indicate an internal structure of the fluid giving the sample a solid-like characteristics, the oils must present some kind of structure, no strong enough for producing a yield point, but sufficient to provoke the need of the equilibration time. Once the equilibration time was determined, the reproducibility of rheological measurements improved considerably. The procedure for determining the equilibration time is based in the report of Mazzeo [32]. Figure 8.12 shows the viscosity of an oil sample of 10.3 °API, from the Gulf of Mexico.

The experimental determination of viscosity was carried out in an Anton Paar MCR 101, with 50 mm parallel plates geometry. The samples were allowed to reach thermal equilibrium for periods of three minutes; also, previous to the actual measurements, a pre-shear step followed by a rest period at least equal to the measured equilibration time was added. Note that this 10.3 °API fluid exhibits a very large viscosity at low values of temperature, this characteristic may be attributed to trapped water, asphaltenes and solids from the formation; also observe, that at it is only at 338 K that the fluid could be considered Newtonian, even at 328 K there is a small degree of shear-thinning; then, if the maximum deformation rate of the transport of oil is below 20 s^{-1}, then it could be safely considered as Newtonian (at 328 K). Another feature to observe, is that even with the very large viscosities at 288 or 298 K, the oils does not show the yield stress phenomenon. According to Mezger [33], the presence of a zero shear-rate viscosity (an almost constant viscosity for very low shear-rates) indicates that there is no yield stress, then, the oil behaves as a fluid, but with high viscosities. A first Newtonian region and the fact that shear thinning is occurring at a wide range of shear rates, rather than a very narrow range, confirms the absence of the gel-like structures of the oils at near rest conditions. Also, when handling this fluid in the laboratory, for any deformation on the surface, the oil levelled out taking some seconds. This self-levelling characteristic is another indication of the absence of the yielding point [33]. This observation will be confirmed in a following section, but now in terms of *Linear Viscoelastic* measurements.

The master curves procedure of Ramírez-González [30] were successfully applied to the untreated samples; however, for some fluids the model can be improved by using one of the well-known models, such as Cross or Carreau-Yasuda, instead of Eq. (8.59). Together with the experimental data, Fig. 8.12a shows the adjustment using the model used in Ramírez-González [30]. It can be gathered that the empirical model fits well the data, yet there is an unrealistic prediction that can be observed in the rectangle drew in Fig. 8.12a. The model slightly overestimates the viscosity but

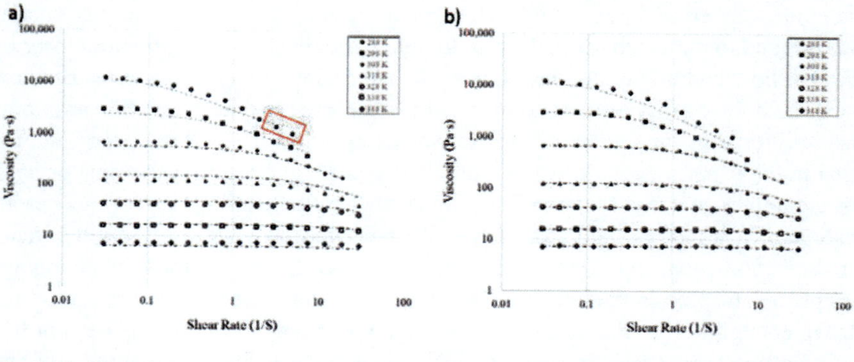

Fig. 8.12 Viscosity curve for the 10.3 °API, **a** empirical model, **b** modified cross model

more important, it also seems to move toward a plateau, which would correspond physically to the second Newtonian region. An analysis of the model indicates a limiting value of 854 Pa·s, a value too large for the second Newtonian plateau. If instead of using Eq. (8.59), the Cross model given by Eq. (8.54) is employed, but following the same procedure described in reference [30], the second Newtonian zone is now about 1 Pa·s, which now is a low value, but seems to be a better approximation considering that even for the largest values of shear-rate, the second plateau is not reached in the experimental determinations. However, as the master curves are of the form (η/η_0) versus $(\dot{\gamma}\eta_0)$, a slight modification to the Cross model was necessary, so that the shear-rate must be multiplied by $\eta_0(T)$, as it can be seen now in Eq. (8.60). The fit of the modified Cross model is in Fig. 8.12b.

$$\frac{\eta(\dot{\gamma}) - \eta_\infty}{\eta_0(T) - \eta_\infty} = \frac{1}{\left(1 + \left[\lambda\eta_0(T)\dot{\gamma}\right]^n\right)} \tag{8.60}$$

In Ramírez-González [30], the zero shear-rates viscosities were measured with a viscometer, while a rheometer was employed for the flow curves. In cases that the only measurements available came from a rheometer, a method to select the most representative value for $\eta_0(T)$ is by adjustment the selected rheological model, for example the ones given by Eq. (8.54) or Eq. (8.59), to each of the temperatures at which the viscosities were determined; but this procedure is only to obtain discrete and representative values of the zero shear-rate at different temperatures. Table 8.5 shows the parameters needed for both models to reproduce the experimental data exhibited in Fig. 8.12.

Figure 8.13 presents viscosity measurements for other two extra heavy oils, an 7 °API and a 11.5 °API fluids, also from the Gulf of México.

Measurements and the adjustment were carried out for other oil samples, Fig. 8.13 shows the experimental data and their curve fitting using the original model of reference [30], which for these 7 °API and 11.5 °API oils, the predictions of the second Newtonian value at the lowest temperatures are 0.13 and 0.08 Pa·s, respectively. For these samples, the model reported by Ramírez-González produced a better approximation than the modified Cross equation.

It can be gathered for the 7 °API and 11.5 °API oils, that at low shear rates and low temperatures both present large values of viscosity. For the 7 °API sample, the

Table 8.5 Parameters for the shift factor, empirical and modified cross models

Shift factor and WLF equation						
	T_0	$\eta_0(T_0)$	d_1	d_2		
	$[K]$	$[Pa \cdot s]$				
	288	13,008	11.712	152.48		
Empirical model, Eq. (8.25) *and modified cross model,* Eq. (8.26)						
h_1	h_2	g_1		η_∞	λ	n
0.93364	0.00040277	0.065655		0.98985	0.00036768	0.88694

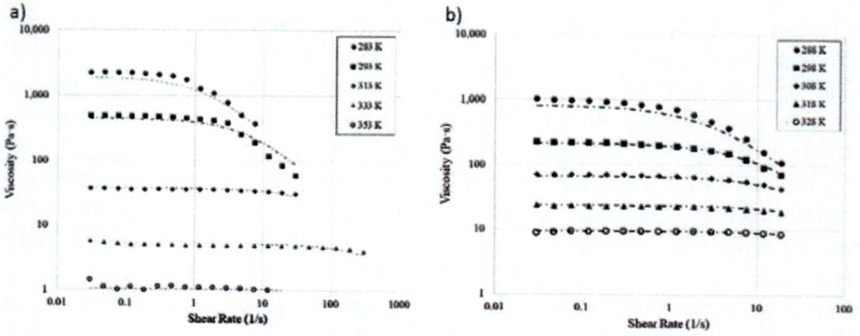

Fig. 8.13 Viscosity curves for two extra heavy oils, **a** 7 °API and, **b** 11.5 °API

Newtonian response can be considered for shear-rates up to 30 s [1]; for measurements at 353 K, the large reduction in viscosity caused difficulties in its measurement. The reduction of viscosity with temperature for 11.5 °API oil seems somehow less significative than for the 7 °API oil. Again, both heavy oils present a zero shear-rate viscosity (i.e. a first Newtonian plateau) which means that the fluids do not yield, as they always behave as fluids. This was again confirmed by their self-leveling nature. An additional note, yield stress phenomena has originated interesting discussion through the years, in part, because its difficulty to measure, i.e. its value depends strongly on the method, but also, because it is difficult to interpret; below the yield point, some consider the material as a solid, or soft-solid but other, consider that the material is only a highly viscous (compared to the viscosity after the sample yields) liquid. For a thorough review on the topic, the manuscripts of Barnes [34] and Coussot [35] are highly recommended.

8.2.2 *Linear Viscoelastic Measurements*

To obtain the flow curves, the rheometer geometry (parallel plates) makes full rotations, which means large degrees of deformation on the fluid sample, breaking down the structure of the fluid, if any. Also, from a flow curve is not distinguishable the level of elasticity of the fluid, this is, a relatively inelastic fluid and a viscoelastic one, could present very similar trends for the η versus $\dot{\gamma}$ plots. Another type of rheometric analysis corresponds *to Linear Viscoelasticity* and, in particular, the so called *Small-Amplitude Oscillatory Shear* (SAOS) is a popular technique to study non-Newtonian characteristic of materials. Linear viscoelastic methods consist on the study of materials at very small deformations, which makes it ideal to study if the sample exhibits any structure, relaxation time, and yield stress. In the SAOS test, the rheometer geometry rotates a just fraction of a turn and then, moves back. The movement is oscillatory (rotates back and forth) and, as normally the fraction of a turn is very small, the deformation imposed on the sample is small enough to keep a linear relationship between stress and deformation, which make this flow suitable to study materials without disturbing its structure, and to characterize viscoelastic materials, see Fig. 8.14 for an schematic representation of such a motion.

Consider an imposed movement on a wall of the rheometer (upper plate, or the cone, or the inner cylinder, depending on the geometry used for the tests). This movement will produce a deformation or strain (γ) in the tested material. The imposed deformation is expressed as follows:

$$\gamma = \gamma_0 \sin(\omega t) \tag{8.61}$$

where γ_0 is the strain amplitude, ω is the frequency of the motion and t is time. The strain rate is then, $d\gamma/dt$, which is given by:

$$\dot{\gamma} = \gamma_0 \omega \cos(\omega t) \tag{8.62}$$

If a sample is subjected to the motion given by Eq. (8.61), the produced shear stress is with the same frequency but quite normally, out-of-phase, this is, the resulting stress wave is shifted by an angle δ, then the stress is expressed as:

$$\tau = \tau_0 \sin(\omega t + \delta) \tag{8.63}$$

The term τ_0 corresponds to the amplitude of the measured stress wave. A representation of the waves of these flow parameters can be seen in Fig. 8.15

Equation (8.63), as is the sine of a sum, can be decomposed in a sine-term and a cosine-term as:

$$\tau = \tau_0' \sin(\omega t) + \tau_0'' \cos(\omega t) \tag{8.64}$$

where τ_0' and τ_0'' are the resulting coefficients of the applied decomposition. It can be seen from Eq. (8.64) that the stress contains a term in phase with the strain (γ)—as in Hooke's law, and the other term is out of phase by 90°, which follows the $\cos(\omega t)$

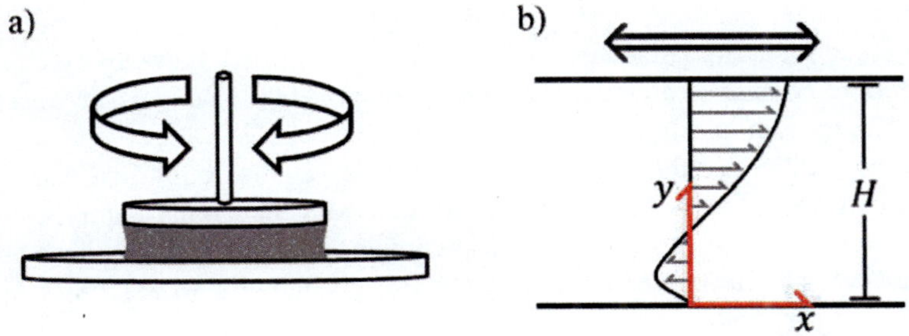

Fig. 8.14 Small-amplitude oscillatory shear; **a** geometry oscillation, **b** representation of the velocity profile

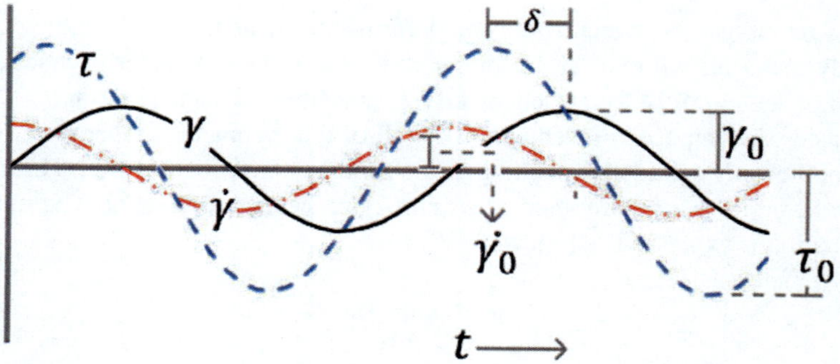

Fig. 8.15 Waves for strain, γ (input), shear-rate, $\dot{\gamma}$, and stress, τ

as the strain-rate does; then, this part of the original stress is proportional to the deformation rate, in the same way as Newton's law of viscosity. In summary, the shift angle δ observable in the stress wave, indicates that this stress can be separated in two contributions, an elastic one and a viscous part. With these in mind, two moduli are defined to represent such contributions: an elastic (or storage) modulus and a viscous (or loss) modulus.

$$G' = \frac{\tau_0'}{\gamma_0} \text{ elastic or in} - \text{phase part,} \tag{8.65}$$

$$G'' = \frac{\tau_0''}{\gamma_0} \text{ viscous or out} - \text{of} - \text{phase part.} \tag{8.66}$$

These moduli have the same dimensions as stress and are positive; these are related to the shift angle by:

$$\tan(\delta) = \frac{G''}{G'} \tag{8.67}$$

The rheometer can make measurements varying the strain amplitude, the frequency and temperature; varying one and keeping the other two fixed. From SAOS measurements, several parameters or properties are reported, however by knowing the values of G' and G'' all the others can be computed [36], as it can be seen from Eq. (8.67), for example.

Another SAOS variable which is frequently reported in published data is the complex viscosity, η^*. As the kinematics imposed in this flow is oscillatory, complex numbers arise in its study, and terms such as complex viscosity and complex modulus (G^*) are defined. The complex viscosity ($|\eta^*|$) is given by Eq. (8.68).

$$|\eta^*| = \sqrt{\left(\frac{G'}{\omega}\right)^2 + \left(\frac{G''}{\omega}\right)^2} \tag{8.68}$$

The complex viscosity can also be separated into a dynamic complex viscosity, $\eta' = G''/\omega$, and an elastic contribution, $\eta'' = G'/\omega$. One of the important aspects of the complex viscosity is that is magnitude, shown in Eq. (8.68), may be related to the viscosity of the flow curves, if the so-called Cox-Merz [37] rule applies. This rule is empirical, but it is expected that many liquids follow it. The rule states that there is an equivalence between the functionality of the complex viscosity with respect to the frequency, and the dynamic viscosity with respect to the shear-rate. This is:

$$\eta(\dot{\gamma}) = \left|\eta^*(\omega)\right|;\ \text{with}\ \dot{\gamma} = \omega \tag{8.69}$$

To use the Cox-Merz rule it is necessary to tested before, by comparing Simple Shear Flow and SAOS experiments at the same temperature.

8.2.2.1 Amplitude Sweep

This is normally the first test needed to perform in SAOS. The main idea is to find the strain limits of the linear viscoelastic or LVE region of the sample. The test may find a minimum deformation, which would appear as data without any trend (noise), meaning that for such low strains, the rheometer is unable to read data properly. The test would also show the maximum strain for the linear deformation range to be valid. The linear viscoelastic region can be found as the range of $\%\gamma$[3] where the moduli G' and G'' remain almost constant. For this, a plot of both moduli, G' and G'' versus $\%\gamma$ is needed. Figure 8.16 is the corresponding plot for the 10.3 °API oil. From this figure it is possible to observe that values of less than $\%\gamma \sim 0.07$, are below the capability of the instrument, at 288.15 K, this is not a characteristic of the fluid, but of the rheometer. It seems that at 348.15 K, the strong reduction in the moduli (more than two decades in the log × log plot) provokes that the noise ends now at $\%\gamma \sim 1$. At 288.15 K the maximum level of strain is about 10 $\%\gamma$, then, for the subsequent measurements in SAOS, it is safe to fix the strain for any value inside $\%\gamma \in [0.07, 10]$ to ensure for the linear viscoelastic region at this low temperature, while for the 348.15 K case. $\%\gamma$ must be inside [1, 100]. Another aspect to note is that, at low temperatures, the values of G' and G'' are closer to each other and, for higher temperatures, as the fluid is becoming 'more' Newtonian, the separation between the loss and storage moduli is much larger, meaning that, as G' is much lower that G'', the fluid is almost inelastic (strictly speaking, G' for a Newtonian fluid must be zero). This is and expected conclusion since, from Fig. 8.12, it has already seen that at 348.15 K, the 10.3 °API oil is practically Newtonian. For Fig. 8.16, a frequency $\omega = 10$ rad/s was used.

Another characteristic of the 10.3 °API oil is the confirmation that the fluid does not show a yield stress. This can be said as that for any value of strain above the sensitivity of the instrument, the viscous modulus (G'') is larger than the elastic one

[3] Usually rheometers report $\%\gamma$ instead of the actual strain. Both γ and $\%\gamma$ are proportional to the angular displacement.

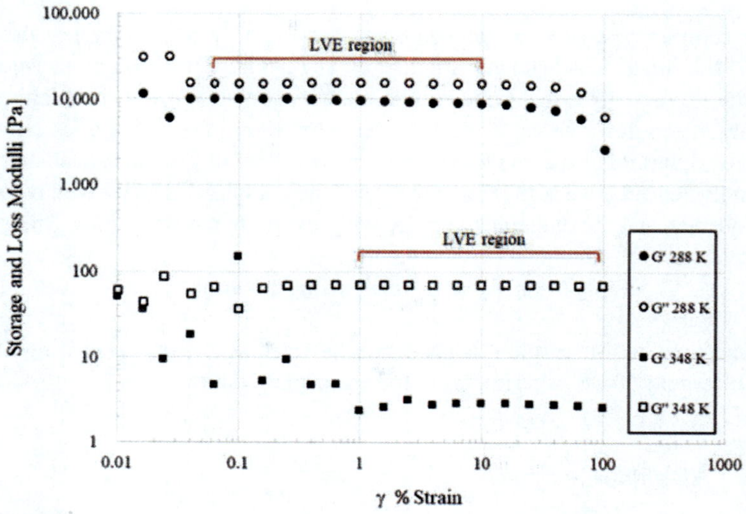

Fig. 8.16 Linear viscoelastic measurements: frequency sweep, 10.3 °API oil

(G'). The viscous or loss modulus corresponds to liquid-like characteristics of the sample, while the elastic or storage one corresponds to solid (elastic) type response; then, if the loss modulus is higher than the storage one, this indicates that the sample is responding as a liquid, unstructured or at least with a structure not 'strong enough' to elicit a solid-like response. A sample with a yield stress would present $G' > G''$ inside de LVE region, and outside the LVE zone, there would be a strain where the moduli crosses, indicating that beyond that strain, the material has yielded and it behaves as a liquid. Figure 8.17 illustrates these ideas.

As mentioned, from an amplitude sweep it is possible to know if a material behaves as fluid or as a gel, depending on the values of G' and G'' relative to each other. Focusing in the upper diagram of Fig. 8.17, Mezger [33] points out, a yield

Fig. 8.17 Characterization of a fluid by amplitude sweep tests

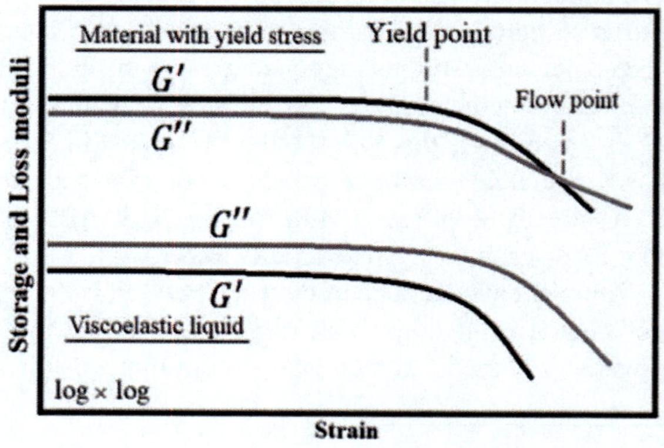

and a flow points may be identified. The yield point is when the storage modulus starts to decay, which means that the elastic (or gel) like structure is beginning to breakdown irreversibly, but at this point, the elastic nature still dominates the liquid one. For larger strains, the crossover of the moduli occurs, and at this point, the material behaves as a fluid and the flow starts. Commonly, the stress corresponding to the flow point is reported as the yield stress of the sample.

For a viscoelastic fluid, curves similar to the lower diagram of Fig. 8.17 are obtained. Even though Fig. 8.16 shows this response, it is still needed to check the degree of elasticity. Strictly speaking, a material with G'-values large enough to be detected by the rheometer, can be considered viscoelastic, at least in some extent: if $G'' > G'$, it is a viscoelastic fluid, if $G' > G''$, the material is a viscoelastic solid. However, for some non-Newtonian materials, the level of elasticity is relatively low, such that it could be considered as an inelastic.

8.2.2.2 Frequency Sweep

Once the LVE region has been determined for the fluid and so, determined whether there is yield point, more information can be obtained by a frequency sweep. For this, the amplitude is fixed to any value inside the LVE region, the temperature is constant, and the frequency is variable. Figure 8.18 shows the moduli plot vs. frequency for the 10.3 °API oil; from this it can be gathered that for all temperatures at all tested frequencies, $G'' > G'$, even at 288 K where the moduli seems very close to each

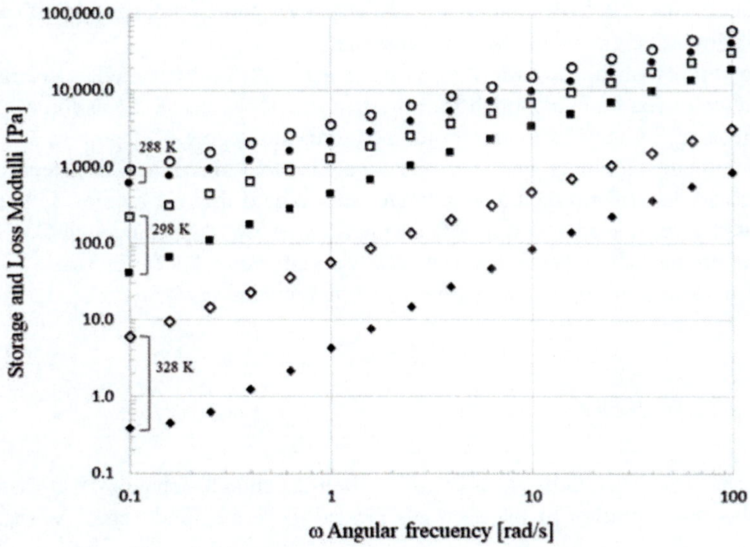

Fig. 8.18 Linear viscoelastic measurements: Frequency sweep for the 10.3 °API oil. G' filled symbols, G" open symbols

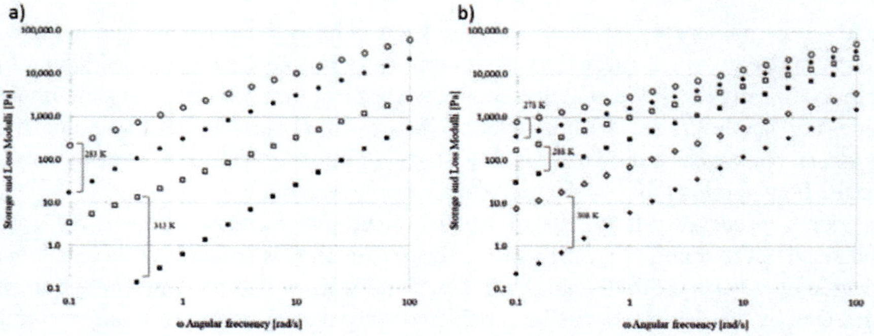

Fig. 8.19 Frequency sweep. G’ filled symbols, G” open symbols, **a** 7 °API and, **b** 11.5 °API

other, there is no crossover. A crossover of the experimental data would mean a change from liquid-like response (at low frequencies) to an elastic (viscoelastic) dominated behavior. Commonly the inverse of the frequency where the crossover occurs, is associated to the largest relaxation time[4] of the fluid [20, 26, 31, 36]. It seems that if the trends of G' and G'' were extrapolated to higher frequencies, the crossover would appear; specifically, data at 288 K looks to be near this point (at $\omega > 100\text{s}^{-1}$, then, assuming that the crossover is at higher frequencies, the largest relaxation time of the oil at such a low temperature would be $\lambda_{\text{max}} < 0.01$ s. The effect of increasing the temperature would be reflected in much lower relaxation times.

Frequency sweeps for the 7 °API and 11.5 °API fluids are in Fig. 8.19. As it was the case for the 10.3 °API fluid, there is no crossover, which indicates that if there is a relaxation, time this would be much less than 0.01 s.

The validity of the Cox-Merz rule for the 10.3 °API oil is tested. It seems that the rule holds reasonable enough for low shear-rates, see Fig. 8.20; as the viscosity is decreasing, by the increments in temperature, the degree of departure seems to be more noticeable. From this, it can be concluded that to use the Cox-Merz rule, it is necessary to perform the comparison to determine if the rule applies. In the case presented here, it seems, that at low frequencies or low shear-rates, the Cox-Merz rule can produce a good estimate of viscosity and its trend, but for increasing values of rate-of-deformation, it significatively overpredicts the viscosity.

8.2.3 *Thixotropy*

The term thixotropy refers to a complex time-dependent behavior of a material, and even the definition of the term has proved to be a difficult task. As reported

[4]The relaxation time is related to the period required by a viscoelastic fluid to vanish any resiudual stress when flow stops. This time is the most important parameter in viscoelasticity.

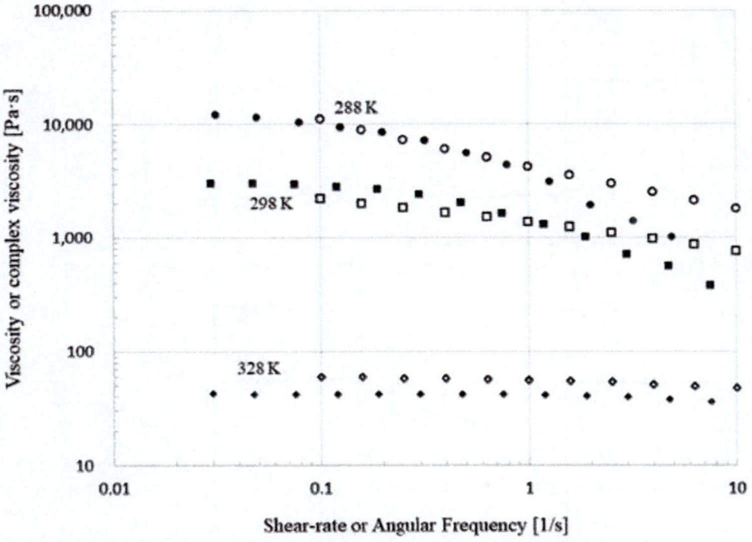

Fig. 8.20 Test of the Cox-Merz rule for the 10.3 °API oil. η filled symbols, |η*| open symbols

by Mewis [38], a general acceptable definition of thixotropy is the continuous decrease of viscosity with time under the influence of shear deformation with a subsequent recovery when the flow is halted. One problem with this definition is that nonlinear viscoelasticity also induces shear-thinning and then it is difficult to distinguish between the two phenomena [38–41]. Thixotropy is related to microstructure of liquids; changes in viscosity are related to microstructural changes produced by breakdown and build-up of such structures. Breakdown is produced by shear forces on the flocculated particles, particle drops, entangled molecules, etc. While the build-up may be the result of in-flow collisions and Brownian motion. In general, the time scales for rebuilding the structures are much larger than the those for breakdown [39]. An important aspect of thixotropy is that the changes in viscosity must be reversible, this is, the fluid must not suffer a permanent structural change after the test [33]. The opposite response in materials is called anti-thixotropy; for such materials, viscosity increases or even become solid-like during and after flow [41].

To distinguish between viscoelasticity and thixotropy, Larson and Wei [41], defined thixotropy as "a reversible, inelastic, time dependence of the viscosity or yield stress during and after flow". The term inelastic is to separate from viscoelastic effects. However, it would be difficult to find a purely "ideal" thixotropic material because, in general, this kind of materials will exhibit a certain degree of elasticity. As noted by Larson and Wei, if elasticity is small, or its time scale is much shorter than that of the thixotropic response, then the sample can be considered as "thixotropic[5]".

[5]Larson and Wei [25], proposed a ratio of relative time scales for viscoelastic and thixotropic process to consider a fluid thixotropic.

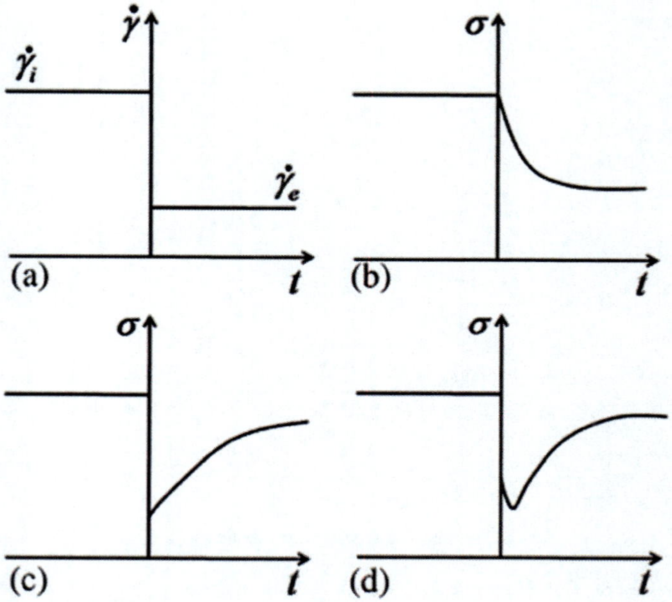

Fig. 8.21 Responses for a step-down in shear-rate [40]: **a** sudden variation in rate-of-deformation, **b** viscoelastic, **c** purely thixotropic, **d** elastic and thixotropic

The loop test are common experiments to study thixotropy; for these protocols, the shear-rate or the shear stress are varied in ramps, which is not appropriate due to the fact that both, stress and rate-of-deformation are time dependent (one as an input and the other as a result), which in turn, complicates the analysis of the measurements (see Barnes [36]). Then to study thixotropy, Mewis and Wagner, in their review [40], stated that a step-down in shear rate is an effective experiment to test for thixotropy. For such a test, viscosity (and stress) are recorded for a sudden reduction (a step) in shear-rate; see Fig. 8.21.

Imposing a sudden change in shear-stress as in Fig. 8.21a, a viscoelastic fluid would suffer a monotonic decrease in the shear stress (σ in the figure) to a new steady value, by viscoelastic (microstructure) stress relaxation as seen in figure Fig. 8.21b. A purely thixotropic fluid, presented Fig. 8.21c would undergo an instantaneous drop in stress, to a lower value followed by a gradual build-up to the new steady stress. A fluid exhibiting both, thixotropic and viscoelastic characteristics, would present the instantaneous drop with a subsequent fast stress relaxation and then, the thixotropic build-up will increase the stress to the new steady-state value, as explained by Mewis and Wagner [40].

The protocol to determine if the oils are thixotropic was similar to the step test consisting of three intervals, proposed by Mezger [33]. The conditions were changed to adjust the particular heavy oils used here, for example, the intermediate shear-rate suggested by Mezger was too large, the flow curves could not achieve that values without oil spilling from the geometry. The actual values here were:

$$\dot{\gamma} = \begin{cases} 0.1\mathrm{s}^{-1} & t \le 60\mathrm{s}, \\ 50\mathrm{s}^{-1} & 60\mathrm{s} < t \le 90\mathrm{s}, \\ 0.1\mathrm{s}^{-1} & t > 90\,\mathrm{s}. \end{cases} \tag{8.70}$$

The resulting stress was monitored by the rheometer, setting the recording to one data per second in the first two intervals, and two data per second for the last, at least for the first five minutes after t = 90 s. Figure 8.22 presents the input shear-rate and the resulting stress for the 10.3 °API fluid at two temperatures (298 and 328 K).

To see the effect of temperature on thixotropy more clearly, no normalization was applied to the stress in Fig. 8.22. By comparing the stress with the cases in Fig. 8.21, it appears that the 10.3 °API oil is a thixotropic fluid with some elastic properties. Mortazavi-Manesh and Shaw [42] mention that the elastic part is normally too fast to be noticed; this seems not the case here. The fluid characteristic and the rheometer allowed to capture that response. It is also noticeable that, at 298 K, even after 500 s, the stress has not reached the previous steady value, but it is very close to it, and still with an increasing trend. The viscoelastic effect last for about 3 s, compared to the 30 s of the high rate-of-deformation of the second step, then, according to the short-term elastic time scale, the oil can be considered just as thixotropic. Now, at 328 K, the viscoelasticity and the thixotropy almost disappear, the difference between the final and initial stress may be within the experimental error. Then, the considerable thixotropic response seen at 298 K, is practically lost at 328 K.

Similar behavior is followed by the 7 °API oil, shown in Fig. 8.23. Here, at 283 K the short-elastic relaxation is hardly seen. At 293 K, just the thixotropic part appears. For the 313 and 333 K cases, thixotropy is not evident at all. It is important to point out that, for the extra heavy oils in here, the recovery time is about 10 min, which could be thought as a fluid with low degree of thixotropy, compared for example, to the model fluid prepared by Moisés et al. [43], showing a recovery of approximately 1 h.

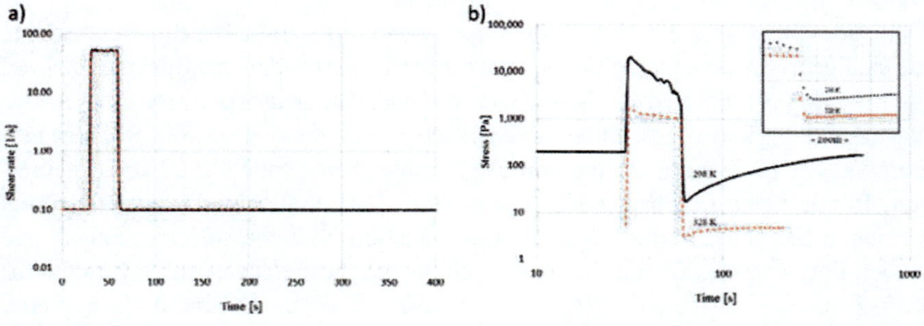

Fig. 8.22 Step test consisting of three intervals for the 10.3 °API at 298 °C and 328 °C: **a** Imposed shear-rate, **b** resulting shear-stress

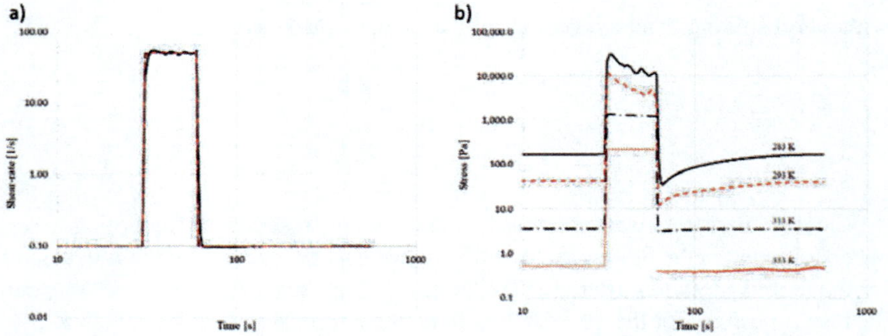

Fig. 8.23 Step test consisting of three intervals for the 7 °API at 283, 293, 313 and 333 °C: **a** Imposed shear-rate, **b** resulting shear-stress

Thixotropy means a time-dependent behavior of stress and viscosity; therefore, modeling of thixotropy involves the evolution of properties with respect to time and a way to quantify the degree of structure at local flow conditions. As mentioned in Chap. 9 of Tropea et al. [44], some of the equations taking into account the time dependence of the degree of structure, present the following form:

$$\frac{d\lambda_{tx}}{dt} = a(1 - \lambda_{tx})^b - c\lambda_{tx}\dot{\gamma}^d \tag{8.71}$$

In this equation, the term $a(1 - \lambda_{tx})^b$ accounts for the build-up of structures and $c\lambda_{tx}\dot{\gamma}^d$ represents their breakdown. The coefficients may be constants or may depend on flow variables. Depending of the values of the b and d parameters, the equations of Moore [45], Cheng and Evans [46] among others, are obtained. The λ_{tx}-parameter is restricted to the interval [0, 1], where $\lambda_{tx} = 1$, corresponds to a fully structured fluid, while $\lambda_{tx} = 0$ to an unstructured one. If the yield-stress phenomenon is present, then, λ_{tx} must take the unit value before yielding, but if the fluid shows not yield, $\lambda_{tx} = 1$ corresponds to the structure at η_0 (the first Newtonian region); while $\lambda_{tx} = 0$ would correspond to η_∞, with no structure at all.

Besides the structure evolution, thixotropy modeling requires an equation to relate the change in stress or viscosity with respect to the rate-of-deformation and the level of structure [44]. Quite frequently, fluids with this characteristic, very often shows plasticity, i.e. yield stress, then, models such as Bingham (Eq. (8.50)), Herschel-Bulkley (Eq. (8.52)) are of common use because their explicit yield stress parameter. It was noted in Chap. 9 of Tropea et al. [44], that there are several forms to couple the chosen structural evolution equation with the selected constitutive model. For example, Tiu and Boger [47] employed the Herschel-Bulkley model for stress $\left(\tau = \tau_y + K\dot{\gamma}^n\right)$ and multiplied the right-hand side by the structural parameter, resulting:

$$\tau = \lambda_{tx}\left(\tau_y + K\dot{\gamma}^n\right) \tag{8.72}$$

In their research, they consider a time evolving equation for λ_{tx} that only accounts for breakdown of structures, for they studied shear degraded food products, which as they do not recover their fully structured condition, they are not properly thixotropic. A very illustrative example can be found on the work of Rubio-Hernández et al. [48] Bingham's model was employed but with a somewhat different coupling of stress-structure variables.

As shown previously, the extra heavy oils analyzed here do not exhibit yield stress; however, the presence of thixotropy and, the closeness between elastic and viscous modules, indicates that there are some kind of structures in the fluid. Considering this, models without an explicit τ_y-parameter can be employed. For example, Fredrickson's [49] model, Cross and Carreau-Yasuda can also be used to describe thixotropy. For more detailed descriptions the manuscripts of Barnes [39], Mujumdar et al. [50], Tropea et al. [44], Mewis and Wagner [40], Larson [51], Larson and Wei [41] should be consulted.

8.2.3.1 Thixotropy and Yield Stress

Modeling the thixotropic behavior of the oils may help to simulate the re-start of pipes or in general, transport processes not at steady state. The studies of the problematic of re-starting pipelines are focused on waxy oils; these oils contain wax crystals which at low temperatures can form a gel phase, with some voids (which makes the gel compressible), some entrapped oil and water, if present. Then, such a gel is not a single phase and also its properties may be not homogeneous inside the pipeline and, as consequence, its rheological nature is hard to describe properly [52]. In fact, rheological models to improve understanding, description, modeling and simulation continue to evolve, for example, Chang et al. [53] explained that the change from solid-like to viscous response is not as simple as a sudden transition. Starting from rest, there is first an elastic response where the stress varies in linear fashion with deformation (strain), up to a point where there the elastic response still dominates, but some creep (flow) can be detected. After this point, the material starts to fracture until a strain where the material completely yields. Chang described the waxy crude oil with three yield stresses, and according to Magda et al. [54] the elastic-limit and the static yield stresses, used by Chang et al. are the ones controlling the restart and also the delay time for the fracture process; this fracture is necessary to reach for the oil to flow. Dimitriou and McKinley [55] developed an elasto-viscoplastic model capable of reproducing many rheological phenomena of waxy oils. Influence of thixotropy to flow restart for a waxy oil was studied by Moisés et al. [43] concluding that thixotropy and yield stress present an important and complex effect on the start-up time but also, a geometric constraint (a solenoid valve, in their study) affect most significantly the flow field in a pipe, due to thixotropy.

The importance of the yield stress in re-start of pipeline flow can be seen in a simple force balance applied to a plugged pipe, $\Delta P \pi R^2 \sim \tau_y \pi D L$, from this, the pressure that must be applied to re-start flow is proportional to the yield stress (τ_y), the length of the plug (L), pipe diameter (D) [55]. Therefore, to predict and design the pumping system for such fluids, the yield stress is necessary, and a better understanding of the phenomena, would imply a safer and economic design. The review of Chala et al. [52] is recommended for a deep study of the problem of transport of waxy oils.

8.2.4 Final Remarks

For practical purposes, it is commonly used to fit stress *vs.* shear rate employing a number of inelastic models, some with an explicit yield stress parameter. This practice may lead to consider a particular oil as having a yield point, although in reality the fluid may not present such phenomenon. For example, in Naiya et al. [56] viscosities are presented for $\dot{\gamma}$ between 10–110 s^{-1}, and with such data, Bingham, power law and Casson's models are fitted. Bingham and Cassson models are used for yield stress fluids and from the stress vs. shear-rate curves it is quite difficult to observe is there is an actual yield point. As mentioned, this is common and it may be quite useful, but if an analysis of flow response at low shear-rates, is needed, then, this procedure may lead to significant errors. The existence of yield stress should be determined by amplitude sweeps, or at least, at very low deformation rates, of about $\dot{\gamma} = 0.1$ or less. For $\dot{\gamma} > 1$, the oil is too far from the rest conditions where the yield stress is defined.

The oils presented here present a significant content of asphaltenes, which gave them their very high viscosities [57] and the almost purely thixotropic behavior [41]. Significant content of asphaltene is common in heavy and extra heavy oils, which may reduce the yield stress [54].

Heavy and extra heavy oils are very complex mixtures, with very complex rheology; then, it seems clear that further research is needed to develop or improve measuring protocols in the rheometer, measuring techniques, constitutive modeling and simulation. There is also a need to corroborate theories and simulations with experiment as close as possible to the actual phenomena occurring in the oil fields and transport ducts.

References

1. Cousins DJ, Enrico MP, Fisher SN, Pickett GR, Shaw NS (1997) Dynamics with a non-Newtonian gas: the force on a body moving through a beam of excitations in superfluid 3He. Phys Rev Lett 79(12):2285–2288. https://doi.org/10.1103/PhysRevLett.79.2285
2. Astillero A (2012) Unsteady non-Newtonian hydrodynamics in granular gases. Phys Rev E 021302:1–16. https://doi.org/10.1103/PhysRevE.85.021302

3. American Petroleum Institute. Technical data book – petroleum refining; 6th ed., Washington, D.C., 1997
4. Chevron. Chevron (online). Available from http://www.chevrontexaco.com/crudemarketing
5. Doherty MF, Perkingd JD (1978) On the dynamics of distillation proceses-I. Chem Eng Sci 33:281–301
6. Doherty MF, Perkins JD (1978) On the dynamics of distillation processes-II. The simple distillation of model solutions. Chem Eng Sci 33(5):569–578. https://doi.org/10.1016/0009-2509(78)80019-0
7. Yarranton HW, Satyro MA (2009) Expanded fluid-based viscosity correlation for hydrocarbons. Ind Eng Chem Res 48(7):3640–3648. https://doi.org/10.1021/ie801698h
8. Satyro MA, Yarranton HW (2010) Expanded fluid-based viscosity correlation for hydrocarbons using an equation of state. Fluid Phase Equilib 298(1):1–11. https://doi.org/10.1016/j.fluid.2010.06.023
9. Motahhari H, Satyro MA, Yarranton HW (2011) Predicting the viscosity of asymmetric hydrocarbon mixtures with the expanded fluid viscosity correlation. Ind Eng Chem Res 50(22):12831–12843. https://doi.org/10.1021/ie201415x
10. Motahhari H, Satyro MA, Yarranton HW (2012) Viscosity prediction for natural gas processing applications. Fluid Phase Equilib 322–323:56–65. https://doi.org/10.1016/j.fluid.2012.03.006
11. Motahhari H, Satyro MA, Taylor SD, Yarranton HW (2013) Extension of the expanded fluid viscosity model to characterized oils. Energy Fuels 27(4):1881–1898. https://doi.org/10.1021/ef301575n
12. Regueira VB, Pereira VJ, Costa GMN, Vieira De Melo SAB (2018) Improvement of the expanded fluid viscosity model for crude oils: effects of the plus-fraction characterization method and density. Energy Fuels 32(2):1624–1633. https://doi.org/10.1021/acs.energyfuels.7b03735
13. Ma M, Chen S, Abedi J (2016) Modeling the density, solubility and viscosity of bitumen/solvent systems using PC-SAFT. J Pet Sci Eng 139:1–12. https://doi.org/10.1016/j.petrol.2015.12.012
14. Yaws CL (1999) Chemical properties handbook. New York
15. Chung T-H, Ajlan M, Lee LL, Starling KE (1988) Generalized multiparameter correlation for nonpolar and polar fluid transport properties. Ind Eng Chem Res 27(4):671–679. https://doi.org/10.1021/ie00076a024
16. Wilke CR (1950) A viscosity equation for gas mixtures. J Chem Phys 18(4):517–519. https://doi.org/10.1063/1.1747673
17. Spencer CF, Danner RP (1972) Improved equation for prediction of saturated liquid density. J Chem Eng Data 17(2):236–241. https://doi.org/10.1021/je60053a012
18. Peng DY, Robinson DB (1976) A new two-constant equation of state. Ind Eng Chem Fundam 15(1):59–64. https://doi.org/10.1021/i160057a011
19. Soreide I (1989) Improved phase behavior predictions of petroluem reservoir fluids from a cubic equation of state. Norwegian Institute of Technology
20. Morrison FA (2001) Understanding rheology. Oxford University Press, New York
21. Middleman S (1977) Fundamentals of polymer processing. McGraw-Hill Education, p 448
22. Sisko AW (1958) The flow of lubricating greases. Ind Eng Chem 50(12):1789–1792. https://doi.org/10.1021/ie50588a042
23. Bingham EC (1916) An investigation of the laws of plastic flow. Bull Bur Stand 13(2):309–353
24. Frigaard IA, Paso KG, de Souza Mendes PR (2017) Bingham's model in the oil and gas industry. Rheol Acta 56(3):259–282. https://doi.org/10.1007/s00397-017-0999-y
25. Tao C, Kutchko BG, Rosenbaum E, Massoudi M (2020) A review of rheological modeling of cement slurry in oil well applications Energies 13. https://doi.org/10.3390/en13030570
26. Barnes HA, Hutton JF, Walters K (1989) An introduction to rheology. Elsevier
27. Yasuda K, Armstrong RC, Cohen RE (1981) Shear flow properties of concentrated solutions of linear and star branched polystyrenes. Rheol Acta 20(2):163–178. https://doi.org/10.1007/BF01513059
28. Bird RB, Armstrong RC, Hassager O (1987) Dynamics of polymeric liquids. Volume 1 fluid mechanics, 2nd edn. Wiley, New York

29. Steffe JF (1992) Rheological methods in food process engineering. Freeman Press
30. Ramírez-González PV (2016) Rheological behavior from light to heavy oils: construction of master curves. Energy Fuels 30(9):7094–7099. https://doi.org/10.1021/acs.energyfuels.6b01340
31. Agassant J-F, Avenas P, Carreau PJ, Vergnes B, Vincent M (2017) Polymer processing principles and modeling. In: Hamilton C (ed) Carl Hanser Verlag
32. Mazzeo FA (2008) Importance of oscillatory time sweeps in rheology. TA Instruments, 1–4
33. Mezger TG (2014) The rheology handbook, 4th edn. Vincentz Network, Hanover, Germany
34. Barnes HA (1999) The yield stress-a review or 'Παντα Ρει'-everything flows? J Nonnewton Fluid Mech 81(1–2):133–178. https://doi.org/10.1016/S0377-0257(98)00094-9
35. Coussot P (2014) Yield stress fluid flows: a review of experimental data. J Nonnewton Fluid Mech 211:31–49. https://doi.org/10.1016/j.jnnfm.2014.05.006
36. Barnes HA (2000) A handbook of elementary rheology. The University of Wales Institute of Non-Newtonian Fluid Mechanics, Aberystwyth, Wales
37. Cox WP, Merz EH (1958) Correlation of dynamic and steady flow viscosities. J Polym Sci 28(118):619–622
38. Mewis J (1979) Thixotropy—a general review. J Nonnewton Fluid Mech 6(1):1–20. https://doi.org/10.1016/0377-0257(79)87001-9
39. Barnes HA (1997) Thixotropy a review. J Nonnewton Fluid Mech 70:1–33
40. Mewis J, Wagner NJ (2009) Thixotropy. Adv Colloid Interface Sci 147–148(C):214–227. https://doi.org/10.1016/j.cis.2008.09.005
41. Larson RG, Wei YA (2019) Review of thixotropy and its rheological modeling. J Rheol 63(3):477–501. https://doi.org/10.1122/1.5055031
42. Mortazavi-Manesh S, Shaw JM (2014) Thixotropic rheological behavior of maya crude oil. Energy Fuels 28(2):972–979. https://doi.org/10.1021/ef4022637
43. Moisés GVL, Alencar LS, Naccache MF, Frigaard IA (2018) The influence of thixotropy in start-up flow of yield stress fluids in a pipe. J Pet Sci Eng 171(August):794–807. https://doi.org/10.1016/j.petrol.2018.07.072
44. Tropea C, Yarin AL, Foss JF (eds) (2007) Springer handbook of experimental fluid mechanics. https://doi.org/10.1007/978-3-540-30299-5
45. Moore F (1959) The rheology of ceramic slips and bodies. Trans Br Ceram Soc 58:470–494
46. Cheng DCH, Evans F (1965) Phenomenological characterization of the rheological behaviour of inelastic reversible thixotropic and antithixotropic fluids. Br J Appl Phys 16(11):1599–1617. https://doi.org/10.1088/0508-3443/16/11/301
47. Tiu C, Boger DV (1974) Complete rheological characterization of time-dependent food products. J Texture Stud 5(3):329–338. https://doi.org/10.1111/j.1745-4603.1974.tb01439.x
48. Rubio-Hernández FJ, Velázquez-Navarro JF, Galindo-Rosales FJ (2007) Thixotropic characterization of fresh cement pastes. Annu Trans Nord Rheol Soc 15
49. Fredrickson AG (1970) A model for the thixotropy of suspensions. AIChE J 16(3):436–441. https://doi.org/10.1002/aic.690160321
50. Mujumdar A, Beris AN, Metzner AB (2002) Transient phenomena in thixotropic systems. J Nonnewton Fluid Mech 102(2):157–178. https://doi.org/10.1016/S0377-0257(01)00176-8
51. Larson RG (2015) Constitutive equations for thixotropic fluids. J Rheol 59(3):595–611. https://doi.org/10.1122/1.4913584
52. Chala GT, Sulaiman SA, Japper-Jaafar A (2017) Flow start-up and transportation of waxy crude oil in Pipelines-a review. J Nonnewton Fluid Mech 2018(251):69–87. https://doi.org/10.1016/j.jnnfm.2017.11.008
53. Chang C, Nguyen QD, Rønningsen HP (1999) Isothermal start-up of pipeline transporting waxy crude oil. J Nonnewton Fluid Mech 87(2–3):127–154. https://doi.org/10.1016/S0377-0257(99)00059-2
54. Magda JJ, El-Gendy H, Oh K, Deo MD, Montesi A, Venkatesan R (2009) Time-dependent rheology of a model waxy crude oil with relevance to gelled pipeline restart. Energy Fuels 23(3):1311–1315. https://doi.org/10.1021/ef800628g

55. Dimitriou CJ, McKinley GH (2014) A comprehensive constitutive law for waxy crude oil: a thixotropic yield stress fluid. Soft Matter 10(35):6619–6644. https://doi.org/10.1039/c4sm00578c
56. Naiya TK, Banerjee S, Kumar R, Mandal A (2015) Heavy crude oil rheology improvement using naturally extracted surfactant. Soc Pet Eng SPE Oil Gas India Conf Exhib
57. Argillier J-F, Barre L, Brucy F, Dournaux J-L, Henaut I, Bouchard R (2007) Influence of asphaltenes content and dilution on heavy oil rheology. Oil Gas Sci Technol Rev IFP. https://doi.org/10.2523/69711-ms

Appendix A
Conversion Factors and the Universal Gas Constant

Table A.1 The Universal Gas constant R

R	Units
8.3144626	J/mol K
1.9858753	Btu/lbmol R
1.9858753	cal/mol K
82.057366	cm^3 atm/mol K

Table A.2 Conversion factors

Magnitude	Conversion Factors
Mass	$1\,kg = 1000\,g = 2.2046226\,lb$
Length	$1\,m = 100\,cm = 0.001\,km = 3.2808299\,ft = 39.370079\,in$
Force	$1\,N = 1 \times 10^5 dyn = 0.1019716\,kgf = 0.2248089\,lbf$
Energy	$1\,J = 1 \times 10^{-3}\,kJ = 1 \times 10^{-6}\,MJ = 0.2388458966\,cal = 9.4782 \times 10^{-4}\,Btu$
Pressure	$1\,bar = 1 \times 10^5\,Pa = 100\,kPa = 0.1\,MPa = 14.503774\,psi = 0.986923267\,atm$
Density	$1\,g/cm^3 = 1000\,kg/m^3 = 62.479606\,lb/ft^3 = 0.0361273\,lb/in^3$
Dynamic viscosity	1 cP =0.01P =0.001 Pa s =1 mPa s =2.4190882 lb/ft hr
Kinematic viscosity	$1\,cSt = 0.01\,St = 0.01\,cm^2/s = 1.0763910 \times 10^{-5}\,ft^2/s = 1.0 \times 10^{-6} m^2/s$
Temperature	$1K = 1.8R;\ K = °C + 273.15;\ °R = °F + 459.67$

B. Carreón-Calderón et al., *Thermophysical Properties of Heavy Petroleum Fluids*, Petroleum Engineering, https://doi.org/10.1007/978-3-030-58831-1

Appendix B
Excel Program for the Thermodynamic Characterization of Unidentifiable Petroleum Fractions

In this appendix, the use of an Excel program for the characterization thermodynamic is explained. The *Fluid Characterization* program intends to be a tool for determining thermodynamic properties of unidentifiable petroleum fractions from the group contribution concept. The functional groups for a given fraction are calculated from a minimization process of the Gibbs free energy as explained in Sect. 3.6.2, where the objective function is given by the logarithm of the fugacity coefficient for a pure component at liquid state. The Peng-Robinson [1] and Soave-Redlich-Kwong [2] equations of state appear as options in order to calculate such an objective function, while the Marreno-Gani [3] and Joback-Reid [4] group contribution methods appear as options in order to calculate thermodynamic properties using the functional groups from the minimization processes. These options lie in the upper left as illustrated below (Fig. B.1).

The minimization process is carried out subject to linear constraints arising from the Molecular Weight, Specific Gravity and Hydrogen-Carbon ratio of the petroleum fraction$(M, S, and HC)$. The first two data are the only ones required, because the third one is estimated as a function of the specific gravity as explained in

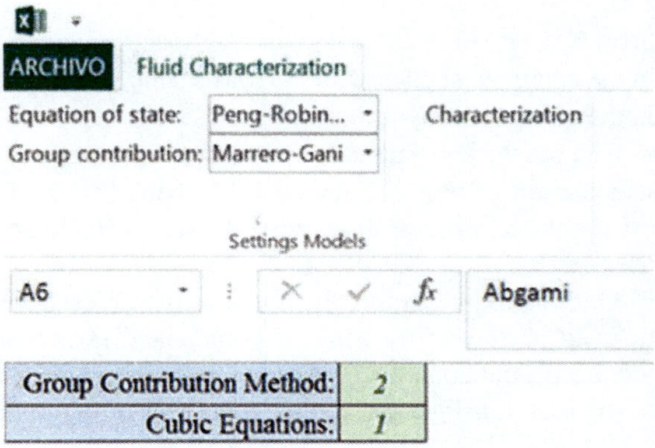

Fig. B.1 Selection of equation of state and group contribution method

B. Carreón-Calderón et al., *Thermophysical Properties of Heavy Petroleum Fluids*, Petroleum Engineering, https://doi.org/10.1007/978-3-030-58831-1

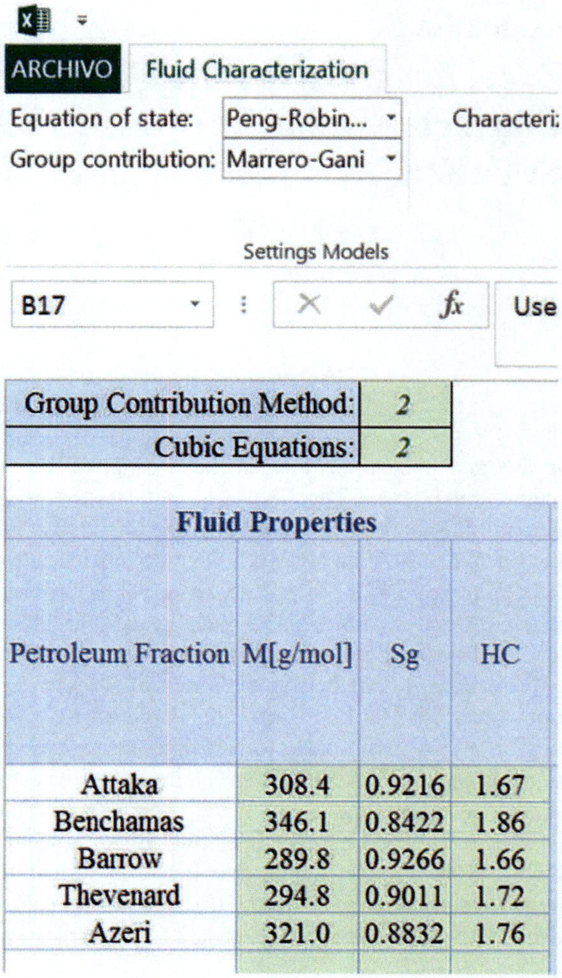

Fig. B.2 Input data: molecular weight and specific gravity

Sect. 6.5. The Molecular Weight and Specific Gravity are introduced as inputs data in the columns labelled with the same name, in front of the petroleum fraction name in the green cells, as shown in Fig. B.2.

The properties of up to ten petroleum fractions can simultaneously calculated, of which eight belong to Chevron's data bank [5] for illustration purposes. However, the input data of these fractions $(M, S, and HC)$ can be modified according to the user needs.

Once the input data for the fractions of interest have been introduced, the calculations are performed by using the command *Characterization*, which lies in the upper left, beside the Equation of state option as shown in Fig. B.1. Finally, the occurrences of the functional groups and the outcomes of the minimization processes appear to

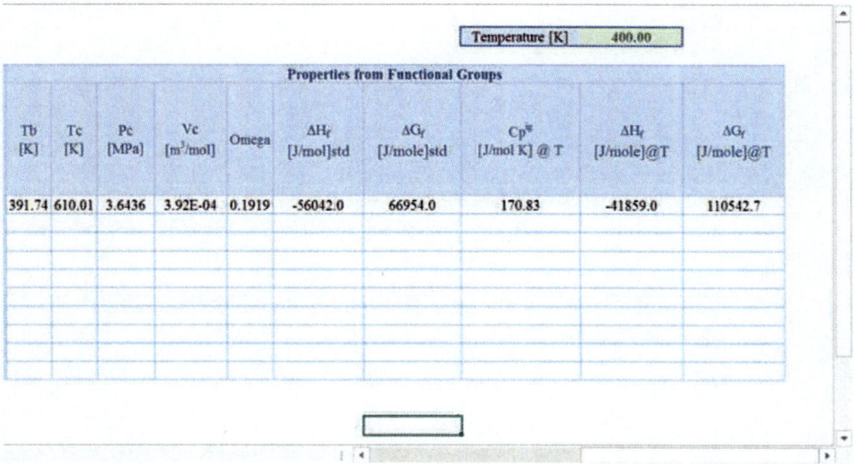

Temperature [K] 400.00

Properties from Functional Groups									
Tb [K]	Tc [K]	Pc [MPa]	Vc [m^3/mol]	Omega	ΔH_f [J/mol]std	ΔG_f [J/mole]std	Cp* [J/mol K] @ T	ΔH_f [J/mole]@T	ΔG_f [J/mole]@T
391.74	610.01	3.6436	3.92E-04	0.1919	-56042.0	66954.0	170.83	-41859.0	110542.7

Fig. B.3 Calculated thermodynamic properties of petroleum fractions

the right of the input data for each characterized fraction. The calculated thermodynamic properties are shown beside these last outcomes. The heat capacity at constant pressure, the molar enthalpy and Gibbs free energy of formation can be calculated any temperature as illustrated below. To save the characterization calculations, they have to be saved as another Excel file as usual so that the original file is always available (Fig. B.3).

References

1. Peng DY, Robinson DB (1976) A new two-constant equation of state. Ind Eng Chem Fundam 15(1):59–64
2. Soave G (1972) Equilibrium constants from a modified Redlich-Kwong equation of state. Chem Eng Sci 27(6):1197–1203
3. Marrero J, Gani R (2001) Group-contribution based estimation of pure component properties. Fluid Phase Equilib 184:183–208
4. Joback KG, Reid RC (1987) Estimation of pure-component properties from group-contributions. Chem Eng Comm, 57(233–243
5. Chevron. Chevron (online) available from http://www.chevrontexaco.com/crudemarketing